REPRESENTATIONS AND
COHOMOLOGY II

D1347609

Already published

Representations and Cohomology

II. Cohomology of groups and modules

D. J. Benson

University of Georgia

PUBLISHED BY THE PRESS SYNDICATE OF THE UNIVERSITY OF CAMBRIDGE
The Pitt Building, Trumpington Street, Cambridge CB2 1RP, United Kingdom

CAMBRIDGE UNIVERSITY PRESS
The Edinburgh Building, Cambridge CB2 2RU, United Kingdom
40 West 20th Street, New York, NY 10011–4211, USA
10 Stamford Road, Oakleigh, Melbourne 3166, Australia

First published 1991
First paperback edition 1998

Printed in the United Kingdom at the University Press, Cambridge

A catalogue record for this book is available from the British Library

ISBN 0 521 36135 4 hardback
ISBN 0 521 63652 3 paperback

Contents

Introduction

This is the second of two volumes which have grown out of about seven years of graduate courses on various aspects of representation theory and cohomology of groups, given at Yale, Northwestern and Oxford. In this second volume, we concentrate on cohomology of groups and modules. We try to develop everything from both an algebraic and a topological viewpoint, and demonstrate the connection between the two approaches. Having in mind the die-hard algebraist who refuses to have anything to do with topology, we have tried to make sure that if the reader omits all sections involving topology, the rest is still a coherent treatment of the subject. But by trying to present the topology with as few prerequisites as possible, we hope to entice such a reader to a more broad-minded point of view. Thus Chapter 1 consists of a predigested summary of the topology required to understand what is happening in Chapter 2.

In Chapter 2, we give an overview of the algebraic topology and K-theory associated with cohomology of groups, and especially the extraordinary work of Quillen which has led to his definition of the higher algebraic K-groups of a ring.

The algebraic side of the cohomology of groups mirrors the topology, and we have always tried to give algebraic proofs of algebraic theorems. For example, in Chapter 3 you will find B. Venkov's topological proof of the finite generation of the cohomology ring of a finite group, while in Chapter 4 you will find L. Evens' algebraic proof. Also in Chapter 4, we give a detailed account of the construction of Steenrod operations in group cohomology using the Evens norm map, a topic usually treated from a topological viewpoint.

One of the most exciting developments in recent years in group cohomology is the theory of varieties for modules, expounded in Chapter 5. In a sense, this is the central chapter of the entire two volumes, since it shows how inextricably intertwined representation theory and cohomology really are.

I would like to record my thanks to the people, too numerous to mention individually, whose insights I have borrowed in order to write these volumes; who have pointed out infelicities and mistakes in the exposition; who have supplied me with quantities of coffee that would kill an average horse; and who have helped me in various other ways. I would especially like to thank Ken Brown for allowing me to explain his approach to induction theorems in I, Chapter 5; Jon Carlson for collaborating with me over a number of years, and without whom these volumes would never have been written; Ralph Cohen for helping me understand the free loop space and its rôle in cyclic homology (Chapter 2 of Volume II); Peter Webb for supplying me with an early copy of the notes for his talk at the 1986 Arcata conference on Representation Theory of Finite Groups, on which Chapter 6 of Volume II is based; David Tranah of Cambridge University Press for sending me a free copy of Tom Körner's wonderful book on Fourier analysis, and being generally helpful in various ways you have no interest in hearing about unless you happen to be David Tranah.

There is a certain amount of overlap between this volume and my Springer lecture notes volume [28]. Wherever I felt it appropriate, I have not hesitated to borrow from the presentation of material there. This applies particularly to parts of Chapters 1, 4 and 5 of Volume I and Chapter 5 of Volume II.

THE SECOND EDITION. In preparing the paperback edition, I have taken the liberty of completely retypesetting the book using the enhanced features of LaTeX 2_ε, $\mathcal{AMS}$-LaTeX 1.2 and XY-pic 3.5. Apart from this, I have corrected those errors of which I am aware. I would like to thank the many people who have sent me lists of errors, particularly Bill Crawley–Boevey, Steve Donkin, Jeremy Rickard and Steve Siegel.

The most extensively changed sections are Section 2.2 and 3.1 of Volume I and Section 5.8 of Volume II, which contained major flaws in the original edition. In addition, in Section 3.1 of Volume I, I have changed to the more usual definition of Hopf algebra in which an antipode is part of the definition, reserving the term bialgebra for the version without an antipode. I have made every effort to preserve the numbering of the sections, theorems, references, and so on from the first edition, in order to avoid reference problems. The only exception is that in Volume I, Definition 3.1.5 has disappeared and there is now a Proposition 3.1.5. I have also updated the bibliography and improved the index. If you find further errors in this edition, please email me at djb@byrd.math.uga.edu.

Dave Benson, Athens, September 1997

CONVENTIONS AND NOTATIONS.

- Maps will usually be written on the left. In particular, we use the left notation for conjugation and commutation: $^g h = ghg^{-1}$, $[g, h] = ghg^{-1}h^{-1}$, and $^g H = gHg^{-1}$.

- We write G/H to denote the action of G as a transitive permutation group on the left cosets of H.

- We write $H \leq_G K$ to denote that "H is G-conjugate to a subgroup of K". Similarly $h \in_G K$ means "h is G-conjugate to an element of K". Thus we write for example $\bigoplus_{g \in_G G}$ to denote a direct sum over conjugacy classes of elements of G.

- The symbol $\square$ denotes the end of a proof.

- We shall use the usual notations $O_p(G)$ for the largest normal p-subgroup of G, $O^p(G)$ for the smallest normal subgroup of G for which the quotient is a p-group, $G^{(\infty)}$ for the smallest normal subgroup of G for which the quotient is soluble, $\Phi(P)$ for the *Frattini subgroup* of a p-group P, i.e., the smallest normal subgroup for which the quotient is elementary abelian, $Z(G)$ for the centre of G, $\Omega_1(G)$ for the subgroup of an abelian p-group G generated by the elements of order p, and so on. The *p-rank* $r_p(G)$ is defined to be the maximal rank of an elementary abelian p-subgroup of G.

- If H and K are subgroups of a group G, then $\sum_{HgK}$ will denote a sum over a set of double coset representatives g of H and K in G.

- We shall write $_\Lambda M_\Gamma$ to denote that M is a Λ-Γ-bimodule, i.e., a left Λ-module which is simultaneously a right Γ-module in such a way that $(\lambda m)\gamma = \lambda(m\gamma)$ for all $\lambda \in \Lambda$, $m \in M$ and $\gamma \in \Gamma$.

- If G is a group of permutations on the set $\{1, \ldots, n\}$ and H is another group, we write $G \wr H$ for the *wreath product*; namely the semidirect product of G with a direct product of n copies of H. Thus elements of $G \wr H$ are of the form $(\pi; h_1, \ldots, h_n)$ with $\pi \in G$, $h_1, \ldots, h_n \in H$ and multiplication given by

$$(\pi'; h_1', \ldots, h_n')(\pi; h_1, \ldots, h_n) = (\pi'\pi; h_{\pi(1)}' h_1, \ldots, h_{\pi(n)}' h_n).$$

- If X is a set with a right G-action and Y is a set with a left G-action, then we write $X \times_G Y$ for the quotient of $X \times Y$ by the equivalence relation $(xg, y) \sim (x, gy)$ for all $x \in X$, $g \in G$, $y \in Y$.

CHAPTER 1

Background from algebraic topology

When we come to give a survey of the cohomology of groups in Chapter 2, we shall need quite a lot of elementary homotopy theory. For the convenience of the reader, we collect in this chapter some of the necessary topological background, and indicate where further details may be found.

1.1. Spaces of maps

First we recall a basic fact from general topology about spaces of maps. If X and Y are topological spaces, we write $\operatorname{Map}(X, Y)$ or Y^X for the space of (continuous) maps from X to Y with the compact-open topology. This is the topology for which the typical sub-basic open set is the set of maps taking a given compact set in X into a given open set in Y.

PROPOSITION 1.1.1. *If X and Y are Hausdorff and Y is locally compact, then the natural map*

$$\operatorname{Map}(X \times Y, Z) \to \operatorname{Map}(X, \operatorname{Map}(Y, Z))$$

sending f to the map f' defined by $f'(x)(y) = f(x, y)$ is a homeomorphism.

PROOF. See for example Hu [**128**, Section V.3]. □

This isomorphism is called the **exponential isomorphism**. This terminology becomes clearer if we write it in the form

$$Z^{Y \times X} \cong (Z^Y)^X.$$

We shall also work with spaces X with a *basepoint* x_0, which we shall denote by (X, x_0), and with pairs of spaces $A \subseteq X$ and a basepoint $x_0 \in A$, which we shall denote by (X, A, x_0). Maps of spaces should take the basepoint to the basepoint.

If (X, x_0) and (Y, y_0) are based spaces, we write $\operatorname{Map}_*(X, Y)$ for the subspace of $\operatorname{Map}(X, Y)$ consisting of those maps taking x_0 to y_0. This is a based space with the constant map as basepoint. If (Z, z_0) is another based space, then under the above correspondence, it is easy to see that the subspace of $\operatorname{Map}(X \times Y, Z)$ corresponding to the subspace

$$\operatorname{Map}_*(X, \operatorname{Map}_*(Y, Z)) \subseteq \operatorname{Map}(X, \operatorname{Map}(Y, Z))$$

consists of those maps sending $(X \times y_0) \cup (x_0 \times Y)$ to z_0.

DEFINITION 1.1.2. *If (X, x_0) and (Y, y_0) are based spaces, we write $X \vee Y$ (X **wedge** Y) for the subspace $(X \times y_0) \cup (x_0 \times Y)$ of $X \times Y$, and $X \wedge Y$ (X **smash** Y) for the quotient space of $X \times Y$ formed by identifying all points of $X \vee Y$ to a single basepoint $*$.*

Thus we have the following:

PROPOSITION 1.1.3. *If (X, x_0) and (Y, y_0) are Hausdorff and (Y, y_0) is locally compact, then the natural map*

$$\mathrm{Map}_*(X \wedge Y, Z) \to \mathrm{Map}_*(X, \mathrm{Map}_*(Y, Z))$$

defined above is a homeomorphism.   □

1.2. Homotopy groups

In this section we give an extremely compressed account of the homotopy groups π_n of a space. The interested reader is advised to refer to a standard source, for example Mosher and Tangora [**195**], Spanier [**247**], Switzer [**258**], Whitehead [**284**], for a more extensive account of this topic.

DEFINITION 1.2.1. *If $f, f' : X \to Y$ are (continuous) maps of topological spaces, then we say f is **homotopic** to f' (written $f \simeq f'$) if there exists a map $F : X \times I \to Y$ (where I is the unit interval $[0, 1]$) such that $F(x, 0) = f(x)$ and $F(x, 1) = f'(x)$. It is clear that homotopy is an equivalence relation on maps. We write $[X; Y]$ for the homotopy classes of maps from X to Y.*

*We say X and Y are **homotopy equivalent** if there are maps $f : X \to Y$ and $f' : Y \to X$ such that the composites are homotopic to the identity maps $f \circ f' \simeq \mathrm{id}_Y$ and $f' \circ f \simeq \mathrm{id}_X$.*

*We say X is **contractible** if it is homotopy equivalent to a single point. In particular, note that a contractible space must be non-empty.*

We write $[X, x_0; Y, y_0]$ for homotopy classes of maps respecting basepoints. The homotopies should respect basepoints in the sense that $F(x_0 \times I) = y_0$. If A is a subspace of X containing x_0 and B is a subspace of Y containing y_0, we write $f : (X, A, x_0) \to (Y, B, y_0)$ to indicate that $f(A) \subseteq B$. We write $[X, A, x_0; Y, B, y_0]$ for homotopy classes of maps $f : X \to Y$ such that $f(A) \subseteq B$ and $f(x_0) = y_0$. Of course, the homotopies $F : X \times I \to Y$ should also have the property that $F(A \times I) \subseteq B$ and $F(x_0 \times I) = y_0$. If F is constant on A, so that f and f' agree on A, we say that f is homotopic to f' **relative to** A.

DEFINITION 1.2.2. *The **homotopy groups** of a based space (X, x_0) are defined to be*

$$\pi_n(X, x_0) = [S^n, s_0; X, x_0]$$

where (S^n, s_0) is an n-sphere with basepoint.

Thus for example $\pi_1(X, x_0)$ is the set of equivalence classes of paths from the basepoint to itself (**loops**), where two such are equivalent if one can be deformed to the other while keeping the two ends fixed. If X is path connected and $\pi_1(X, x_0) = 0$ then we say (X, x_0) is **simply connected**.

Since S^0 consists of two points, one of which is the basepoint, $\pi_0(X, x_0)$ is the set of path components of X.

GROUP STRUCTURE. The set $\pi_n(X, x_0)$ $(n \geq 1)$ can be given the structure of a group as follows. We regard (S^n, s_0) as an n-cube with its boundary identified to a single point, $(I^n, \dot{I}^n)$. Now to compose two elements $[f]$ and $[g]$ of $\pi_n(X, x_0)$, we "divide and stretch" along the first coordinate:

$$f * g : I^n \to X$$

$$(f * g)(s_1, \ldots, s_n) = \left\{ \begin{array}{ll} f(2s_1, s_2, \ldots, s_n) & 0 \leq s_1 \leq \frac{1}{2} \\ g(2s_1 - 1, s_2, \ldots, s_n) & \frac{1}{2} \leq s_1 \leq 1. \end{array} \right. \tag{1}$$

It is easy to check that if $f \simeq f'$ and $g \simeq g'$ then $f * g \simeq f' * g'$, so that this induces a well defined multiplication on $\pi_n(X, x_0)$.

PROPOSITION 1.2.3. *The above defined multiplication makes $\pi_n(X, x_0)$ into a group for $n \geq 1$, and an abelian group for $n \geq 2$.*

PROOF. The identity element is given by the constant map. The inverse in $\pi_n(X, x_0)$ of an element $[f]$ is given by $[f']$, where

$$f'(s_1, s_2, \ldots, s_n) = f(1 - s_1, s_2, \ldots, s_n).$$

The associative law corresponds to the "obvious" homotopy from $(f_1 * f_2) * f_3$ to $f_1 * (f_2 * f_3)$ given by

$$F((s_1, \ldots, s_n), t) = \left\{ \begin{array}{ll} f_1(\frac{4s_1}{1+t}, s_2, \ldots, s_n) & 0 \leq s_1 \leq \frac{1+t}{4} \\ f_2(4s_1 - 1 - t, s_2, \ldots, s_n) & \frac{1+t}{4} \leq s_1 \leq \frac{2+t}{4} \\ f_3(\frac{4s_1 - 2 - t}{2-t}, s_2, \ldots, s_n) & \frac{2+t}{4} \leq s_1 \leq 1 \end{array} \right.$$

(Draw a diagram!) Similarly for $n \geq 2$, the commutative law corresponds to the following diagram:

$$\begin{array}{|c|c|} \hline f_1 & f_2 \\ \hline \end{array} \simeq \begin{array}{|c|c|} \hline f_1 & * \\ \hline * & f_2 \\ \hline \end{array} \simeq \begin{array}{|c|c|} \hline * & f_1 \\ \hline f_2 & * \\ \hline \end{array} \simeq \begin{array}{|c|c|} \hline f_2 & f_1 \\ \hline \end{array}$$

Here, the horizontal and vertical directions are the first and second coordinates in I^n, and the areas marked with an asterisk all go to the basepoint in X. $\square$

REMARK. It is sometimes convenient to work with the (homotopy equivalent) space of maps from $[0, a] \times I^{n-1}$ to X (sending the boundary to x_0) rather than I^n to X, where a is regarded as a variable. Composition of a map from $[0, a] \times I^{n-1}$ and a map from $[0, b] \times I^{n-1}$ gives a map from $[0, a+b] \times I^{n-1}$. This composition has the advantage of being strictly associative rather than just homotopy associative. For $n = 1$, such maps are called **Moore loops** on X.

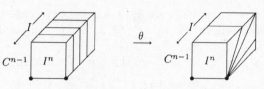

FIGURE 1. The map θ

DEFINITION 1.2.4. *The group $\pi_1(X, x_0)$ is called the* **fundamental group** *of the space (X, x_0).*

REMARK. The set of homotopy classes of paths in X with possibly different ends, where the homotopies are required to fix both ends of the path, form a groupoid. Namely, we can only compose one path with another if the endpoint of the first equals the starting point of the second. This groupoid is called the **fundamental groupoid** of X, and does not depend on choice of basepoint.

RELATIVE HOMOTOPY GROUPS. If (X, A, x_0) is a based pair, we define its **relative homotopy groups** to be

$$\pi_n(X, A, x_0) = [D^n, S^{n-1}, s_0; X, A, x_0] = [I^n, \dot{I}^n, C^{n-1}; X, A, x_0].$$

Here D^n is the n dimensional disc, with boundary S^{n-1}, and $C^{n-1} = \dot{I}^n \setminus \overset{\circ}{I}^{n-1}$ is the boundary of I^n with the interior of the face I^{n-1} (with the *last* coordinate zero) removed. This subspace is identified to a point and regarded as the basepoint.

For example if $n = 2$ then

$$C^{n-1} =$$

PROPOSITION 1.2.5. *Suppose $\pi_n(X, A, x_0) = 0$. Then any map*

$$f : (D^n, S^{n-1}, s_0) \to (X, A, x_0)$$

is homotopic relative to S^{n-1} (i.e., via a homotopy which is constant on S^{n-1}) to a map f' with $f'(D^n) \subseteq A$.

PROOF. Regard f as a map $(I^n, \dot{I}^n, C^{n-1}) \to (X, A, x_0)$. There is a homotopy $F : I^n \times I \to X$ with $F(u, 0) = f(u)$ and $F(u, 1) = x_0$ for all $u \in I^n$, $F(u, t) \in A$ for all $u \in \dot{I}^n$, and $F(u, t) = x_0$ for all $u \in C^{n-1}$. Composing F with the map $\theta : I^n \times I \to I^n \times I$ given in Figure 1 we obtain the required homotopy from f to a suitable f'. $\qquad\square$

For $n \geq 2$, we can give $\pi_n(X, A, x_0)$ a multiplication according to Formula 1.

PROPOSITION 1.2.6. *The set $\pi_n(X, A, x_0)$ is a pointed set (i.e., a set with a distinguished element, namely the constant map at the basepoint) for $n = 1$, a group for $n \geq 2$, and an abelian group for $n \geq 3$.* ☐

WARNING. There is no excision in homotopy, so that in general
$$\pi_n(X, A, x_0) \neq \pi_n(X/A, A/A).$$

BOUNDARY MAP. If $f : (I^n, \dot{I}^n, C^{n-1}) \to (X, A, x_0)$ then we define
$$\partial_n(f) = f|_{I^{n-1}} : (I^{n-1}, \dot{I}^{n-1}) \to (A, x_0)$$
(note that $I^{n-1} \subseteq \dot{I}^n$ and $C^{n-1} \cap I^{n-1} = \dot{I}^{n-1}$). It is easy to check that if $f \simeq f'$ as maps of based pairs then $\partial_n(f) \simeq \partial_n(f')$ as maps of based spaces, so that ∂_n induces a well defined map
$$\partial_n : \pi_n(X, A, x_0) \to \pi_{n-1}(A, x_0) \qquad (n \geq 1).$$

LONG EXACT SEQUENCE.

PROPOSITION 1.2.7. *Given a based pair (X, A, x_0), the natural maps $i : (A, x_0) \hookrightarrow (X, x_0)$ and $j : (X, x_0) \to (X, A, x_0)$ together with the boundary map defined above give rise to a long exact sequence*
$$\cdots \xrightarrow{j_*} \pi_{n+1}(X, A, x_0) \xrightarrow{\partial_*} \pi_n(A, x_0) \xrightarrow{i_*} \pi_n(X, x_0) \xrightarrow{j_*} \pi_n(X, A, x_0) \xrightarrow{\partial_*} \cdots$$

REMARK. Exactness of the above sequence should be interpreted as follows. For $n \geq 1$, it is an exact sequence of groups, and the image of $\pi_2(X, A, x_0)$ lies in the centre of $\pi_1(A, x_0)$. For $n = 0$, these are only pointed sets (sets with a distinguished basepoint). The sequence is exact everywhere in the sense that the kernel of each map (pre-image of the basepoint) is the image of the previous map. Also, the map $\pi_1(X, A, x_0) \to \pi_0(A, x_0)$ extends to an action of $\pi_1(X, A, x_0)$ on $\pi_0(A, x_0)$, and elements are in the same orbit if and only if they have the same image in $\pi_0(X, x_0)$.

PROOF. We shall assume that $n \geq 2$ for the proof, and leave the reader to make the necessary changes for $n = 0$ and 1. There are six separate checks to be made here.

$j_* \circ i_* = 0$: If $f : (I^n, \dot{I}^n, C^{n-1}) \to (X, A, x_0)$ with $f(I^n) \subseteq A$ then we have a homotopy $F : I^n \times I \to X$ given by $F(t_1, \ldots, t_n, t) = f(t_1, \ldots, t_{n-1}, t + (1 - t)t_n)$, showing that $[f] = 0$ in $\pi_n(X, A, x_0)$.

$\mathrm{Ker}(j_*) \subseteq \mathrm{Im}(i_*)$: If $f : (I^n, \dot{I}^n) \to (X, x_0)$ and $j_*[f] = 0$ then there is a homotopy $F : I^n \times I \to X$ such that $F(u, 0) = f(u)$ and $F(u, 1) = x_0$ for all $u \in I^n$, $F(u, t) \in A$ for all $u \in \dot{I}^n$, and $F(u, t) = x_0$ for all $u \in C^{n-1}$. Compose F with the map $\theta : I^n \times I \to I^n \times I$ given in Figure 1, to obtain a homotopy from f to a map $(I^n, \dot{I}^n) \to (A, x_0)$.

$i_* \circ \partial_* = 0$: If $f : (I^{n+1}, \dot{I}^{n+1}, C^n) \to (X, A, x_0)$ then $i_* \partial_*[f] = [f|_{I^n}]$. The map f provides a homotopy from $f|_{I^n}$ to the constant map at x_0.

$\mathrm{Ker}(i_*) \subseteq \mathrm{Im}(\partial_*)$: If $f : (I^n, \dot{I}^n) \to (A, x_0)$ with $i_*[f] = 0$ in $\pi_n(X, x_0)$, then there is a homotopy $F : (I^n \times I, \dot{I}^n \times I) \to (X, x_0)$ with $F(u, 0) = f(u)$ and $F(u, 1) = x_0$. Regarding F as a map $I^{n+1} \to X$, $[F]$ is an element of $\pi_{n+1}(X, A, x_0)$ with $\partial_*[F] = [f]$.

$\partial_* \circ j_* = 0$: If $f : (I^n, \dot{I}^n) \to (X, x_0)$ then $\partial_* j_*[f] = [f|_{I^{n-1}}]$. Since f sends I^{n-1} to x_0, $\partial_* j_*[f] = 0$.

$\mathrm{Ker}(\partial_*) \subseteq \mathrm{Im}(j_*)$: For this, we need the following fact. Given any map from $\dot{I}^n \times I \cup I^n \times \{0\}$ to X, we may extend it to a map from $I^n \times I$ to X. In other words, any partial homotopy may be extended to a homotopy. We express this by saying that the pair $(I^n, \dot{I}^n)$ has the **homotopy extension property** (HEP) with respect to X. A space which has the homotopy extension property with respect to all spaces X is called a **cofibration**. The reason why $(I^n, \dot{I}^n)$ is a cofibration is because there is a **retraction** $I^n \times I \to \dot{I}^n \times I \cup I^n \times \{0\}$, i.e., a map whose composite with the inclusion is the identity map on $\dot{I}^n \times I \cup I^n \times \{0\}$.

Now if $f : (I^n, \dot{I}^n, C^{n-1}) \to (X, A, x_0)$ with $\partial_*[f] = [f|_{I^{n-1}}] = 0$, then there is a homotopy $F : (I^{n-1} \times I, \dot{I}^{n-1} \times I) \to (A, x_0)$ with $F(u, 0) = f(u)$ and $F(u, 1) = x_0$. Thus there is a partially defined homotopy

$$\dot{I}^n \times I \cup I^n \times 0 \to X$$
$$(u, t) \in I^{n-1} \times I \mapsto F(u, t)$$
$$(u, t) \in (\dot{I}^n \setminus I^{n-1}) \times I \mapsto x_0$$
$$(u, t) \in I^n \times 0 \mapsto f(u).$$

Extend to a homotopy $I^n \times I \to X$, and restrict to $I^n \times 1$ to obtain a map $\alpha : I^n \to X$ with $f \simeq j(\alpha)$, so that $[f] = j_*[\alpha]$. $\square$

ACTION OF π_1 ON π_n, AND INVARIANCE OF BASEPOINT. If x_0 and x_1 are two different basepoints in X, and $\omega : I \to X$ is a path with $\omega(0) = x_0$ and $\omega(1) = x_1$, we define a map

$$\omega^* : \pi_n(X, x_1) \to \pi_n(X, x_0)$$

as follows. Given a map $f : (S^n, s_0) \cong (D^n, S^{n-1}) \to (X, x_1)$ we define

$$\omega^*(f) : (D^n, S^{n-1}) \to (X, x_0)$$

$$u \mapsto \begin{cases} f(2u) & \text{if } 0 \le |u| \le \frac{1}{2} \\ \omega(2 - 2|u|) & \text{if } \frac{1}{2} \le |u| \le 1 \end{cases}$$

where $|u|$ is the distance from the origin to u in $D^n \subseteq \mathbb{R}^n$, and $2u$ is calculated inside the vector space $\mathbb{R}^n$. In words, the part of the disk from radius 1 to radius $\frac{1}{2}$ is sent along the path ω. The part of the disk with radius at most $\frac{1}{2}$ is homeomorphic to the whole disk, and is sent into X by composing the original map with this homeomorphism.

If $f \simeq f'$ then $\omega^*(f) \simeq \omega^*(f')$ so that this map is well defined.

Similarly if x_0, $x_1 \in A \subseteq X$ are two different basepoints in (X, A), and $\omega : I \to A$ is a path with $\omega(0) = x_0$ and $\omega(1) = x_1$, we define an isomorphism

$$\omega^* : \pi_n(X, A, x_1) \to \pi_n(X, A, x_0)$$

as follows. Given a map $f : (D^n, S^{n-1}, (1, 0, \dots, 0)) \to (X, A, x_0)$, the map $\omega^*(f) : (D^n, S^{n-1}, (1, 0, \dots, 0)) \to (X, A, x_1)$ is given by sending $(x_0, \dots, x_n)$ to $\omega(x_0)$ if $0 \leq x_0 \leq 1$. We choose a homeomorphism between the half of the disk with $x_0 \leq 0$ with the points with $x_0 = 0$ identified to a single point $*$, and the entire disk D^n with basepoint $(1, 0, \dots, 0)$. We compose this homeomorphism with the original map f to define $\omega^*(f)$ for $x_0 \leq 0$.

Again if $f \simeq f'$ then $\omega^*(f) \simeq \omega^*(f')$, so that this map is well defined.

PROPOSITION 1.2.8. *If ω, $\omega' : I \to X$ are homotopic relative to $\{0, 1\}$, then $\omega^* = (\omega')^* : \pi_n(X, x_1) \to \pi_n(X, x_0)$. For $x_0 = x_1$, this constitutes a group action of $\pi_1(X, x_0)$ on $\pi_n(X, x_0)$.*

If $\omega \simeq \omega' : I \to A$ then $\omega^* = (\omega')^* : \pi_n(X, A, x_1) \to \pi_n(X, A, x_0)$. For $x_0 = x_1$, this constitutes a group action of $\pi_1(A, x_0)$ on $\pi_n(X, A, x_0)$.

The long exact homotopy sequence

$$\cdots \to \pi_{n+1}(X, A, x_0) \to \pi_n(A, x_0) \to \pi_n(X, x_0) \to \pi_n(X, A, x_0) \to \cdots$$

is a long exact sequence of $\pi_1(A, x_0)$-modules.

PROOF. This is a long but straightforward exercise in keeping your head. See Spanier [247, Section 7.3] for the details. □

NOTATION. We write $\pi'_n(X, x_0)$ for the quotient of $\pi_n(X, x_0)$ obtained by identifying $\omega^*(f)$ with f for each $[\omega] \in \pi_1(X, x_0)$, and $\pi'_n(X, A, x_0)$ for the quotient of $\pi_n(X, A, x_0)$ obtained by identifying $\omega^*(f)$ with f for each $[\omega] \in \pi_1(A, x_0)$. Note that the action of $\pi_1(X, x_0)$ on itself is by conjugation, so that $\pi'_1(X, x_0)$ is its abelianisation.

We say that (X, x_0) is a **simple space** if $\pi_1(X, x_0)$ acts trivially on $\pi_n(X, x_0)$ for all n, so that $\pi'_n(X, x_0) = \pi_n(X, x_0)$, and in particular $\pi_1(X, x_0)$ is abelian.

Elements of $\pi_n(X, x_0)$ may be thought of as obstructions to extending maps $S^n \to X$ to maps $D^{n+1} \to X$ as shown by the following proposition:

PROPOSITION 1.2.9. *Suppose X is path connected and $\pi_n(X, x_0) = 0$. Then any map $f : S^n \to X$ may be extended to a map $\tilde{f} : D^{n+1} \to X$.*

PROOF. Since X is path connected, $\pi_n(X, f(s_0)) = 0$, so the map $f : (S^n, s_0) \to (X, f(s_0))$ is homotopic to the constant map at $f(s_0)$. Such a homotopy is a map $S^n \times I \to X$ sending $S^n \times 0 \cup s_0 \times I$ to s_0 and equal to f on $S^n \times 1$. Since $(S^n \times I, S^n \times 0 \cup S^n \times 1 \cup s_0 \times I, S^n \times 0 \cup s_0 \times I) \cong (D^{n+1}, S^n, s_0)$ the proposition follows. □

The theory of obstructions was developed by Eilenberg and others, and an account of it may be found, for example, in Mosher and Tangora [195, Chapter 1].

LOOP SPACES.   There is an alternative approach to homotopy groups based on the concept of a loop space.

DEFINITION 1.2.10. *We define the* **loop space** ΩX *of a based space* X *to be*

$$\Omega X = \mathrm{Map}_*(S^1, X),$$

with the constant path at x_0 *(written* $*$*) as basepoint.*

Now a path in ΩX is the same thing as a homotopy between loops in X (by the correspondence given in Proposition 1.1.1), and so we have

$$\pi_1(X, x_0) \cong \pi_0(\Omega X, *).$$

By Proposition 1.1.3, we have

$$\Omega^2 X = \mathrm{Map}_*(S^1, \mathrm{Map}_*(S^1, X)) \cong \mathrm{Map}_*(S^1 \wedge S^1, X) \cong \mathrm{Map}_*(S^2, X).$$

Continuing in this way, and using the fact that $S^{n-1} \wedge S^1 \cong S^n$, we have

$$\Omega^n X \cong \mathrm{Map}_*(S^n, X).$$

A path in $\Omega^n X$ is the same as a homotopy between maps $S^n \to X$, and so we have proved the following:

PROPOSITION 1.2.11. $\pi_n(X, x_0) \cong \pi_0(\Omega^n X, *)$. $\qquad\qquad$ $\square$

In a similar way, if (X, A, x_0) is a based pair, we define $P(X, A, x_0)$ to be the space of paths in X beginning at x_0 and ending in A. By a similar argument to the above, we have the following:

PROPOSITION 1.2.12. $\pi_n(X, A, x_0) \cong \pi_0(\Omega^{n-1} P(X, A, x_0), *)$. $\qquad$ $\square$

For further details see for example Switzer [**258**, Chapter 3], where the long exact sequence (Proposition 1.2.7) is developed from this point of view.

EXERCISE. The (reduced) **suspension** SX of a space X is defined to be $S^1 \wedge X$. The nth suspension of X is $S^n X = S \cdots S X = S^n \wedge X$. Show that if X is Hausdorff then there is a natural homeomorphism

$$\mathrm{Map}_*(SX, Y) \to \mathrm{Map}_*(X, \Omega Y).$$

Deduce that there is a natural bijection

$$[S^n X; Y] \cong [X; \Omega^n Y].$$

Thus S and Ω are adjoint functors.

1.3. The Hurewicz theorem

For an arbitrary topological space, we have the **singular homology** $H_p(X; R)$ with coefficients in a commutative ring R, defined as the homology of the singular chain complex $\Delta_*(X, A; R) = \Delta_*(X; R)/\Delta_*(A; R)$, where $\Delta_p(X; R) = \Delta_p(X) \otimes R$ is the free R-module on the singular p-simplices. A singular p-simplex in X is a (continuous) map $\Delta^p \to X$ where Δ^p is a standard p-simplex

$$\Delta^p = \{(x_0, \ldots, x_p) \in \mathbb{R}^{p+1} \mid \sum x_i = 1 \text{ and each } x_i \geq 0\}.$$

For more details see for example Spanier [**247**]. Similarly **singular cohomology** $H^p(X; R)$ is the cohomology of the cochain complex

$$\Delta^p(X, A; R) = \Delta^p(X, R)/\Delta^p(A; R),$$

where $\Delta^p(X; R) = \text{Hom}(\Delta_p(X), R)$. In case $R = \mathbb{Z}$, we write $H_p(X)$ and $H^p(X)$.

Recall that if (X, x_0) is a based space then $\pi_n(X, x_0)$ is defined to be the group of homotopy classes of maps from (S^n, s_0) to (X, x_0). If $f : (S^n, s_0) \to (X, x_0)$ is such a map, then we have an induced map in homology

$$f_* : H_n(S^n) \to H_n(X)$$

which only depends on the homotopy class $[f] \in \pi_n(X, x_0)$. Since $H_n(S^n) \cong \mathbb{Z}$, we can define

$$h_n([f]) = f_*(1) \in H_n(X)$$

to obtain a well defined map

$$h_n : \pi_n(X, x_0) \to H_n(X)$$

called the **Hurewicz map**.

Similarly if $[f] \in \pi_n(X, A, x_0)$, then $[f]$ is represented by a map

$$f : (D^n, S^{n-1}, s_0) \to (X, A, x_0).$$

This induces a map

$$f_* : H_n(D^n, S^{n-1}) \to H_n(X, A),$$

and since $H_n(D^n, S^{n-1}) \cong \mathbb{Z}$ we can define

$$h_n([f]) = f_*(1) \in H_n(X, A)$$

to obtain a well defined Hurewicz map

$$h_n : \pi_n(X, A, x_0) \to H_n(X, A).$$

PROPOSITION 1.3.1. *If $n \geq 1$ then $h_n : \pi_n(X, x_0) \to H_n(X)$ is a group homomorphism. If ω is a path in X then $h_n(\omega^*(f)) = h_n(f)$ (see Section 1.2 for notation), so that we have a well defined map $h_n : \pi'_n(X, x_0) \to H_n(X)$.*

If $n \geq 2$ then $h_* : \pi_n(X, A, x_0) \to H_n(X, A)$ is a group homomorphism. If ω is a path in A then $h_n(\omega^*(f)) = h_n(f)$, so that we have an induced map $h_n : \pi'_n(X, A, x_0) \to H_n(X, A)$.

PROOF. See Spanier [**247**, Section 7.4].   □

If we wish to compare the maps h_n for different values of n, we must choose the identifications $H_n(S^n) \cong \mathbb{Z}$ and $H_n(D^n, S^{n-1}) \cong \mathbb{Z}$ in a consistent way, since there are two possible choices for each n. Since $(S^n, s_0) \cong (I^n, \dot{I}^n)$ and $(D^n, S^{n-1}, s_0) \cong (I^n, \dot{I}^n, C^{n-1})$, we may do this inductively as follows.

$$H_{n+1}(I^{n+1}, C^n) \longrightarrow H_{n+1}(I^{n+1}, \dot{I}^{n+1}) \xrightarrow{\partial_*} H_n(\dot{I}^{n+1}, C^n) \longrightarrow H_n(I^{n+1}, C^n)$$

$$\parallel \qquad\qquad\qquad\qquad\qquad \cong \uparrow j_* \qquad\qquad\qquad \parallel$$

$$0 \qquad\qquad\qquad\qquad\qquad H_n(I^n, \dot{I}^n) \qquad\qquad\qquad 0$$

If z_n is our choice of generator for $H_n(I^n, \dot{I}^n)$ then we let $z_{n+1} = \partial_*^{-1} j_*(z_n)$.

With these choices of identifications, we have the following theorem:

THEOREM 1.3.2. *Given a based pair* (X, A, x_0) *we have a commutative diagram*

$$\cdots \longrightarrow \pi_{n+1}(X, A, x_0) \longrightarrow \pi_n(A, x_0) \longrightarrow \pi_n(X, x_0) \longrightarrow \pi_n(X, A, x_0) \longrightarrow \cdots$$
$$\downarrow{h_{n+1}} \qquad\qquad \downarrow{h_n} \qquad\qquad \downarrow{h_n} \qquad\qquad \downarrow{h_n}$$
$$\cdots \longrightarrow H_{n+1}(X, A) \longrightarrow H_n(A) \longrightarrow H_n(X) \longrightarrow H_n(X, A) \longrightarrow \cdots$$

PROOF. See Spanier [**247**, Section 7.4].   □

THEOREM 1.3.3 (Absolute Hurewicz theorem). *Suppose that* $\pi_i(X, x_0) = 0$ *for* $i < n$. *Then* $H_i(X) = 0$ *for* $0 < i < n$, *and* $h_n : \pi'_n(X, x_0) \to H_n(X)$ *is an isomorphism. Note that if* $n > 1$, $\pi'_n(X, x_0) = \pi_n(X, x_0)$, *while* $\pi'_1(X, x_0) = \pi_1(X, x_0)/[\pi_1(X, x_0), \pi_1(X, x_0)]$, *the abelianisation of* $\pi_1(X, x_0)$.

THEOREM 1.3.4 (Relative Hurewicz theorem). *Suppose that* A *is path connected and*

$$\pi_i(X, A, x_0) = 0 \text{ for } i < n.$$

Then $H_i(X, A) = 0$ *for* $i < n$, *and* $h_n : \pi'_n(X, A, x_0) \to H_n(X, A)$ *is an isomorphism.*

PROOF. The above two theorems are proved simultaneously by induction, together with a third theorem called the "homotopy addition theorem". For the details, see Spanier [**247**, Section 7.5].   □

Note that the converse of the Hurewicz theorem is false. It is not hard to construct spaces (X, x_0) with $\pi_i(X, x_0)$ non-zero for infinitely many different $i > 0$, but with $H_i(X) = 0$ for all $i > 0$. However, the following theorem is easy to deduce from the absolute Hurewicz theorem.

THEOREM 1.3.5. *Suppose* $\pi_1(X, x_0) = 0$ *and* $H_i(X) = 0$ *for* $0 < i < n$. *Then* $\pi_i(X, x_0) = 0$ *for* $0 < i < n$, *and* $h_n : \pi_n(X, x_0) \to H_n(X)$ *is an isomorphism.*   □

EXAMPLE. Let $n \geq 2$. Since $H_i(S^n) = 0$ for $0 < i < n$, $H_n(S^n) \cong \mathbb{Z}$, and $\pi_1(S^n, s_0) = 0$, we have $\pi_i(S^n, s_0) = 0$ for $i < n$ and $\pi_n(S^n, s_0) \cong \mathbb{Z}$. One of the hardest problems in algebraic topology, and one which has motivated many developments in the subject, is the problem of determining $\pi_i(S^n, s_0)$ for $i > n$. See for example Ravenel [219] for a detailed survey of this still unsolved problem.

There is also a "local" version of the Hurewicz theorem at a prime.

THEOREM 1.3.6. *Suppose that k is a field (of characteristic zero or p), and that $k \otimes_{\mathbb{Z}} \pi_i(X, x_0) = 0$ for $i < n$. Then $H_i(X; k) = 0$ for $0 < i < n$ and $h_n : k \otimes_{\mathbb{Z}} \pi_i'(X, x_0) \to H_i(X; k)$ is an isomorphism.*

PROOF. See Mosher and Tangora [195] Chapter 10. $\qquad\square$

1.4. The Whitehead theorem

There is a way of turning any map into an inclusion, as follows:

DEFINITION 1.4.1. *If $f : X \to Y$, we define the* **mapping cylinder** M_f *of f as the quotient space*

$$M_f = \frac{(X \times I) \,\dot{\cup}\, Y}{(x, 1) \sim f(x)}$$

(with the quotient topology).

We have maps $i : X \to M_f$ sending $x \mapsto (x, 0)$ and $j : Y \to M_f$ sending $y \mapsto (y)$.

LEMMA 1.4.2. *The inclusion $j : Y \to M_f$ is a homotopy equivalence.*

PROOF. Define a map $\pi : M_f \to Y$ via $(x, t) \mapsto f(x)$, $y \mapsto (y)$. Then $\pi \circ j = \mathrm{id}_Y$, and there is a homotopy $F : j \circ \pi \simeq \mathrm{id}_{M_f}$ given by $F((y), t') = (y)$, $F((x, t), t') = (x, 1 - t'(1 - t))$. $\qquad\square$

Thus we may regard $i : X \to M_f$ as an inclusion which is "f up to homotopy" in the sense of the following diagram:

$$
\begin{array}{ccc}
X & \xrightarrow{\;f\;} & Y \\
\| & & \simeq \downarrow j \\
X & \xrightarrow{\;i\;} & M_f
\end{array}
$$

We define the **mapping cone** C_f to be the space obtained from M_f by identifying the image of i to a single point. Thus for example $\tilde{H}_n(C_f) \cong H_n(M_f, X)$.

Since $j_* : \pi_n(Y, y_0) \cong \pi_n(M_f, y_0) \cong \pi_n(M_f, x_0)$, we may use this construction to obtain long exact sequences in homotopy and homology from any map as follows:

$$\cdots \to \pi_{n+1}(M_f, X, x_0) \to \pi_n(X, x_0) \xrightarrow{\;f_*\;} \pi_n(Y, y_0) \to \pi_n(M_f, X, x_0) \to \cdots$$

$$\cdots \longrightarrow H_{n+1}(M_f, X) \longrightarrow H_n(X) \xrightarrow{f_*} H_n(Y) \longrightarrow H_n(M_f, X) \longrightarrow \cdots$$

The Hurewicz map connects these long exact sequences to make a commutative diagram as in Theorem 1.3.2.

THEOREM 1.4.3 (Whitehead). *Suppose X and Y are path connected, and $f : (X, x_0) \to (Y, y_0)$. Suppose that for some $n \geq 1$,*

(A) $f_* : \pi_i(X, x_0) \to \pi_i(Y, y_0)$ *is* $\begin{cases} iso & i < n \\ epi & i = n. \end{cases}$

Then

(B) $f_* : H_i(X) \to H_i(Y)$ *is* $\begin{cases} iso & i < n \\ epi & i = n. \end{cases}$

Conversely if X and Y are simply connected, or more generally simple spaces, then (B) $\Rightarrow$ (A).

PROOF. By the above long exact sequence, (A) is equivalent to the statement that $\pi_i(M_f, X, x_0) = 0$ for $i < n$, and (B) is equivalent to the statement that $H_i(M_f, X) = 0$ for $i < n$. Thus the theorem follows from the relative Hurewicz theorem.   □

1.5. CW-complexes and cellular homology

For many purposes in algebraic topology one needs to have a "good" category of spaces to work with, because an arbitrary topological space is too badly behaved. For the most part we shall be interested in homotopy properties of spaces, and it turns out that an appropriate category of spaces to work in is that of CW-complexes. These are built up by an inductive procedure of adding on cells, which we now describe.

DEFINITION 1.5.1. *Let D^n denote an n-cell, i.e., a closed unit ball in $\mathbb{R}^n$, and let $S^{n-1} = \dot{D}^n$ be its boundary, an $(n-1)$-sphere. Suppose A is a Hausdorff topological space and $\{f_\alpha : S^{n-1} \to A\}$ is a family of maps* (**attaching maps**). *Let $Z = \bigcup_\alpha D_\alpha^n$ be a disjoint union of n-cells indexed by α. Then $\{f_\alpha\}$ induces a map $f : \dot{Z} \to A$. Let*

$$X = A \cup_f Z = (A \,\dot{\cup}\, Z)/(z \sim f(z) \text{ for } z \in \dot{Z})$$

with the quotient topology. Then we say X is **obtained from A by attaching n-cells**.

Let σ_α^n be the image of D_α^n in X. Then we have a map

$$g_\alpha : (D^n, S^{n-1}) \to (X, A)$$

whose image is σ_α^n. This is called the **characteristic map** *of the cell σ_α^n.*

Note that the space automatically has the *weak topology*: $K \subseteq X$ is closed if and only if $K \cap A$ is closed and $K \cap \sigma_\alpha^n$ is closed for each α.

DEFINITION 1.5.2. *A* **relative CW-complex** *(X, A) is a pair $A \subseteq X$ with A Hausdorff, and a filtration $A = X^{(-1)} \subseteq X^{(0)} \subseteq X^{(1)} \subseteq \cdots \subseteq X$ such that*

(i) $X^{(k)}$ *is obtained from* $X^{(k-1)}$ *by attaching* k-*cells,*

(ii) $X = \bigcup_k X^{(k)}$, *and*

(iii) X *has the* **weak topology**: $K \subseteq X$ *is closed if and only if* $K \cap X^{(k)}$ *is closed in* $X^{(k)}$ *for all* k.

The subspace $X^{(k)} \subseteq X$ *is called the* k-**skeleton** *of* (X, A).

CLOSURE FINITENESS. Note that the boundary of a cell $\dot{\sigma}_\alpha^k$ is *compact*, and is hence contained in the union of A and a *finite number* of the attached cells.

REMARK. "CW" stands for "Closure finite, Weak topology".

DEFINITION 1.5.3. *A* **CW-complex** *is as above with* $A = \emptyset$. *A CW-complex is* **finite** *if it has only finitely many cells, and* **of finite type** *if it has only finitely many cells in each dimension.*

EXAMPLE. A simplicial complex is a CW-complex in which each cell is a simplex and the attaching maps take the boundary of a cell (linearly) to the union of the faces of the simplex. In particular the boundary maps here are injective, which sometimes simplifies arguments. It turns out that every CW-complex is homotopy equivalent to a simplicial complex.

REMARK. According to a theorem of Milnor [**189**], if X is a compact topological space and Y has the homotopy type of a CW-complex then the space of maps from X to Y (with the compact-open topology) has the homotopy type of a CW-complex.

Note that with the usual product topology, the product of two CW-complexes is not necessarily a CW-complex, unless one of them is finite, or both of them are countable (see Milnor [**186**], p. 272). The problem is that the product does not have the weak topology with respect to its finite subcomplexes. To solve this problem, we take products in the category of compactly generated spaces. This is the same as using the weak topology with respect to the products of finite subcomplexes.

CELLULAR CHAINS. In the case of a CW-complex, there is a very efficient way of calculating homology and cohomology, via *cellular chains*. If (X, A) is a relative CW-complex, we define

$$C_p(X, A; R) = H_p(X^p, X^{p-1}; R) = \bigoplus_{p\text{-cells } \alpha} R_\alpha$$

the free R-module on the p-cells. Then the short exact sequence of chain complexes

$$0 \to \Delta_*(X^{(p-1)}, X^{(p-2)}; R) \to \Delta_*(X^{(p)}, X^{(p-2)}; R)$$
$$\to \Delta_*(X^{(p)}, X^{(p-1)}; R) \to 0$$

gives rise to the *boundary map of the triple*

$$\partial : H_p(X^{(p)}, X^{(p-1)}; R) \to H_{p-1}(X^{(p-1)}, X^{(p-2)}; R).$$

It is easy to check that $\partial^2 = 0$, so we use this map

$$\partial \, : \, C_p(X, A; R) \to C_{p-1}(X, A; R)$$

to make a chain complex $(C_*(X, A; R), \partial)$, the *complex of cellular chains*.

Similarly in cohomology the coboundary map

$$\delta \, : \, H^{p-1}(X^{(p-1)}, X^{(p-2)}; R) \to H^p(X^{(p)}, X^{(p-1)}; R)$$

gives us a cochain complex $(C^*(X, A; R), \delta)$. This is the same as the complex obtained by dualising the complex of cellular chains. It is called the *complex of cellular cochains*.

Note that for a CW-complex of finite type, the cellular chains and cellular cochains form complexes of *finitely generated* R-modules.

Since a sphere only has reduced homology in one degree, we have

$$H_q(X^{(p)}, X^{(p-1)}) = 0$$

for $q \neq p$. It is easy to deduce by induction from this that $H_q(X^{(p)}, A) = 0$ for $q > p$, and that $H_q(X^{(p)}, A) = H_q(X, A)$ for $q < p$.

It is worth reading the proof of the following theorem before trying to read the section on spectral sequences, since in some sense the argument is a thinly disguised collapsing spectral sequence argument. We shall reprove the theorem as an easy application of the spectral sequence of a filtered chain complex in Section 3.2.

THEOREM 1.5.4. *We have natural isomorphisms*

$$H_p(X, A; R) \cong H_p(C_*(X, A; R), \partial), \quad H^p(X, A; R) \cong H^p(C^*(X, A; R), \delta).$$

PROOF. We have various long exact sequences which interlock as in the following diagram.

$$
\begin{array}{ccc}
H_{p+1}(X^{(p+1)}, X^{(p)}) & & 0 = H_{p-1}(X^{(p-2)}, A) \\
\downarrow{\scriptstyle \partial'} \quad \searrow^{\partial} & & \downarrow \\
H_p(X^{(p-1)}, A) \longrightarrow H_p(X^{(p)}, A) \longrightarrow H_p(X^{(p)}, X^{(p-1)}) \xrightarrow{i_*} H_{p-1}(X^{(p-1)}, A) \\
\| \qquad\qquad\qquad \downarrow{\scriptstyle j_*} \qquad\qquad\qquad \searrow^{\partial} \qquad\qquad \downarrow \\
0 \qquad\qquad H_p(X^{(p+1)}, A) \cong H_p(X, A) \qquad H_{p-1}(X^{(p-1)}, X^{(p-2)}) \\
\downarrow \\
H_p(X^{(p+1)}, X^{(p)}) = 0
\end{array}
$$

In this diagram, the rows and columns are exact. Thus we have

$$H_p(C_*(X, A; R), \partial) = \mathrm{Ker}(\partial)/\mathrm{Im}(\partial) = \mathrm{Ker}(i_*)/\mathrm{Im}(\partial)$$

$$\cong H_p(X^{(p)}, A)/\mathrm{Im}(\partial') = H_p(X^{(p)}, A)/\mathrm{Ker}(j_*)$$

$$\cong H_p(X^{(p+1)}, A) \cong H_p(X, A).$$

Similarly for cohomology we use the dual of the above argument. $\square$

CELLULAR APPROXIMATION.

DEFINITION 1.5.5. *A map* $f : (X, A) \to (Y, B)$ *of relative CW-complexes is said to be* **cellular** *if it maps the k-skeleton of* (X, A) *into the k-skeleton of* (Y, B) *for each* $k \geq 0$.

THEOREM 1.5.6 (Cellular Approximation Theorem). *Every map*

$$f : (X, A) \to (Y, B)$$

of relative CW-complexes is homotopic (relative to A) to a cellular map.

PROOF. See Spanier [**247**, Section 7.6, Cor. 18]. □

WEAK HOMOTOPY EQUIVALENCES.

DEFINITION 1.5.7. *A map* $f : X \to Y$ *is a* **weak homotopy equivalence** *if*

$$f_* : \pi_n(X, x) \to \pi_n(Y, f(x))$$

is an isomorphism for all $n \geq 0$, *and for all* $x \in X$. *Note that if* X *is path connected then it is sufficient to check this for a single choice of basepoint, by Section 1.2.*

A space X *is* **weakly contractible** *if* $\pi_n(X, x) = 0$ *for all* $n \geq 0$ *and for all* $x \in X$.

It is clear that a homotopy equivalence is a weak homotopy equivalence. We shall show that for CW-complexes, the converse is true.

THEOREM 1.5.8 (Whitehead). *If* X *and* Y *are CW-complexes and* $f : X \to Y$ *is a weak homotopy equivalence, then* f *is a homotopy equivalence. In particular if* X *is a weakly contractible CW-complex then* X *is contractible.*

PROOF. By the cellular approximation theorem, we may assume that f is cellular. This has the effect that the mapping cylinder M_f is a CW-complex. Since f is a weak homotopy equivalence, the long exact sequence

$$\cdots \to \pi_{n+1}(M_f, X, x_0) \to \pi_n(X, x_0) \xrightarrow{f_*} \pi_n(Y, y_0) \to \pi_n(M_f, X, x_0) \to \cdots$$

of Section 1.4 shows that $\pi_n(M_f, X, x_0) = 0$ for all $n \geq 0$.

Now suppose (K, L) is a relative CW-complex and $g : (K, L) \to (M_f, X)$ is a map. By induction, it is clear that the restriction of g to the n-skeleton of (K, L) is homotopic relative to L to a map sending $K^{(n)}$ into X. Namely, once the $(n-1)$-skeleton is in, the obstruction to pushing the n-skeleton in lies in $\pi_n(M_f, X, x_0) = 0$. We may fit these homotopies together by doing the first skeleton in the interval $[0, \frac{1}{2}]$, the second in $[\frac{1}{2}, \frac{3}{4}]$, and so on, to show that g is homotopic relative to L to a map sending K into X.

Applying the above statement to the identity map from (M_f, X) to itself, we see that the inclusion $X \subseteq M_f$ is a homotopy equivalence. Hence by the discussion in Section 1.4, so is f. □

1.6. Fibrations and fibre bundles

A fibre bundle is a sort of twisted product of one space by another, like for example a Möbius band. A fibration is a homotopy theoretic generalisation of this concept. The main homotopy theoretic properties of fibrations are the long exact sequence in homotopy and the spectral sequence in homology.

DEFINITION 1.6.1. *A map $p : E \to B$ has the* **homotopy lifting property** *(HLP) with respect to a space X if given the following commutative diagram of spaces, a map H can always be found.*

$$
\begin{array}{ccc}
X \times 0 & \xrightarrow{\ f\ } & E \\
\downarrow & {\scriptstyle H}\nearrow & \downarrow {\scriptstyle p} \\
X \times I & \xrightarrow{\ h\ } & B
\end{array}
$$

REMARK. You should think of this as saying that a lift f of one end of the homotopy h can be extended to a lift H of the entire homotopy.

DEFINITION 1.6.2. *A map $p : E \to B$ is a* **Hurewicz fibration** *if it has HLP with respect to all spaces X. It is a* **Serre fibration** *if it has HLP with respect to all simplicial complexes.*

REMARKS. (i) The reason why it is convenient to work with the larger class of Serre fibrations rather than Hurewicz fibrations is that in general fibre bundles are only Serre fibrations, and this condition is strong enough for homotopy theoretic constructions. Note however that fibre bundles with B paracompact are known to be Hurewicz fibrations. The proof is quite involved, and is due to Hurewicz.

(ii) Note that taking X to be a single point in the definition shows that the image of p is a union of path components of B.

(iii) The space B is called the **base space** of the fibration, while E is called the **total space**. The inverse image in E of a point in B is called a **fibre** and is usually denoted F. We shall see that all fibres in a path component of B are homotopy equivalent.

The definition of fibration is easier to use in a slightly stronger form, as given in the next lemma.

DEFINITION 1.6.3. *A subspace $A \subseteq X$ is called a* **strong deformation retract** *of X if there is a homotopy $h : X \times I \to X$ with $h(x,0) = x$, $h(a,t) = a$ and $h(x,1) \in A$ for all $x \in X$, $a \in A$ and $t \in I$.*

LEMMA 1.6.4. *If (X, A) is a simplicial pair with A a strong deformation retract of X and $p : E \to B$ is a Serre fibration then we can lift maps as in the following diagram:*

$$
\begin{array}{ccc}
A & \longrightarrow & E \\
\downarrow & \nearrow & \downarrow \\
X & \longrightarrow & B
\end{array}
$$

PROOF. This follows from the diagram

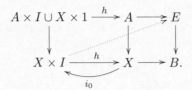

The following is one of the important properties of fibrations:

THEOREM 1.6.5 (Fibring property). *Suppose $p : E \to B$ is a Serre fibration with basepoints $p(e_0) = b_0$. If $B' \subseteq B$ and $E' = p^{-1}(B')$, then for $n \geq 1$, $p_* : \pi_n(E, E', e_0) \to \pi_n(B, B', b_0)$ is an isomorphism.*

PROOF. (i) p_* is onto:
Elements of π_n lift as in the following diagram, in which $p_*(\gamma) = \alpha$:

(ii) p_* is one–one:
If $h : p_*(\alpha) \simeq p_*(\beta)$ is a homotopy then we may lift it to a homotopy $H : \alpha \simeq \beta$ as in the following diagram:

$$(I^n \times 0) \cup (C^{n-1} \times I) \cup (I^n \times 1) \xrightarrow{\alpha \cup (e_0) \cup \beta} E$$

$$I^n \times I \xrightarrow{\quad h \quad} B$$

(with diagonal map H)

where (e_0) is the map sending the whole of $C^{n-1} \times I$ to the basepoint e_0. □

In the particular case where $B' = \{b_0\}$ we have $E' = F = p^{-1}(b_0)$ and

$$p_* : \pi_n(E, F, e_0) \xrightarrow{\cong} \pi_n(B, b_0, b_0) = \pi_n(B, b_0).$$

Thus the long exact sequence (Proposition 1.2.7) of the pair (E, F) and Proposition 1.2.8 give us the following theorem.

THEOREM 1.6.6. *If $F \xrightarrow{i} E \xrightarrow{p} B$ is a Serre fibration then there is a long exact sequence of homotopy groups*

$$\cdots \to \pi_n(F, e_0) \xrightarrow{i_*} \pi_n(E, e_0) \xrightarrow{p_*} \pi_n(B, b_0) \xrightarrow{d_*} \pi_{n-1}(F, e_0) \to \cdots$$

where the map d_* is given by the composite

$$\pi_n(B, b_0) \xrightarrow{p_*^{-1}} \pi_n(E, F, e_0) \xrightarrow{\partial_*} \pi_{n-1}(F, e_0).$$

The above is a long exact sequence of modules for $\pi_1(F, e_0)$. □

Note that in homology, one does not obtain a long exact sequence, but rather a spectral sequence. This is the subject of Section 3.3.

PROPOSITION 1.6.7. *Suppose $p : E \to B$ is a fibration, and $\omega : I \to B$ is a path in B. Then the fibres $p^{-1}(\omega(0))$ and $p^{-1}(\omega(1))$ at the two ends of the paths are homotopy equivalent by a map whose homotopy class only depends on the homotopy class of ω (relative to its ends).*

PROOF. Let $F = p^{-1}(\omega(0))$ and $F' = p^{-1}(\omega(1))$ be the fibres. Then the diagram

$$
\begin{array}{ccc}
F \times 0 & \xrightarrow{} & E \\
\downarrow & {\scriptstyle \alpha} \nearrow & \downarrow \\
F \times I & \longrightarrow I \xrightarrow{\;\omega\;} & B
\end{array}
$$

gives us a map $F \cong F \times 1 \xrightarrow{\alpha_1} F'$, and the diagram

$$
\begin{array}{ccc}
F \times (I \times 0 \cup 0 \times I \cup I \times 1) & \xrightarrow{\;\alpha \cup \pi_F \cup \alpha'\;} & E \\
\downarrow & & \downarrow \\
F \times I \times I & \longrightarrow I \times I \xrightarrow{\;\omega \simeq \omega'\;} & B
\end{array}
$$

shows that any two maps α, α' obtained in this way from homotopic paths ω, ω' are homotopic. If we do the same for the opposite path $\bar{\omega}(t) = \omega(1-t)$ we obtain a map $F' \to F$, again well defined up to homotopy. Since $\omega * \bar{\omega}$ is homotopic to the identity path, the composite of these maps $F \to F' \to F$ is homotopic to the identity map. The same applies to $F' \to F \to F'$, and so $F \simeq F'$. □

Note that the proof of the above proposition gives us an action of the fundamental groupoid of B on the fibres, up to homotopy. In particular, for example, it shows that $\pi_1(B, b_0)$ acts on homotopy invariants such as the homology and cohomology groups of F.

DEFINITION 1.6.8. *A* **fibre bundle** *with base B, total space E and fibre F consists of a surjective map $p : E \to B$ such that there is an open covering $\{U_\alpha\}$ of B and homeomorphisms ϕ_α making the following square commute:*

$$
\begin{array}{ccc}
U_\alpha \times F & \xrightarrow{\;\phi_\alpha\;} & p^{-1}(U_\alpha) \\
{\scriptstyle \pi_1} \downarrow & & \downarrow {\scriptstyle p} \\
U_\alpha & =\!\!=\!\!=\!\!= & U_\alpha
\end{array}
$$

In other words, a fibre bundle is locally like a product of B and F, but may be globally twisted. Examples are given by the Möbius band, and by tangent and cotangent bundles on a smooth manifold.

If $\xi = (p : E \to B)$ is a fibre bundle, we write $E(\xi)$ for the total space E, and $B(\xi)$ for the base space B.

LEMMA 1.6.9. *If $E = B \times F$, $p = \pi_B : E \to B$ then $p : E \to B$ has HLP with respect to all X. Thus $p : E \to B$ is a Hurewicz fibration.*

PROOF.

$$
\begin{array}{ccc}
X \times 0 & \xrightarrow{\;f\;} & B \times F \\
\pi_X \big\uparrow \big\downarrow i & {\scriptstyle H} \nearrow & \big\downarrow \pi_B \\
X \times I & \xrightarrow{\;h\;} & B
\end{array}
$$

Take $H = h \times (\pi_F \circ f \circ \pi_X)$.   □

In order to show that fibre bundles are Serre fibrations, we need the following lemma.

LEMMA 1.6.10. *If $p : E \to B$ has the homotopy lifting property with respect to all n-cubes I^n, then $p : E \to B$ has the **homotopy extension lifting property** (HELP) with respect to simplicial pairs (X, A). In other words, partial lifts of homotopies can be extended to lifts as in the following diagram.*

$$
\begin{array}{ccc}
X \times 0 \cup A \times I & \xrightarrow{\;f\;} & E \\
\big\downarrow & {\scriptstyle H} \nearrow & \big\downarrow \\
X \times I & \xrightarrow{\;h\;} & B
\end{array}
$$

In particular, $p : E \to B$ is a Serre fibration.

PROOF. We proceed by induction on the skeleta of X, the zero skeleton being easy to handle. Suppose we have extended the lift to $X \times 0 \cup X^{(n-1)} \times I$ and suppose Δ is an n-simplex of X, so that the boundary $\dot{\Delta} \subseteq X^{(n-1)}$. Then we can extend h to $X \times 0 \cup (X^{(n-1)} \cup \Delta) \times I$ as in the following diagram:

$$
\begin{array}{ccc}
\Delta \times 0 \cup \dot{\Delta} \times I & \xrightarrow{\;g\;} & E \\
\big\downarrow & {\scriptstyle H} \nearrow & \big\downarrow \\
\Delta \times I & \xrightarrow{\;h\;} & B
\end{array}
$$

Fitting together these extensions for the various n-simplices of X, we have extended to $X \times 0 \cup X^{(n)} \times I$. By the weak topology we may continue and extend to $X \times I$.   □

THEOREM 1.6.11. *If $f : E \to B$ is a fibre bundle then it is a Serre fibration.*

PROOF. By Lemma 1.6.10, we only need show that p has HLP for the n-cube I^n.

$$
\begin{array}{ccc}
I^n \times 0 & \xrightarrow{\;g\;} & E \\
\big\downarrow & {\scriptstyle H} \nearrow & \big\downarrow \\
I^n \times I & \xrightarrow{\;h\;} & B
\end{array}
$$

Let $\{U_\alpha\}$ be as in the definition of fibre bundle. Then $\{h^{-1}(U_\alpha)\}$ is an open covering of $I^n \times I$, and so by compactness we can choose $\varepsilon > 0$ such that the image under h of every ball of diameter less than ε in $I^n \times I$ lies in some U_α (ε is called the **Lebesgue covering number** of this open covering). Cut $I^n \times I$ into subcubes $I_{i_1,\ldots,i_n} \times I_j$ each small enough to fit into a ball of radius ε. We order these subcubes lexicographically and define H on them by induction. At each stage, we need to know that products $U_\alpha \times F$ have HELP with respect to pairs of finite simplicial complexes made up out of cubes. This follows from the above two lemmas. $\square$

COVERING SPACES.

DEFINITION 1.6.12. *If $p : E \to B$ is a fibre bundle in which the fibres are discrete sets, then E is called a* **covering space** *of B. By Proposition 1.6.7, there is an action of $\pi_1(B, b_0)$ on $F = p^{-1}(b_0)$. If E and B are path connected then the homotopy sequence*

$$(0 =)\pi_n(F, e_0) \to \pi_n(E, e_0) \to \pi_n(B, b_0) \to \pi_{n-1}(F, e_0)(= 0) \qquad (n \geq 2)$$

$$(0 =)\pi_1(F, e_0) \to \pi_1(E, e_0) \to \pi_1(B, b_0) \to \pi_0(F, e_0)(= F) \to \pi_0(E, e_0)(= 0)$$

shows that $\pi_n(E, e_0) \cong \pi_n(B, b_0)$ for $n \geq 2$, $\pi_1(B, b_0)$ acts transitively on F, and $\pi_1(E, e_0)$ is the stabiliser of the point $e_0 \in F$. If E and B are path connected and $\pi_1(E, e_0) = 0$, then E is called the **universal cover** *of B. Under these conditions F is the regular permutation representation of $\pi_1(B, b_0)$.*

THEOREM 1.6.13. *Suppose B is path connected, and every neighbourhood of a point in B contains a simply connected neighbourhood. Then B has a universal cover $p : (E, e_0) \to (B, b_0)$. In particular, every connected CW-complex has a universal cover.*

If $p' : (E', e_0') \to (B, b_0)$ is any other cover then there is a unique map $(E, e_0) \to (E', e_0')$ whose composite with p' gives p. In particular, the universal cover is unique up to isomorphisms of fibre bundles over B.

There is a one–one correspondence between isomorphism classes of path connected covering spaces $p' : (E', e_0') \to (B, b_0)$ and conjugacy classes of subgroups $H \leq \pi_1(B, b_0)$, given by $(E', e_0') \cong (E, e_0)/H$.

PROOF. See Spanier [247, Sec. 2.5], where this theorem is proved under slightly weaker hypotheses. The idea of the proof is as follows. We construct E as the set of ordered pairs $(x, [\omega])$ consisting of a point x in B and a homotopy class $[\omega]$ of paths from b_0 to x. This set is topologised by taking as a base for the topology sets of the following form. If U is open in B, and $[\omega]$ is a path from b_0 to $x \in U$, let $\langle U, \omega \rangle$ be the set consisting of the pairs $(u, [\omega * \omega'])$ where ω' is a path inside U. It is easy to check that this gives a fibre bundle, and hence a Serre fibration.

If $p' : (E', e_0') \to (B, b_0)$ is another cover, then we construct a map $(E, e_0) \to (E', e_0')$ using the homotopy lifting property with respect to the unit interval. Such a lift is, of course, unique, since otherwise at the point of bifurcation there is no neighbourhood in B whose pre-image in E' is a product.

If E' is path connected then the map $(E, e_0) \to (E', e_0')$ is surjective. Letting $H = \text{Im}(\pi_1(E', e_0') \to \pi_1(B, b_0))$, the long exact homotopy sequence shows that H is the stabiliser of a point F' and so $(E', e_0') \cong (E, e_0)/H$. $\square$

EXAMPLES. (i) The universal cover of the circle S^1 is the real line $\mathbb{R}$. The pre-image of the basepoint is $\pi_1(S^1, s_0) \cong \mathbb{Z}$.

(ii) The universal cover of the projective space $\mathbb{R}P^n$ is the sphere S^n. The pre-image of the basepoint is $\pi_1(\mathbb{R}P^n, x_0) \cong \mathbb{Z}/2$.

(iii) The universal cover of the torus T^2 is the plane $\mathbb{R}^2$. The pre-image of the basepoint is $\pi_1(T^2, x_0) \cong \mathbb{Z}^2$.

EXERCISE: THE PATH FIBRATION. If X is a based space, denote by PX the path space $\text{Map}_*((I, 0), (X, x_0))$ of paths in X starting at the basepoint x_0. Show that the map $PX \to X$ sending a path $\omega : (I, 0) \to (X, x_0)$ to $\omega(1)$ is a Hurewicz fibration, with fibre ΩX. Show that PX is contractible.

REPLACING A MAP BY A FIBRATION. If $f : Y \to X$ is a map of based spaces, denote by P_f the space of ordered pairs (y, ω) consisting of a point y in Y and an unbased map $\omega : I \to X$ with $\omega(1) = f(y)$; in other words, the pullback

$$
\begin{array}{ccc}
P_f & \longrightarrow & X^I \\
\downarrow & & \downarrow \\
Y & \longrightarrow & X.
\end{array}
$$

Show that P_f is homotopy equivalent to Y via the first projection $(y, \omega) \mapsto y$. Show that the composite map $P_f \to Y \to X$ is homotopic to the map $P_f \to X$ sending (y, ω) to $\omega(0)$, and that the latter map is a Hurewicz fibration. In this way we have replaced the original map $Y \to X$ by the fibration $P_f \to X$. The fibre F_f of the map $P_f \to X$ is called the **homotopy fibre** of f, and consists of ordered pairs (y, ω) as above with $\omega(1) = f(y)$ and $\omega(0) = x_0$; in other words, the pullback

$$
\begin{array}{ccc}
F_f & \longrightarrow & PX \\
\downarrow & & \downarrow \\
Y & \longrightarrow & X.
\end{array}
$$

1.7. Paracompact spaces

In the next chapter, we shall be discussing a particular class of fibre bundles, called principal G-bundles, where G is a topological group. We shall classify principal G-bundles over a paracompact base space. In preparation for this, we include here a brief discussion of paracompact spaces.

DEFINITION 1.7.1. *Suppose $\{U_\alpha\}$ is an open cover of a space X. A (locally finite)* **partition of unity** *subordinate to $\{U_\alpha\}$ consists of functions $f_\alpha : X \to \mathbb{R}$ such that*

(i) $f_\alpha(x) \geq 0$ *for all α and x.*

(ii) $f_\alpha(x) = 0$ *if $x \notin U_\alpha$.*

(iii) *For each $x \in X$, $\{\alpha \mid f_\alpha(x) \neq 0\}$ is a finite set.*

(iv) *For each $x \in X$, $\sum_\alpha f_\alpha(x) = 1$.*

If $\{U_\alpha\}$ and $\{V_\beta\}$ are two open covers of X, we say $\{U_\alpha\}$ is a **refinement** *of $\{V_\beta\}$ if each U_α is contained in some V_β.*

An open cover $\{U_\alpha\}$ is said to be **locally finite** *if every point has a neighbourhood which intersects only a finite number of the U_α non-trivially.*

A Hausdorff space X is **paracompact** *if every open cover has a locally finite refinement.*

EXAMPLE. It is shown in Lundell and Weingram [**168**, p. 54–55] that every CW-complex is paracompact.

The main facts we shall need about paracompact spaces are contained in the following two theorems.

THEOREM 1.7.2. *A Hausdorff space X is paracompact if and only if there is a partition of unity subordinate to any given open cover of X.*

PROOF. This is proved in Appendix 1 of Lundell and Weingram [**168**]. In fact they use the above property as the definition of a paracompact space, and show that it is equivalent to the definition given above.   □

THEOREM 1.7.3. *Suppose X is a paracompact space and $\{U_\alpha\}$ is an open cover of X. Then there is a countable open cover $\{W_j\}$ of X such that each W_j is a disjoint union of open sets, each of which is contained in some U_α.*

If each $x \in X$ is in at most n of the U_α then we may choose a finite open cover $\{W_j, \ 1 \leq j \leq n\}$ with the above property.

PROOF. Let $\{f_\alpha\}$ be a partition of unity subordinate to $\{U_\alpha\}$. For each $x \in X$, let $S(x)$ be the set of indices α such that $f_\alpha(x) \neq 0$ (so that $S(x)$ is finite for all $x \in X$). If S is a subset of the set of indices α, let

$$W(S) = \{x \in X \mid f_\alpha(x) > f_\beta(x) \text{ for all } \alpha \in S, \beta \notin S\}.$$

It is clear that $W(S)$ is open in X, and that if $S \not\subseteq S'$, $S' \not\subseteq S$ then $W(S) \cap W(S') = \emptyset$. Thus if we let

$$W_j = \bigcup_{|S(x)|=j} W(S(x))$$

then W_j is a disjoint union of sets $W(S(x))$, and each $W(S(x)) \subseteq U_\alpha$ whenever $\alpha \in S(x)$.

Under the last hypothesis, $W_j = \emptyset$ for $j > n$.   □

COROLLARY 1.7.4. *Suppose $p : E \to B$ is a fibre bundle with paracompact base space B. Then there is a countable open cover $\{U_j\}$ of B such that $p : p^{-1}(U_j) \to U_j$ is a product bundle for each U_j.*   □

1.8. Simplicial sets

In this book, I have tried wherever possible to avoid the language of simplicial sets, in order to make the discussion as accessible as possible to a wide audience. However, at times they are indispensable; and at times discussions are greatly clarified when described in this language. This will be especially true when we discuss cyclic homology in Section 2.12.

The idea of the theory of simplicial sets is that homotopy theory really should not depend on properties of point set topology, and ought to admit an algebraic, or at least combinatorial description. The appropriate combinatorial framework was first formulated by Eilenberg and Zilber [99] and developed by D. M. Kan [143, 145]; other good references are Bousfield and Kan [50], May [180], Lamotke [163] and Section VIII.5 of MacLane [170]. We shall content ourselves with a brief summary of the salient features of the theory, and the reader is referred to these references for further details.

Let $\triangle$ be the category whose objects are the finite totally ordered sets, and whose arrows are the monotonic functions. A **simplicial object** X in a category $\mathcal{C}$ is a contravariant functor $\triangle \to \mathcal{C}$. A **simplicial set** is a simplicial object in the category of sets. If X is a simplicial object, we write X_n for the image of $\{0,\ldots,n\}$ in $\mathcal{C}$ (the "set of n-simplices"), $d_i : X_n \to X_{n-1}$ $(0 \le i \le n)$ for the image of the ith "face map"

$$\{0,\ldots,n-1\} \to \{0,\ldots,n\}$$

$$j \mapsto \begin{cases} j & \text{if } j < i \\ j+1 & \text{if } j \ge i \end{cases}$$

and $s_i : X_n \to X_{n+1}$ $(0 \le i \le n)$ for the image of the ith "degeneracy map"

$$\{0,\ldots,n+1\} \to \{0,\ldots,n\}$$

$$j \mapsto \begin{cases} j & \text{if } j \le i \\ j-1 & \text{if } j > i \end{cases}$$

These maps satisfy the relations

$$\begin{aligned}
d_i d_j &= d_{j-1} d_i & i < j \\
d_i s_j &= s_{j-1} d_i & i < j \\
d_i s_j &= 1 & i = j \text{ or } j+1 \\
d_i s_j &= s_j d_{i-1} & i > j+1 \\
s_i s_j &= s_{j+1} s_i & i \le j
\end{aligned}$$

and all maps and relations coming from $\triangle$ follow from these. So giving a simplicial object X in $\mathcal{C}$ is the same as giving objects X_n in $\mathcal{C}$, $0 \le n < \infty$, and maps $d_i : X_n \to X_{n-1}$ and $s_i : X_n \to X_{n+1}$ satisfying the above relations.

A **simplicial map** $f : X \to Y$ between simplicial objects is a natural transformation of functors; this amounts to giving maps $f_n : X_n \to Y_n$ satisfying $f_{n-1} d_i = d_i f_n$ and $f_{n+1} s_i = s_i f_n$ for all $0 \le i \le n$. Thus simplicial objects in a category $\mathcal{C}$ form a category **Simp** $\mathcal{C}$.

If X is a simplicial set, an n-simplex $x \in X_n$ is said to be **degenerate** if $x = s_i y$ for some $y \in X_{n-1}$ and some i. The remaining simplices are said to be **non-degenerate**. Note that because of the above relations, the maps s_i are injective, so that there are in general lots of degenerate simplices.

The relationship with topological simplices is as follows. We denote by Δ^n the standard n-simplex

$$\Delta^n = \{(x_0, \dots, x_n) \in \mathbb{R}^{n+1} \mid \sum x_i = 1 \text{ and each } x_i \geq 0\}.$$

The maps $\mathbf{d}^i : \Delta^{n-1} \to \Delta^n$ and $\mathbf{s}^i : \Delta^{n+1} \to \Delta^n$ are defined by

$$\mathbf{d}^i(x_0, \dots, x_{n-1}) = (x_0, \dots, x_i, 0, x_{i+1}, \dots, x_{n-1})$$

$$\mathbf{s}^i(x_0, \dots, x_{n+1}) = (x_0, \dots, x_i + x_{i+1}, \dots, x_{n+1}).$$

These satisfy the above relations given for the d_i and s_i, but with the order of composition reversed. If X is a simplicial set, we define its **topological realisation** $|X|$ to be the quotient of

$$\bigsqcup_{n \geq 0} X_n \times \Delta^n$$

by the equivalence relation given by identifying $(d_i x, y)$ with $(x, \mathbf{d}^i y)$ and $(s_i x, y)$ with $(x, \mathbf{s}^i y)$. This gives a CW-complex (not in general a simplicial complex since faces of a simplex may get identified) with one n-cell for each non-degenerate n-simplex in X. The topological standard simplex itself is the topological realisation of the simplicial set $\Delta[n]$ whose r-simplices are the sequences of integers $0 \leq a_0 \leq \dots \leq a_r \leq n$ with

$$d_i(a_0, \dots, a_r) = (a_0, \dots, a_{i-1}, a_{i+1}, \dots, a_r)$$

$$s_i(a_0, \dots, a_r) = (a_0, \dots, a_i, a_i, \dots, a_r).$$

The maps $\mathbf{d}^i$ and $\mathbf{s}^i$ correspond to obvious maps $\mathbf{d}^i : \Delta[n-1] \to \Delta[n]$ and $\mathbf{s}^i : \Delta[n+1] \to \Delta[n]$. There is also the obvious corresponding notion of the topological realisation of a map of simplicial sets, so that topological realisation is a functor from simplicial sets to CW-complexes.

If Y is a topological space, then the **singular simplices** on Y form a simplicial set $\mathrm{Sing}(Y)$ with $\mathrm{Sing}(Y)_n = \mathrm{Map}(\Delta^n, Y)$ and with maps d_i and s_i induced by the maps $\mathbf{d}^i$ and $\mathbf{s}^i$. There is an obvious adjunction between topological realisation and singular simplices

$$\mathrm{Map}(|X|, Y) \overset{\mathrm{nat}}{\cong} \mathrm{Hom}_{\mathbf{SimpSet}}(X, \mathrm{Sing}(Y)).$$

If X and Y are simplicial sets then $X \times Y$ is the simplicial set with $(X \times Y)_n = X_n \times Y_n$ with the obvious face and degeneracy maps. This definition takes a bit of getting used to, because a product of degenerate simplices may be non-degenerate. This is related to the topological product by

$$|X \times Y| = |X| \times |Y|,$$

where the product on the right is taken in the category of compactly generated spaces (the usual topology of the product is wrong for infinite dimensional spaces). For example, taking $X = Y = \Delta[1]$ (the "simplicial unit
interval"), $X \times Y$ has four zero-simplices $(0,0)$, $(0,1)$, $(1,0)$, $(1,1)$, five non-
degenerate one-simplices $(00,01)$, $(01,00)$, $(01,01)$, $(01,11)$ and $(11,01)$, and
two non-degenerate two-simplices $(001,011)$ and $(011,001)$. Thus $|X \times Y|$ is
a unit square, triangulated by cutting into two triangles using the diagonal.
A proof of the fact that products commute with topological realisation can
be found in Chapter III of May [180].

The (reduced) **suspension** SX of a pointed simplicial set (X, x_0) is the
simplicial set with one zero-simplex $*$, and whose n-simplices for $n > 0$ are
the pairs (i, x), $i \geq 1$, $x \in X_{n-i}$, modulo the equivalence relation $(i, s_0^n x_0) \sim
s_0^{n+i} *$. The face and degeneracy maps are determined by the relations

$$d_0(1, x) = s_0^{n-1} * \qquad s_0(i, x) = (i + 1, x)$$
$$d_{i+1}(1, x) = (1, d_i x) \qquad s_{i+1}(1, x) = (1, s_i x).$$

It is easy to check that this is related to the usual reduced suspension (Sec-
tion 1.2) by $|SX| = S|X|$.

The simplicial version of the space of maps from one space to another goes
as follows. If X and Y are simplicial sets, then $\mathrm{Hom}(X, Y)$ is the simplicial
set in which an n-simplex is a simplicial map $f : X \times \Delta[n] \to Y$, with
face and degeneracy maps given by composition $d_i(f) = f \circ (\mathrm{id}_X \times \mathbf{d}^i)$ and
$s_i(f) = f \circ (\mathrm{id}_X \times \mathbf{s}^i)$. The realisation of an n-simplex $f : X \times \Delta[n] \to Y$
gives a map $|X| \times \Delta^n \to |Y|$ and hence a map $\Delta^n \to \mathrm{Map}(|X|, |Y|)$. Putting
these together gives a map

$$|\mathrm{Hom}(X, Y)| \to \mathrm{Map}(|X|, |Y|)$$

which is a homotopy equivalence.

The notion of a fibration in **SimpSet** is similar to that of a Serre fibration in the topological category. Let $\Delta[n, i]$ be the simplicial subset of
$\Delta[n]$ generated by all except the ith face; in other words, if σ is the unique
non-degenerate n-simplex in $\Delta[n]$ then $\Delta[n, i]$ is generated by $d_1 \sigma, \ldots, d_{i-1}\sigma$,
$d_{i+1}\sigma, \ldots, d_n \sigma$. Thus the topological realisation $\Delta^{n,i}$ of $\Delta[n, i]$ is the subset
of Δ^n consisting of those $(x_0, \ldots, x_n)$ such that $x_j = 0$ for some $j \neq i$. One
can show that a map of topological spaces is a Serre fibration if and only if
it has the lifting property with respect to the pair $(\Delta^n, \Delta^{n,i})$. So we define a
map $E \to B$ of simplicial sets to be a **Kan fibration** if the diagonal arrow
can always be filled in, as in the following diagram:

A number of remarks need to be made at this point. First of all, in contrast to the situation in topology, if we denote by $*$ the simplicial set with one n-simplex for each n (so that the topological realisation of $*$ is a single point), given a simplicial set X, the (unique) map $X \to *$ is not necessarily a Kan fibration. A simplicial set is called a **Kan complex** if $X \to *$ is a fibration. However, if X is (the underlying simplicial set of) a simplicial group, then X is necessarily a Kan complex (see for example May [**180**], Theorem 17.1). If $E \to B$ is a Serre fibration of topological spaces, then $\mathrm{Sing}(E) \to \mathrm{Sing}(B)$ is a Kan fibration, as can easily be seen using the adjunction between singular simplices and topological realisation. Thus for any topological space Y, $\mathrm{Sing}(Y)$ is a Kan complex. In particular, the functor $X \mapsto \mathrm{Sing}(|X|)$ provides a canonical way of turning simplicial sets into Kan complexes. There is also a combinatorial construction of such a functor $X \mapsto Ex^{\infty}(X)$ constructed by Kan [**144**].

If $f, f' : X \to Y$ are simplicial maps from one simplicial set to another, then a homotopy from f to f' is a simplicial map $F : X \times \Delta[1] \to Y$ with the property that $F \circ (\mathrm{id}_X \times d_0) = f$ and $F \circ (\mathrm{id}_X \times d_1) = f'$. As long as Y is a Kan complex, homotopy is an equivalence relation on maps from X to Y. The topological realisation of a homotopy between simplicial maps is a homotopy between maps of CW-complexes. We write $[X; Y]$ for the homotopy classes of maps from X to Y when Y is a Kan complex. There are also obvious notions of homotopy classes of pointed maps $[X, x_0; Y, y_0]$, where x_0 and y_0 are zero-simplices, and of maps between pairs $[X, A, x_0; Y, B, y_0]$.

THEOREM 1.8.1. *The adjunction between the singular simplices functor and the topological realisation functor passes down to an equivalence of categories between CW-complexes and homotopy classes of maps, and Kan complexes and homotopy classes of simplicial maps.*

PROOF. See May [**180**] Chapter III §16, or Bousfield and Kan [**50**] Chapter VIII §4.   □

This theorem says that we can do homotopy theory in the category of simplicial sets, without reference to topological spaces. For example, we give the simplicial definition of homotopy groups.

The simplicial n-sphere $S[n]$ is the unique simplicial set with one non-degenerate zero-simplex $*$, one non-degenerate n-simplex, and no other non-degenerate simplices. Thus $|S[n]| = S^n$ (the n-sphere is obtained from Δ^n by identifying the boundary to a point). So if X is a Kan complex, we define $\pi_n(X, x_0) = [S[n], *; X, x_0]$. Now a map from $(S[n], *)$ to (X, x_0) consists of an n-simplex in X with the property that each face map gives the fully degenerate $(n-1)$-simplex $s_0^{n-1}(x_0)$. Therefore $\pi_n(X, x_0)$ consists of equivalence classes of certain n-simplices in X under an equivalence relation which the reader is invited to write down explicitly in terms of the existence of an $(n+1)$-simplex with certain properties.

Finally, we discuss the homology of a simplicial set with coefficients in a commutative ring R. If X is a simplicial set, regarded as a contravariant functor $\triangle \to \textbf{Set}$, then we may compose with the free module functor $\textbf{Set} \to {}_R\textbf{Mod}$ to get a simplicial R-module RX. The chain complex of a simplicial R-module M is the chain complex of R-modules with M_n in degree n and boundary map $\partial_n = \sum_{i=0}^{n}(-1)^i d_i : M_n \to M_{n-1}$. Note that the subcomplex of degenerate simplices is contractible, since the degeneracies can be interpreted as contracting homotopies (see Mac Lane [**170**]), so one may as well only use the non-degenerate simplices in forming this chain complex. It is easy to see that this reduced complex is just the complex of cellular chains on $|X|$ regarded as a CW-complex, so we use the notation $C_*(X; R)$. The homology groups $H_n(X; R)$ are defined to be the homology groups of $C_*(X; R)$, and the cohomology groups $H^n(X; R)$ are defined to be the cohomology groups of $C^*(X; R) = \text{Hom}_R(C_*(X; R), R)$. If Y is a topological space, then the above construction on $\text{Sing}(Y)$ yields the usual singular homology groups $H_n(Y; R)$.

The **Eilenberg–Zilber Theorem** says that there is a natural chain homotopy equivalence

$$C_*(X; R) \otimes C_*(Y; R) \to C_*(X \times Y; R).$$

given as follows. If $p + q = n$ and $x \in X_p$, $y \in Y_p$, then

$$x \otimes y \mapsto \sum_{(\mu,\nu)} (-1)^{\varepsilon(\mu,\nu)} (s_{\nu_q} \ldots s_{\nu_1} x, s_{\mu_p} \ldots s_{\mu_1} y).$$

Here, the sum runs over the **Eilenberg–Mac Lane** (p, q)-**shuffles** (μ, ν) of n. Namely, $\mu = (\mu_1, \ldots, \mu_p)$ and $\nu = (\nu_1, \ldots, \nu_q)$ with $\mu_1 < \cdots < \mu_p$ and $\nu_1 < \cdots < \nu_q$, in such a way that $\{\mu_1, \ldots, \mu_p, \nu_1, \ldots, \nu_q\}$ is a permutation of $\{1, \ldots, n\}$, and the sign of the permutation is $\varepsilon(\mu, \nu)$. For further details, see Mac Lane [**170**], Section VIII.8.

EXERCISE. If $\mathcal{C}$ is a category, the **nerve** of $\mathcal{C}$ is the simplicial set $N\mathcal{C}$ whose n-simplices are the chains of n consecutive maps $\bullet \to \bullet \to \cdots \to \bullet$ in $\mathcal{C}$. Write down the face and degeneracy maps, and check that this is a simplicial set. The **classifying space** $B\mathcal{C}$ of $\mathcal{C}$ is the topological realisation of the nerve $|N\mathcal{C}|$.

If X is a CW-complex, let $\mathcal{C}(X)$ be the category formed from the partially ordered set of singular simplices on X. Show that $B\mathcal{C}(X)$ is the "barycentric subdivision" of $|\text{Sing}(X)|$ (cf. Section 6.1), and hence using Theorem 1.8.1 that there is a homotopy equivalence between X and $B\mathcal{C}(X)$. Thus every CW-complex is homotopy equivalent to the classifying space of a category.

1.9. The Milnor exact sequence

The n-**skeleton** of a simplicial set is the simplicial subset generated by the i-simplices for $i \leq n$. When studying homotopy classes of maps $X \to Y$ from one simplicial set to another, it is often easy to see what is going on on the n-skeleton of X for each n, and one would like to pass to the limit.

However, it is not true in general that if $X_0 \leq X_1 \leq \cdots$ is a sequence of inclusions with $X = \bigcup X_n$ then $[X; Y] = \varprojlim [X_n; Y]$. There is also a $\varprojlim^1$ term, which we now describe. In Chapter 4 of Volume I, we described $\varprojlim^1$ for an inverse system of abelian groups; we now need this notion for inverse systems of groups in general. If

$$\cdots \to G_2 \xrightarrow{\alpha_2} G_1 \xrightarrow{\alpha_1} G_0$$

are groups and homomorphisms, then the group $\prod G_n$ acts on the set $\prod G_n$ via

$$(g_0, g_1, \dots)(x_0, x_1, \dots) = (g_0 x_0 \alpha_1(g_1)^{-1}, g_1 x_1 \alpha_2(g_2)^{-1}, \dots)$$

and $\varprojlim^1 G_n$ is the quotient of the set $\prod G_n$ by the equivalence relation defined by this action. It is only a pointed set, and not a group in general, but if the G_n are abelian then $\varprojlim^1 G_n$ is an abelian group. An exact sequence

$$\{1_n\} \to \{G'_n\} \to \{G_n\} \to \{G''_n\} \to \{1_n\}$$

of inverse systems of groups gives rise to a six term exact sequence of groups and pointed sets

$$1 \to \varprojlim G'_n \to \varprojlim G_n \to \varprojlim G''_n \to \varprojlim^1 G'_n \to \varprojlim^1 G_n \to \varprojlim^1 G''_n \to 1.$$

THEOREM 1.9.1 (Milnor; Bousfield–Kan [**50**] Chapter IX §3).
Suppose that Y is a Kan complex and $X_0 \leq X_1 \leq \cdots$ is a sequence of inclusions of simplicial sets with union X. Then there is an exact sequence of pointed sets

$$1 \to \varprojlim^1 [SX_n; Y] \to [X; Y] \to \varprojlim [X_n; Y] \to 1. \qquad \square$$

CHAPTER 2

Cohomology of groups

2.1. Overview of group cohomology

A good introduction to the history of the cohomology of groups can be found in Mac Lane [171]. He traces the history back to the works of Hurewicz (1936) on aspherical spaces and Hopf (1942) on the relationship between the fundamental group and the second homology group of a space. We shall not dwell here on the historical development, but refer the reader to Mac Lane's article for further information and comments.

The purpose of this chapter is to give a survey of group cohomology and how it is connected to various other parts of mathematics, and in particular to topological and algebraic K-theory. In the first few sections, where we provide several definitions of group cohomology and show how they are related, we give fairly complete proofs. Later on, we lapse into description and give enough references so that the interested reader may chase up the proofs (we trust that the reader will also excuse some forward references to Chapter 3 on spectral sequences during the later sections of this chapter). We hope that this romp through large chunks of mathematics will be taken as a joy ride, and not as an indigestible pill.

The first approach to group cohomology, which we have already examined in some detail in Volume I, is the algebraic approach. Recall from Section 2.4 of Volume I that if Λ is an augmented algebra over a commutative ring of coefficients R, then we define homology and cohomology groups

$$H_n(\Lambda, R) = \operatorname{Tor}_n^\Lambda(R, R), \quad H^n(\Lambda, R) = \operatorname{Ext}_\Lambda^n(R, R).$$

More generally, if M is a right Λ-module, we define

$$H_n(\Lambda, M) = \operatorname{Tor}_n^\Lambda(M, R)$$

while if M is a left Λ-module we define

$$H^n(\Lambda, M) = \operatorname{Ext}_\Lambda^n(R, M).$$

These are both covariant functors in M, and from Section 2.5 of Volume I we have long exact sequences

$$\cdots \to H_n(\Lambda, M') \to H_n(\Lambda, M) \to H_n(\Lambda, M'') \to H_{n-1}(\Lambda, M') \to \cdots$$
$$\cdots \to H^n(\Lambda, M') \to H^n(\Lambda, M) \to H^n(\Lambda, M'') \to H^{n+1}(\Lambda, M') \to \cdots$$

associated to a short exact sequence of modules (right in the first case and left in the second)

$$0 \to M' \to M \to M'' \to 0.$$

In the case where $\Lambda = RG$, the group algebra of G over R, we write $H_n(G, M)$ and $H^n(G, M)$ for $H_n(RG, M)$ and $H^n(RG, M)$ respectively. Of course, by the comments in Section 3.1 of Volume I, we may regard right RG-modules as left RG-modules via $gm = mg^{-1}$ and vice-versa.

Recall also from Section 2.6 of Volume I that Yoneda composition defines a product structure on cohomology

$$
\begin{array}{ccc}
H^m(G, R) \times H^n(G, R) & \longrightarrow & H^{m+n}(G, R) \\
\| & & \| \\
\mathrm{Ext}^m_{RG}(R, R) \times \mathrm{Ext}^n_{RG}(R, R) & \longrightarrow & \mathrm{Ext}^{m+n}_{RG}(R, R).
\end{array}
$$

We saw in Section 3.1 of Volume I that in the case of a group algebra, this product is *graded commutative* in the sense that if $a \in H^m(G, R)$ and $b \in H^n(G, R)$ then $ab = (-1)^{mn}ba$. The reader is warned that the same is not true for more general augmented algebras, even for augmented symmetric algebras. It is also not true that the ring $\mathrm{Ext}^*_{RG}(M, M)$ is graded commutative for an arbitrary (even simple) RG-module M.

It is often convenient for theoretical purposes, although seldom for practical computation, to use a standard projective resolution of R as an RG-module. In Section 3.3 of Volume I, we introduced the *standard resolution* or *bar resolution* and in Section 3.4 of Volume I we used it to interpret low degree cohomology in terms of group extensions.

The second approach to cohomology we shall examine in Section 2.2 is that of Eilenberg–MacLane spaces. If G is a group, then there is a CW-complex $K(G, 1)$ whose fundamental group is G and whose higher homotopy vanishes. This space is unique up to homotopy equivalence, and is characterised by the contractibility of its universal cover. Our second definition of group cohomology is

$$H_n(G, R) = H_n(K(G, 1); R), \quad H^n(G, R) = H^n(K(G, 1); R)$$

for R a commutative ring of coefficients. We shall show that this agrees with the algebraic definition by regarding the cellular chains of the universal cover of $K(G, 1)$ as a projective resolution of R as an RG-module.

The third definition, which we shall examine in Section 2.4, comes via the theory of principal G-bundles. Roughly speaking, a principal G-bundle is a fibre bundle where each fibre is a copy of the group G. This definition makes sense for an arbitrary *topological group* G, and it turns out that the functor $\mathrm{Princ}_G(X)$ assigning to a CW-complex X the set of principal G-bundles over

X is a representable contravariant functor. In other words, there is a CW-complex written BG with the property that there is a natural isomorphism

$$[X; BG] \overset{\text{nat}}{\cong} \text{Princ}_G(X).$$

This space comes equipped with a *universal* principal G-bundle, written $\xi_G :$ $EG \to BG$, with the property that the above correspondence is given by

$$[f] \mapsto f^*(\xi_G).$$

The universal bundle is unique up to homotopy, and we give Milnor's construction to provide a concrete model of this space. For Milnor's model, it turns out that EG is contractible, and so if G has the discrete topology then BG is an Eilenberg–Mac Lane space $K(G, 1)$. Thus if we define

$$H_n(G, R) = H_n(BG; R), \quad H^n(G, R) = H^n(BG; R)$$

then this definition will agree with our previous definitions in case G is discrete. In some sense, Milnor's model for EG corresponds to the bar resolution in the algebraic approach.

In case G is the unitary group $U(n)$ (or the orthogonal group $O(n)$), we shall see that principal G-bundles correspond naturally to complex (or real) vector bundles, and in this way we shall see that the Grassmannians $Gr_n(\mathbb{C})$ and $Gr_n(\mathbb{R})$ are models for $BU(n)$ and $BO(n)$.

One of the benefits of the topological approach is that we have a natural definition of *characteristic classes* of representations. Namely if M is a $\mathbb{C}G$-module (resp. $\mathbb{R}G$-module) then $EG \times_G M$ is a vector bundle over BG and hence corresponds to a map from BG to $BU(n)$ (resp. $BO(n)$). Thus pulling back cohomology classes of $BU(n)$ (resp. $BO(n)$) gives naturally defined Chern classes (resp. Stiefel–Whitney classes) of representations. In Chapter 3, we shall see that these characteristic classes are related to the Atiyah spectral sequence, which is a device for comparing the ordinary representation ring with the cohomology ring.

It is also worth knowing about the Kan–Thurston Theorem [**146**], which we now state without proof. Roughly speaking, the Kan–Thurston theorem says that every topological space is homologically the same as some group (usually infinitely generated, even if the space has only finitely many cells). More precisely, given any path connected space X with base point, there exists an Eilenberg–Mac Lane space $K(\pi, 1)$ and a map $K(\pi, 1) \to X$ which induces an isomorphism on (singular) homology and cohomology with any local coefficients. In fact, the space X may be recovered up to (weak) homotopy from the group π and the kernel of the map $\pi \to \pi_1(X)$, which is a perfect normal subgroup of π. Namely, there is a construction due to Quillen, called the plus construction, which we describe in Section 2.10, which kills a perfect normal subgroup of π_1 of a space without affecting the homology or cohomology. The map $K(\pi, 1) \to X$ gives rise to a map $K(\pi, 1)^+ \to X$ which is a weak homotopy equivalence.

It follows from the Kan–Thurston theorem that any operations one can define on the cohomology of groups, such as for example the Steenrod operations discussed in Chapter 4, are automatically operations on the cohomology of all topological spaces. Of course, Steenrod operations were first defined for spaces.

2.2. Eilenberg–Mac Lane spaces

DEFINITION 2.2.1. *A space (X, x_0) is of* **type** (π, n) *if* $\pi_n(X, x_0) \cong \pi$, *and* $\pi_i(X, x_0) = 0$ *for* $i \neq n$. *If X has the homotopy type of a CW-complex then it is an* **Eilenberg–Mac Lane space** $K(\pi, n)$.

REMARK. By Theorem 1.5.8, a CW-complex X is a $K(\pi, 1)$ if and only if $\pi_1(X, x_0) \cong \pi$ and the universal cover of X is contractible.

EXAMPLES. Take $X = \mathbb{R}P^\infty$, the union of the real projective spaces $\mathbb{R}P^n \subseteq \mathbb{R}P^{n+1} \subseteq \cdots$ with the weak topology. The sphere is a 2-fold cover of $\mathbb{R}P^n$, and so S^∞ is a 2-fold cover of $\mathbb{R}P^\infty$. We claim that $\pi_i(S^\infty) = 0$ for all i, so that S^∞ is contractible. This is because any map from an i-sphere to S^∞ has image lying in some finite sphere S^n, by compactness. Taking n greater than i, we see that the map is homotopic to zero already in S^n. It follows that $\mathbb{R}P^\infty$ is a $K(\mathbb{Z}/2, 1)$.

In a similar way, we see from the long exact homotopy sequence of the fibration $S^1 \to S^{2n+1} \to \mathbb{C}P^n$ that $\pi_2(\mathbb{C}P^n, x_0) \cong \mathbb{Z}$, and $\pi_i(\mathbb{C}P^n, x_0) = 0$ for $i = 1$ and $2 < i < n$. Thus $\pi_2(\mathbb{C}P^\infty, x_0) \cong \mathbb{Z}$ and $\pi_i(\mathbb{C}P^\infty, x_0) = 0$ for $i \neq 2$, so that $\mathbb{C}P^\infty$ is a $K(\mathbb{Z}, 2)$.

Another easy example is that S^1 is a $K(\mathbb{Z}, 1)$.

THEOREM 2.2.2. *Given $n \geq 1$ and π a group, with π abelian if $n \geq 2$, then there exists an Eilenberg–Mac Lane space $K(\pi, n)$.*

PROOF. We start with the case $n = 1$. Choose a free presentation

$$1 \to R \to F \to \pi \to 1$$

with F free and R a normal subgroup of F. We construct a CW-complex X as follows. We take as the 1-skeleton X^1 of X a wedge sum of copies of the circle S^1_α, indexed over generators α of F as a free group. Thus we have $\pi_1(X^1, x_0) \cong F$. Now choose elements $\beta \in R$ generating R as a normal subgroup of F. Each element β corresponds to an element of $\pi_1(X^1, x_0)$ and is hence represented by a map

$$\hat{\beta} : (S^1, s_0) \to (X^1, x_0).$$

We take for the 2 skeleton of X

$$X^2 = X^1 \cup_{\hat{\beta}} \bigcup (\sigma^2_\beta),$$

the space formed by attaching 2-cells to X^1, using the $\hat{\beta}$ as attaching maps.

We have $\pi_1(X^2, X^1, x_0) = 0$, and so the homotopy exact sequence is

$$\cdots \longrightarrow \pi_2(X^2, X^1, x_0) \longrightarrow \pi_1(X^1, x_0) \longrightarrow \pi_1(X^2, x_0) \longrightarrow 0.$$

$$\parallel$$
$$F$$

For each generator β of R there is an obvious corresponding element in the group $\pi_2(X^2, X^1, x_0)$, which goes to $\beta \in \pi_1(X^1, x_0)$. Thus the kernel of $\pi_1(X^1, x_0) \to \pi_1(X^2, x_0)$ contains R. To show that it is exactly R, we notice that the covering $\tilde{X}^1$ of X^1 corresponding to the subgroup R of F (as in Section 1.6) has $\pi_1(\tilde{X}^1, x_0) \cong R$, and the fibre of $\tilde{X}^1 \to X^1$ is F/R. Thus for each 2-cell σ_β^2 of X^2 and each element of F/R we attach a 2-cell to $\tilde{X}^1$ in the obvious way, we obtain a space $\tilde{X}^2$ covering X^2, with fibre F/R. The commutative diagram

$$
\begin{array}{ccccc}
R & = & \pi_1(\tilde{X}^1, x_0) & \longrightarrow & \pi_1(\tilde{X}^2, x_0) & \longrightarrow & 0 \\
& & \downarrow & & \downarrow & & \\
F & = & \pi_1(X^1, x_0) & \longrightarrow & \pi_1(X^2, x_0) & \longrightarrow & 0 \\
& & & & \downarrow & & \\
& & & & F/R & &
\end{array}
$$

now shows that $\pi_1(X^2, x_0) = F/R$ and $\pi_1(\tilde{X}^2, x_0) = 0$.

We now kill the higher homotopy groups by induction. Suppose we have constructed the m-skeleton X^m with

$$\pi_r(X^m, x_0) = \begin{cases} \pi & r = 1 \\ 0 & 1 < r < m. \end{cases}$$

We choose generators γ for $\pi_m(X^m, x_0)$ and representatives

$$\hat{\gamma} : (S^m, s_0) \to (X^m, x_0)$$

and form

$$X^{m+1} = X^m \cup_{\hat{\gamma}} \bigcup \sigma_\gamma^{m+1}$$

by using the $\hat{\gamma}$ as attaching maps. We have

$$\pi_{m+1}(X^{m+1}, X^m, x_0) \cong \bigoplus_\gamma \mathbb{Z}_\gamma,$$

a free abelian group on generators corresponding to the elements γ. The boundary map

$$\pi_{m+1}(X^{m+1}, X^m, x_0) \xrightarrow{\partial} \pi_m(X^m, x_0)$$

sends the generator of $\mathbb{Z}_\gamma$ to γ and is hence surjective, so that $\pi_m(X^{m+1}, x_0) = 0$. The lower homotopy groups are unaltered, and so we have constructed

X^{m+1}. We now set $X = \bigcup_{m \geq 1} X^m$ with the weak topology, so that X is a $K(\pi, 1)$.

The construction of a $K(\pi, n)$ for $n \geq 2$ is the same only easier. Begin with a presentation

$$1 \to R \to F \to \pi \to 1$$

of π as an abelian group, so that F is free abelian and $R \leq F$. Choose generators α for F as a free abelian group and take as the n-skeleton X^n a wedge sum of copies of a sphere S^n_α indexed over the generators α. Then $\pi_i(X^n, x_0) = 0$ for $i < n$ and $\pi_n(X^n) \cong F$. Choose generators β for R and attach $(n+1)$-cells corresponding to the generators β. Kill higher homotopy exactly as before. $\square$

REMARK. If π is abelian then the set of homotopy classes of maps from X to $K(\pi, n)$] is in natural bijection with $H^n(X; \pi)$ via the map

$$[X; K(\pi, n)] \to H^n(X; \pi)$$
$$f \mapsto f^*(i_n)$$

where $i_n \in H^n(K(\pi, n); \pi) \cong \operatorname{Hom}(H_n(K(\pi, n)), \pi)$ corresponds to the inverse of the Hurewicz homomorphism

$$\pi = \pi_n(K(\pi, n), x_0) \to H_n(K(\pi, n)).$$

If $n = 1$ and π is not necessarily abelian, then $[X; K(\pi, 1)]$ is in natural bijection with $\operatorname{Hom}(\pi_1(X, x_0), \pi)$ via the map $f \to f_*$. These statements are proved by obstruction theory, and may be found for example in Mosher and Tangora [195].

In either case it follows that $K(\pi, n)$ is unique up to homotopy, since if X and Y are both $K(\pi, n)$'s, then there are maps $X \to Y$ and $Y \to X$ corresponding to the isomorphism $\pi_n(X) \cong \pi_n(Y)$). The composite in either direction induces the identity map on π_n and is hence homotopic to it by again applying the bijection. It also follows that $[K(\pi, n); K(\pi', n)] \cong \operatorname{Hom}(\pi, \pi')$.

The above statements may be phrased by saying that $H^n(-; \pi)$ is a *representable functor* (see Section 2.1 of Volume I) represented by $K(\pi, n)$. Thus for example Yoneda's lemma shows that the natural transformations

$$H^n(-; \pi) \rightsquigarrow H^m(-; \pi')$$

are in one–one correspondence with homotopy classes of maps

$$[K(\pi, n); K(\pi', m)]$$

or equivalently with elements of

$$H^m(K(\pi, n); \pi').$$

In particular the Hurewicz theorem shows that there can be no non-zero natural transformations if $m < n$, and if $m = n$ the natural transformations are in one–one correspondence with $\operatorname{Hom}(\pi, \pi')$. For $m > n$ the natural transformations can be quite interesting. For example if $\pi = \mathbb{Z}/p$ then there are the Steenrod operations, which we shall discuss from a much more algebraic point of view in Chapter 4.

THEOREM 2.2.3. (i) $H_n(G, R) \cong H_n(K(G, 1); R)$
(ii) $H^n(G, R) \cong H^n(K(G, 1); R)$.

PROOF. Let X be a $K(G, 1)$ which is a CW-complex, and let $\tilde{X}$ be its universal cover, so that $\tilde{X}$ is contractible (see the remark after the definition). G acts freely on $\tilde{X}$ permuting the cells. So the cellular chains $C_*(\tilde{X}; R)$ on $\tilde{X}$ form an exact sequence of free RG-modules, and the quotient by the G-action is $C_*(X; R)$. Together with the augmentation $C_0(\tilde{X}; R) \to R \to 0$ we obtain a free resolution of R as an RG-module.

We have

$$C_i(\tilde{X}; R) \otimes_{RG} R \cong C_i(X; R)$$

and so using this resolution to calculate Tor, we have

$$H_i(G, R) = \mathrm{Tor}_i^{RG}(R, R) \cong H_i(X; R).$$

Similarly

$$\mathrm{Hom}_{RG}(C_i(\tilde{X}; R), R) \cong \mathrm{Hom}_R(C_i(X; R), R) = C^i(X; R)$$

and so

$$H^i(G, R) = \mathrm{Ext}_{RG}^i(R, R) \cong H^i(X; R). \qquad \square$$

EXERCISE. If X is a $K(\pi, n)$, show that the loop space ΩX is a $K(\pi, n-1)$.

2.3. Principal G-bundles

DEFINITION 2.3.1. *A* **topological group** G *is a group which is simultaneously a Hausdorff topological space, in such a way that the multiplication map $G \times G \to G$ and the inverse map $G \to G$ are continuous.*

Examples of topological groups are the general linear group $GL_n(\mathbb{R})$, the unitary group $U(n)$, the orthogonal group $O(n)$, the group $\mathrm{Diff}^+(M)$ of orientation preserving self-diffeomorphisms of a manifold M (with the compact-open topology), any group with the discrete topology, and so on.

DEFINITION 2.3.2. *A (locally trivial)* **principal G-bundle** ξ *is a fibre bundle $p : E \to B$ with fibre homeomorphic to G, and a* **right** *G-action $E \times G \to E$ such that there exists an open cover $\{U_\alpha\}$ of B and homeomorphisms*

$$
\begin{array}{ccc}
U_\alpha \times G & \xrightarrow{\phi_\alpha} & p^{-1}(U_\alpha) \\
\downarrow & & \downarrow{\scriptstyle p} \\
U_\alpha & =\!=\!=\!= & U_\alpha
\end{array}
$$

EXAMPLES. (i) The covering $S^n \to \mathbb{R}P^n$ discussed in the last section is a principal $\mathbb{Z}/2$-bundle.

(ii) According to a theorem of Chevalley, if G is a Lie group and H is a closed subgroup, then $G \to G/H$ is a principal H-bundle.

(iii) Let $V^n(\mathbb{R}^m)$ denote the Stiefel variety; namely the set whose elements are ordered n-tuples of orthonormal vectors in $\mathbb{R}^m$ (these are called n-**frames** in $\mathbb{R}^m$), topologised as a subset of $(\mathbb{R}^m)^n$. Let $G^n(\mathbb{R}^m)$ denote the Grassmann variety; namely the set whose elements are n-dimensional subspaces of $\mathbb{R}^m$, topologised as a subset of the projective space corresponding to $\Lambda^n(\mathbb{R}^m)$. It is not hard to check that the natural map $V^n(\mathbb{R}^m) \to G^n(\mathbb{R}^m)$ associating to each n-frame its linear span, is a fibre bundle. There is a right action of the orthogonal group $O(n)$ on $V^n(\mathbb{R}^m)$, namely an $n \times n$ orthogonal matrix tells you how to go from one n-frame in $\mathbb{R}^m$ to another, by linear substitutions. The orbits of $O(n)$ are the fibres of the map $V^n(\mathbb{R}^m) \to G^n(\mathbb{R}^m)$, making it a principal $O(n)$-bundle.

Similarly, we write $V^n(\mathbb{C}^m) \to G^n(\mathbb{C}^m)$ for the principal $U(n)$-bundle formed by performing the same construction with orthonormal n-frames in $\mathbb{C}^m$.

There is an inclusion $V^n(\mathbb{R}^m) \hookrightarrow V^n(\mathbb{R}^{m+1})$ given by the inclusion $\mathbb{R}^m \hookrightarrow \mathbb{R}^{m+1}$, and we write $V^n(\mathbb{R}^\infty)$ for the union of the $V^n(\mathbb{R}^m)$, with the weak topology with respect to the union. We construct $G^n(\mathbb{R}^\infty)$ in a similar way, so that $V^n(\mathbb{R}^\infty) \to G^n(\mathbb{R}^\infty)$ is again a principal $O(n)$-bundle.

DEFINITION 2.3.3. *If $\xi = (p : E \to B)$ is a principal G-bundle and F is a left G-space (i.e., G acts on F in such a way that the action is given by a continuous map $G \times F \to F$), we form a bundle $\xi[F]$ as follows. The total space of $\xi[F]$ is $E \times_G F = (E \times F)/\sim$, where $\sim$ is the equivalence relation given by $(xg, y) \sim (x, gy)$ for all $x \in E$, $y \in F$, $g \in G$. The base space is again B, and the map $p_F : E \times_G F \to B$ given by $p_F(x,y) = p(x)$ makes $\xi[F]$ a fibre bundle with fibre F. We say $\xi[F]$ is a bundle with* **structure group** *G and fibre F.*

Thus for example if $G = GL_n(\mathbb{C})$ (resp. $GL_n(\mathbb{R})$) and $F = \mathbb{C}^n$ (resp. $\mathbb{R}^n$) then $\xi[F]$ is a fibre bundle whose fibres are complex (resp. real) vector spaces and where the maps $\phi_\alpha : p^{-1}(U_\alpha) \to U_\alpha \times F$ are linear on each fibre. Such a fibre bundle is called a rank n **vector bundle**. Conversely, given a vector bundle $p : E \to B$ with fibre $F \cong \mathbb{C}^n$ (resp. $\mathbb{R}^n$), we may recover ξ as follows. We may form in a canonical way a vector bundle whose fibres are $\mathrm{Hom}(F,F) \cong \mathrm{Mat}_n(\mathbb{C})$ (resp. $\mathrm{Mat}_n(\mathbb{R})$). The subbundle whose fibres are the isomorphisms from F to F is the principal $GL_n(\mathbb{C})$-bundle (resp. $GL_n(\mathbb{R})$-bundle) ξ. Thus there is a one–one correspondence between rank n complex (resp. real) vector bundles over B and principal $GL_n(\mathbb{C})$-bundles (resp. $GL_n(\mathbb{R})$-bundles) over B.

DEFINITION 2.3.4. *If $f : B' \to B$ is a map of base spaces and $\xi = (p : E \to B)$ is a fibre bundle over B, we define $f^*(\xi)$ to be the* **pullback bundle** *over B' with total space*

$$E' = \{(x, y) \in B' \times E \mid f(x) = p(x)\}$$

and with $p' : E' \to B'$ given by $p'(x, y) = x$.

It is easy to see that $f^*(\xi)$ is a fibre bundle with the same fibre as ξ. If ξ is a principal G-bundle then so is $f^*(\xi)$.

EXERCISES. 1. If $E \to B$ and $E' \to B$ are n and m dimensional vector bundles, show that the pullback

$$
\begin{array}{ccc}
E'' & \longrightarrow & E' \\
\downarrow & & \downarrow \\
E & \longrightarrow & B
\end{array}
$$

constructs an $(n + m)$-dimensional vector bundle $E'' \to B$, called the **Whitney sum** of the two bundles.

2. Tensor products of vector bundles are harder to define. We begin with product bundles. If V and W are vector spaces then the tensor product of the bundles $B \times V$ and $B \times W$ is the bundle $B \times (V \otimes W)$. Using the fact that the map $\mathrm{Hom}(V, V') \otimes \mathrm{Hom}(W, W') \to \mathrm{Hom}(V \otimes W, V' \otimes W')$ is continuous, show that if $B \times V \to B \times V'$ and $B \times W \to B \times W'$ are maps of vector bundles over B then we obtain a map of vector bundles $B \times (V \otimes W) \to B \times (V' \otimes W')$. Deduce that the topology on $B \times (V \otimes W)$ does not depend on the choice of product structure. Now define tensor products of bundles $E \otimes E' \to B$ using an open cover on which both bundles are products.

3. In a similar way, if $E \to B$ and $E' \to B$ are vector bundles of dimension m and n, construct a vector bundle $\mathrm{Hom}(E, E') \to B$ of dimension mn. Construct exterior power bundles $\Lambda^i(E) \to B$ and symmetric powers $S^i(E) \to B$ and show that $E \otimes E \cong \Lambda^2(E) \oplus S^2(E)$.

2.4. Classifying spaces

We now describe the classification of principal G-bundles over a paracompact base space. Fuller details may be found in Husemoller [137].

PROPOSITION 2.4.1. *If B' is paracompact, $f, g : B' \to B$ are homotopic, and ξ is a principal G-bundle over B then $f^*(\xi) \cong g^*(\xi)$ as principal G-bundles over B'.*

PROOF. Let $\rho : B' \times I \to B' \times I$ be defined by $\rho(x, t) = (x, 1)$. We claim that given any bundle $\xi' = (E' \to B' \times I)$, there is a map $\theta : E' \to E'$ such that the diagram

$$
\begin{array}{ccc}
E' & \xrightarrow{\ \theta\ } & E' \\
\downarrow{\scriptstyle p} & & \downarrow{\scriptstyle p} \\
B' \times I & \xrightarrow{\ \rho\ } & B' \times I
\end{array}
$$

commutes, and gives a bundle isomorphism $\xi' \cong \rho^*(\xi')$. If $h : B' \times I \to B$ is a homotopy from f to g, then applying this claim with $\xi' = h^*(\xi)$ proves the proposition.

To prove the claim, we argue as follows. Since I is compact we may choose an open cover $\{U_\alpha\}$ of B' so that the restriction of ξ' to each $U_\alpha \times I$ is a product bundle. Since B' is paracompact, we may assume that $\{U_\alpha\}$ is locally finite. Choose a partition of unity $\{f_\alpha\}$ subordinate to $\{U_\alpha\}$. Set $f'_\alpha(x) = f_\alpha(x)/\max_\beta f_\beta(x)$, so that the equation $\sum_\alpha f_\alpha(x) = 1$ is replaced by $\max_\alpha f_\alpha(x) = 1$. Choose isomorphisms

$$\phi_\alpha : U_\alpha \times I \times G \to p^{-1}(U_\alpha \times I).$$

For each α, define maps $\theta_\alpha : E' \to E'$, $\rho_\alpha : B' \times I \to B' \times I$ fitting into a commutative square as above, by the formulae

$$\theta_\alpha(\phi_\alpha(x,t,g)) = \phi_\alpha(x, \max(f'_\alpha(x),t),g)$$
$$\rho_\alpha(x,t) = (x, \max(f'_\alpha(x),t))$$

inside U_α, and the identity outside. Since $\{U_\alpha\}$ is locally finite, if we totally order the α's, then the infinite composites of these maps have only finitely many non-identity maps in a neighbourhood of each point. So the infinite composites make sense, and give the required maps θ and ρ. $\square$

COROLLARY 2.4.2. *If B is paracompact and contractible then every principal G-bundle over B is isomorphic to the product bundle $B \times G$.* $\square$

DEFINITION 2.4.3. *A **universal G-bundle** ξ_G is a principal G-bundle $EG \to BG$ such that for all paracompact spaces B the map*

$$[B; BG] \to \mathrm{Princ}_G(B)$$
$$f \mapsto f^*(\xi_G)$$

from homotopy classes of maps $B \to BG$ to principal G-bundles over B is a bijection.

LEMMA 2.4.4. *If $\xi_G = (E \to B)$ and $\xi'_G = (E' \to B')$ are universal G-bundles with B and B' paracompact then there is a homotopy equivalence $f : B \to B'$ with $\xi_G = f^*(\xi'_G)$. In particular E is homotopy equivalent to E' by a map commuting with the G-action.*

PROOF. Let $f : B \to B'$ correspond to ξ_G under the bijection $[B; B'] \to \mathrm{Princ}_G(B)$, and $f' : B' \to B$ correspond to ξ'_G under the bijection $[B'; B] \to \mathrm{Princ}_G(B')$. Then $f \circ f' \simeq \mathrm{id}_{B'}$ and $f' \circ f \simeq \mathrm{id}_B$. $\square$

To prove existence of a classifying space, we use Milnor's construction.

DEFINITION 2.4.5. *If X and Y are topological spaces then the **join** $X * Y$ is the quotient of the product space $X \times I \times Y$ by the equivalence relation*

$$(x,0,y) \sim (x',0,y) \quad \forall\, x,x' \in X,\ y \in Y$$
$$(x,1,y) \sim (x,1,y') \quad \forall\, x \in X,\ y,y' \in Y.$$

We now define EG to be the infinite join $G * G * \cdots$. Thus EG consists of formal elements

$$(t_1 g_1, t_2 g_2, \dots)$$

with $t_i \in [0,1]$, $t_i = 0$ for all but finitely many i, and $\sum t_i = 1$, modulo the equivalence relation given by

$$(t_1 g_1, t_2 g_2, \dots) \sim (t_1 g_1', t_2 g_2', \dots)$$

provided $g_i = g_i'$ whenever $t_i \neq 0$. In case G is a CW-complex, we may give the infinite join the weak topology with respect to the union of the finite joins, so that it is again a CW-complex, and hence paracompact. The free right G-action on EG is given by

$$(t_1 g_1, t_2 g_2, \dots)g = (t_1 g_1 g, t_2 g_2 g, \dots).$$

The base space BG is simply the quotient space $BG = EG/G$. It is easy to check that with the above definitions $\xi_G = (p_G : EG \to BG)$ is a (locally trivial) principal G-bundle.

REMARK. All the topological groups we are interested in are CW-complexes. If the group G is not a CW-complex, then the weak topology on EG is not necessarily the right one, as it is not clear that the action of G is continuous. One needs to use the **strong topology**, namely use as few open sets as possible so that the coordinate functions $t_i : EG \to [0,1]$ and $g_i : t_i^{-1}(0,1] \to G$ are continuous. Since BG is no longer necessarily paracompact, one is forced to discuss numerable bundles over an arbitrary base space, since otherwise one loses the uniqueness argument of Lemma 2.4.4. One also loses local triviality of the bundle (see Segal [**229**] for a discussion of this point). In Husemoller [**137**], you will find the theory developed using numerable bundles, and without using the local triviality condition.

THEOREM 2.4.6 (Milnor [**187**]). *The above bundle* $\xi_G = (p_G : EG \to BG)$ *is a universal G-bundle.*

PROOF. If $\xi = (p : E \to B)$ is a principal G-bundle over a paracompact base space B, then by Corollary 1.7.4, we may choose a countable open cover $\{U_j\}$ such that the restriction to each U_j is a product bundle. Choose a partition of unity $\{f_j\}$ subordinate to $\{U_j\}$. Choose isomorphisms

$$\phi_j : p^{-1}(U_j) \to U_j \times G$$

and denote by $q_j : U_j \times G \to G$ the second projection. Define a map $\tilde{f} : E \to EG$ via

$$\tilde{f}(x) = (f_1(p(x))q_1(\phi_1(x)), f_2(p(x))q_2(\phi_2(x)), \dots).$$

Then $\tilde{f}$ commutes with the G-action, and so we get a map $f : B \to BG$. The square

$$
\begin{array}{ccc}
E & \xrightarrow{\tilde{f}} & EG \\
{\scriptstyle p}\downarrow & & \downarrow{\scriptstyle p_G} \\
B & \xrightarrow{f} & BG
\end{array}
$$

is a pullback square, and so $\xi \cong f^*(\xi_G)$.

Next, we must show that if $f, f' : B \to BG$ are maps with $f^*(\xi_G) \cong (f')^*(\xi_G)$ then $f \simeq f'$. Define inclusions

$$EG \xrightarrow{i_{\mathrm{odd}}} EG$$
$$(t_1 g_1, t_2 g_2, \dots) \mapsto (t_1 g_1, 0, t_2 g_2, 0, \dots)$$

and

$$EG \xrightarrow{i_{\mathrm{even}}} EG$$
$$(t_1 g_1, t_2 g_2, \dots) \mapsto (0, t_1 g_1, 0, t_2 g_2, 0, \dots).$$

The inclusion i_{odd} is homotopic to the identity map via the G-map

$$t.i_{\mathrm{odd}} + (1-t).\mathrm{id}_{EG} : EG \times I \to EG$$

and similarly for i_{even}. Thus we may pass down to maps $i_{\mathrm{odd}}, i_{\mathrm{even}} : BG \to BG$ homotopic to the identity map.

Let $\xi = f^*(\xi_G) \cong (f')^*(\xi_G)$. Then we have maps $\alpha, \beta : E(\xi) \to EG$ and lying over $f \circ i_{\mathrm{odd}}$ and $f' \circ i_{\mathrm{even}}$ respectively. Then the map

$$\gamma : E(\xi) \times I \to EG$$
$$(x, t) \mapsto t\alpha(x) + (1-t)\beta(x)$$

lies over a homotopy h from $f \circ i_{\mathrm{odd}}$ to $f' \circ i_{\mathrm{even}}$

$$
\begin{array}{ccc}
E(\xi) \times I & \xrightarrow{\gamma} & EG \\
\downarrow & & \downarrow \\
B \times I & \xrightarrow{h} & BG
\end{array}
$$

So we have $f \simeq f \circ i_{\mathrm{odd}} \simeq f' \circ i_{\mathrm{even}} \simeq f'$. $\square$

LEMMA 2.4.7. *The total space EG of Milnor's construction is weakly contractible. In particular, if G has the homotopy type of a CW-complex then EG is contractible.*

PROOF. If $f : S^n \to EG = G * G * \cdots$ then since S^n is compact, the image is contained in some finite join $G * \cdots * G$ (r times). But $G * \cdots * G$ ($r+1$ times) contains the cone $G * \cdots * G * (\mathrm{id}_G)$, which is contractible. Hence f is homotopic to a constant map in $G * \cdots * G$ ($r+1$ times). The result now follows from Whitehead's Theorem 1.5.8. $\square$

THEOREM 2.4.8. *If G has the homotopy type of a CW-complex then a principal G-bundle ξ (over a base space $B(\xi)$ having the homotopy type of a CW-complex) is universal if and only if the total space $E(\xi)$ is contractible.*

PROOF. If ξ is universal then by Lemma 2.4.4 $E(\xi)$ is homotopy equivalent to Milnor's EG, and is hence contractible. Conversely, if $E(\xi)$ is contractible then we have a diagram

$$
\begin{array}{ccccc}
E(\xi) & \longleftarrow & E(\xi) \times EG & \longrightarrow & EG \\
\downarrow & & \downarrow & & \downarrow \\
B(\xi) & \longleftarrow & (E(\xi) \times EG)/G & \longrightarrow & BG.
\end{array}
$$

The horizontal arrows are fibrations with contractible fibres, so the long exact homotopy sequence (Theorem 1.6.6) shows that they are weak homotopy equivalences, and hence homotopy equivalences by Theorem 1.5.8. Hence ξ is universal. $\qquad\square$

EXERCISES. (i) If $G = \mathbb{Z}/2$ then $G * \cdots * G$ (r times) is homeomorphic to the $(r-1)$-sphere S^{r-1} with G acting as the map taking each point to the antipodal point. Thus we have $EG \cong S^\infty$ and $BG \cong \mathbb{R}P^\infty$.

(ii) Similarly if $G = S^1 = \{z \in \mathbb{C} \mid |z| = 1\}$ is the circle group then $G * \cdots * G$ (r times) is the unit sphere S^{2r-1} in $\mathbb{C}^r$ with G acting by scalar multiplication. Thus we have $EG = S^\infty$ and $BG = \mathbb{C}P^\infty = K(\mathbb{Z}, 2)$.

(iii) If G is the orthogonal group $O(n)$ then we claim that the principal $O(n)$-bundle $V^n(\mathbb{R}^\infty) \to G^n(\mathbb{R}^\infty)$ is universal. Since $G^n(\mathbb{R}^\infty)$ and $V^n(\mathbb{R}^\infty)$ are CW-complexes, by the theorem it suffices to prove that $V^n(\mathbb{R}^\infty)$ is weakly contractible. If $f : S^r \to V^n(\mathbb{R}^\infty)$ then the image of f lies in some $V^n(\mathbb{R}^m)$. But $V^n(\mathbb{R}^{n+m})$ contains the cone on $V^n(\mathbb{R}^m)$, and so f is homotopic to a constant map inside $V^n(\mathbb{R}^{n+m})$. We thus write $BO(n)$ for the space $G^n(\mathbb{R}^\infty)$ and $EO(n)$ for the space $V^n(\mathbb{R}^\infty)$.

If in the definition of the Stiefel variety $V^n(\mathbb{R}^\infty)$ we replace the orthogonal n-frames by *all* (linearly independent) n-frames, we still find that we have a contractible space $\hat{V}^n(\mathbb{R}^\infty)$, by the same argument, and hence $\hat{V}^n(\mathbb{R}^\infty) \to G^n(\mathbb{R}^\infty)$ is a universal $GL_n(\mathbb{R})$-bundle. So we have $BGL_n(\mathbb{R}) = G^n(\mathbb{R}^\infty) = BO(n)$.

Now recall that there is a one–one correspondence between principal $GL_n(\mathbb{R})$-bundles over B and rank n vector bundles over B (see Section 2.3), given by $\xi \mapsto \xi[\mathbb{R}^n]$. If we apply this to the bundle $\hat{V}^n(\mathbb{R}^\infty) \to G^n(\mathbb{R}^\infty)$, we obtain a canonical rank n vector bundle over $G^n(\mathbb{R}^\infty)$ whose total space $(\hat{V}^n(\mathbb{R}^\infty) \times \mathbb{R}^n)/GL_n(\mathbb{R}) = (V^n(\mathbb{R}^\infty) \times \mathbb{R}^n)/O(n)$ has as its points the ordered pairs consisting of an n-dimensional subspace of $\mathbb{R}^\infty$ and a point in it. We write $\hat{G}^n(\mathbb{R}^\infty) \to G^n(\mathbb{R}^\infty)$ for this **canonical bundle**. We have thus proved the following:

THEOREM 2.4.9. *There is a natural one–one correspondence between rank n real vector bundles over a paracompact base space B and homotopy classes* *of maps $f : B \to G^n(\mathbb{R}^\infty)$. The vector bundle corresponding to f is given by pulling back the canonical bundle*

$$\hat{G}^n(\mathbb{R}^\infty) \to G^n(\mathbb{R}^\infty). \qquad \square$$

In exactly the same way, the space $V^n(\mathbb{C}^\infty)$ is also contractible, so that $V^n(\mathbb{C}^\infty) \to G^n(\mathbb{C}^\infty)$ is a universal bundle for the unitary group $U(n)$. The space $\hat{V}^n(\mathbb{C}^\infty)$ is again contractible. We thus write $BU(n) = BGL_n(\mathbb{C})$ for the space $G^n(\mathbb{C}^\infty)$ and $EU(n)$ for the space $V^n(\mathbb{C}^\infty)$. We write $\hat{G}^n(\mathbb{C}^\infty)$ for the total space

$$(\hat{V}^n(\mathbb{C}^\infty) \times \mathbb{C}^n)/GL_n(\mathbb{C}) = (V^n(\mathbb{C}^\infty) \times \mathbb{C}^n)/U(n)$$

of the canonical rank n vector bundle over $G^n(\mathbb{C}^\infty)$, whose points are ordered pairs consisting of an n-dimensional subspace of $\mathbb{C}^\infty$ and a point in it. The corresponding theorem is as follows.

THEOREM 2.4.10. *There is a natural one–one correspondence between rank n complex vector bundles over a paracompact base space B and homotopy classes of maps $f : B \to G^n(\mathbb{C}^\infty)$. The vector bundle corresponding to f is given by pulling back the canonical bundle*

$$\hat{G}^n(\mathbb{C}^\infty) \to G^n(\mathbb{C}^\infty). \qquad \square$$

We now return to the general situation.

THEOREM 2.4.11. (i) *There is a homotopy equivalence $\Omega BG \simeq G$.*
(ii) $\pi_i(BG) \cong \pi_{i-1}(G)$.
(iii) *If G is discrete then Milnor's BG is an Eilenberg–Mac Lane space $K(G, 1)$ with universal cover EG.*

PROOF. (i) We form a pullback diagram

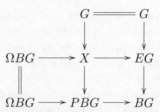

where $PBG \to BG$ is the path fibration on BG, see the exercise in Section 1.6. We thus have a diagram of fibrations

$$
\begin{array}{ccc}
G & =\!\!=\!\!= & G \\
\downarrow & & \downarrow \\
\Omega BG \longrightarrow X & \longrightarrow & EG \\
\| & \downarrow & \downarrow \\
\Omega BG \longrightarrow PBG & \longrightarrow & BG
\end{array}
$$

and since PBG and EG are contractible we have

$$\Omega BG \simeq \Omega BG \times EG \simeq X \simeq G \times PBG \simeq G.$$

(ii) This follows from the long exact sequence of the fibration $EG \to BG$.

(iii) This follows from the fact that EG is contractible and $\pi_1(BG) \cong G$, by the remark after the definition of Eilenberg–Mac Lane space. $\quad\square$

It follows from the above theorem that cohomology of the classifying space $H^n(BG; R)$ is a generalisation to arbitrary topological groups of group cohomology $H^n(G, R)$ of discrete groups. In the discrete case, there is an obvious CW decomposition of Milnor's EG for which the cellular chains form the standard (bar) resolution of the trivial RG-module R. Thus for a topological group Milnor's EG may be thought of as a sort of continuous bar resolution.

THEOREM 2.4.12. (i) *If N is a closed normal subgroup of G, then there is a fibration $BG \to B(G/N)$ with fibre BN.*

(ii) *If H is a closed subgroup of G there is a fibration $BH \to BG$ with fibre the coset space G/H.*

PROOF. (i) Let $E(G/N)$ and EG be contractible spaces on which G/N and G act freely. Then N also acts freely on EG, and we may take for BN the quotient space $(EG)/N$. Thus G/N acts on BN and we may take for BG the space $(E(G/N) \times EG)/G = (E(G/N) \times BN)/(G/N)$. There is then an obvious fibration $BG \to B(G/N) = E(G/N)/(G/N)$ given by the first projection, and the fibre is BN.

(ii) Let EG be a contractible space on which G acts freely. Then H also acts freely on EG and so we take $BG = (EG)/G$, $BH = (EG)/H$. The fibre of the obvious map $BH \to BG$ is the coset space G/H. $\quad\square$

More generally, if $\rho : G \to G'$ is any group homomorphism, we may take for BG the quotient space $(EG \times EG')/G$, where G acts on EG' through ρ, and for BG' the quotient space EG'/G'. Then we write $B\rho$ for the obvious map from BG to BG'.

EXERCISES. 1. Show that every real vector bundle on a paracompact space may be given an orthogonal inner product, and that every complex vector bundle on a paracompact space may be given a unitary inner product.

2. Using Yoneda's lemma, show that there is a map $BU(n) \times BU(m) \to BU(n+m)$ (or $BO(n) \times BO(m) \to BO(n+m)$ in the real case) corresponding to the Whitney sum of vector bundles, and a map $BU(n) \times BU(m) \to BU(nm)$ (resp. $BO(n) \times BO(m) \to BO(nm)$) corresponding to tensor product of vector bundles.

3. If G is a (discrete) group, define the **Cayley category** of G to be the category $\mathcal{C}(G)$ whose objects are the elements of G, and whose arrows are the left multiplications by elements of G, so that there is a unique arrow between each pair of objects. G acts on $\mathcal{C}(G)$ by right multiplication (which commutes with left multiplication). Show that the nerve $N\mathcal{C}(G)$ of this category (see the exercise at the end of Section 1.8) is a simplicial set whose topological realisation is Milnor's model of EG.

If we regard the group G as a category with one object, in which every arrow is an isomorphism, show that $NG = N\mathcal{C}(G)/G$ so that $BG = |NG|$ is Milnor's model of the classifying space BG.

Define the notion of a topological category (with the topology on the arrows, not the objects) and show that its nerve is a simplicial (topological) space. Define the topological realisation of a simplicial space. Show that if G is a topological group regarded as a topological category with one object, then the topological realisation of its nerve is Milnor's model for BG.

For further remarks on these points, see Segal [**229**].

2.5. K-theory

In the last section, we saw that there is a natural one–one correspondence between complex vector bundles of dimension n over a paracompact space B and homotopy classes of maps from B to $BU(n) = G^n(\mathbb{C}^\infty)$. Now the obvious maps $G^n(\mathbb{C}^m) \hookrightarrow G^{n+1}(\mathbb{C}^{m+1})$ give us a map $BU(n) \hookrightarrow BU(n+1)$, which corresponds to adding a trivial one dimensional summand to an n dimensional vector bundle to give an $(n+1)$ dimensional bundle. We set $BU = \bigcup_{n \geq 1} BU(n)$ via the above maps, with the weak topology with respect to the union. Thus a homotopy class of maps from B to BU corresponds to an equivalence class of vector bundles, where two vector bundles are regarded as equivalent if they can be made equal by adding trivial bundles on both sides.

If we restrict to the case where B is compact, then given any vector bundle over B, the corresponding map $B \to G^n(\mathbb{C}^\infty)$ has image lying in some $G^n(\mathbb{C}^m)$. So there is some $(n-m)$ dimensional vector bundle the addition of which forms an m dimensional trivial bundle.

DEFINITION 2.5.1. *If B is a compact space, we write $K(B)$ for the additive group whose generators $[E]$ correspond to the complex vector bundles $E \to B$, and whose relators say that $[E] + [E'] = [E'']$ whenever $E'' \to B$ is the Whitney sum of $E \to B$ and $E' \to B$.*

Note that in $K(B)$, a typical element is of the form $[E_1] - [E_2]$, and that $[E_1] - [E_2]$ is equal to $[E_3] - [E_4]$ if and only if for some E_5 the Whitney sums $E_1 \oplus E_4 \oplus E_5$ and $E_2 \oplus E_3 \oplus E_5$ are isomorphic. By the above remarks, we may always assume E_2, E_4 and E_5 are trivial bundles.

PROPOSITION 2.5.2. *If B is a compact space, then $K(B) \cong [B; BU \times \mathbb{Z}]$.*

PROOF. Since both sides are additive on connected components, we may assume that B is connected. Thus a map $B \to BU \times \mathbb{Z}$ corresponds to a map $B \to BU$ and an integer, which we think of as giving the dimension. The map $B \to BU$ has image lying in some $BU(n)$, and hence corresponds to some vector bundle $E \to B$ of dimension n. We add or subtract a suitable trivial bundle to make the dimension correspond to the image of $B \to \mathbb{Z}$, and hence obtain an element of $K(B)$. It is easy to check that the element of $K(B)$ obtained in this way is well defined. There is an obvious map the other way, from $K(B)$ to $[B; BU \times \mathbb{Z}]$, taking a generator $[E]$ of dimension n to the map given by $B \to BU(n) \to BU$ and the constant map $B \to \mathbb{Z}$ with image n. □

If $f : B' \to B$ is a map of compact spaces, the pullback of vector bundles gives a map $f^* : K(B) \to K(B')$. If B has a basepoint x, then the maps $\{x\} \hookrightarrow B \twoheadrightarrow \{x\}$ gives us maps $\mathbb{Z} = K(\{x\}) \to K(B) \to K(\{x\}) = \mathbb{Z}$, whose composite is the identity. Thus if we write $\tilde{K}(B)$ for the kernel of $K(B) \to \mathbb{Z}$ then $K(B) \cong \tilde{K}(B) \oplus \mathbb{Z}$. Note that every element of $\tilde{K}(B)$ is represented by an actual vector bundle.

The tensor product of vector bundles induces a multiplication which makes $K(B)$ (or $KO(B)$) into a commutative ring. $\tilde{K}(B)$ is a ring without unit, and the canonical process of adjoining a unit yields $K(B)$.

Using the action of a cyclic group of order n on the tensor nth power of a vector bundle, one may copy the construction described in Section 5.9 of Volume I to obtain operations ψ^n on $K(B)$ or $\tilde{K}(B)$. Exactly the same proof as given in Proposition 5.9.1 of Volume I shows that these ψ^n are ring homomorphisms from $K(B)$ or $\tilde{K}(B)$ to itself. In this context, the operations ψ^n are called the **Adams operations**.

We think of $\tilde{K}(B)$ as the zeroth part of a (reduced) **generalised cohomology theory** as follows. A (reduced) generalised cohomology theory consists of contravariant functors $\tilde{h}^n$, $n \in \mathbb{Z}$, and natural isomorphisms $\tilde{h}^n \cong \tilde{h}^{n+1}S$ (S denotes suspension) from a suitable category of spaces and maps to abelian groups, satisfying the usual Eilenberg–Steenrod axioms for (reduced) cohomology, except the axiom giving the value on a point. These axioms are the **homotopy axiom**: each $\tilde{h}^n$ takes homotopic maps to the same map; and the **exactness axiom**: if $f : X \to Y$ is a map with mapping cone C_f, then the sequence

$$\tilde{h}^n(C_f) \xrightarrow{j^*} \tilde{h}^n(Y) \xrightarrow{f^*} \tilde{h}^n(X)$$

is exact. Note that the mapping cone of the inclusion $Y \to C_f$ is homotopy equivalent to the suspension SX, so that this implies we have a long exact sequence

$$\cdots \to \tilde{h}^{n-1}(X) = \tilde{h}^n(SX) \to \tilde{h}^n(C_f) \to \tilde{h}^n(Y) \to \tilde{h}^n(X) \to \cdots$$

Note also that corresponding to a reduced generalised cohomology theory, there is a corresponding unreduced theory $h^n(X) = \tilde{h}^n(X \overset{\cdot}{\cup} \text{(point)})$ and relative theory $h^n(X, Y) = \tilde{h}^n(C_i)$, where C_i is the mapping cone of the inclusion $i : Y \to X$ of a subspace. This relative theory satisfies the **excision axiom**: if Z is a subspace of Y whose closure is in the interior of Y, then the mapping cone of the inclusion $Y \backslash Z \to X \backslash Z$ is homotopy equivalent to C_i, and so $h^n(X \backslash Z, Y \backslash Z) \to h^n(X, Y)$ is an isomorphism. For further information about generalised cohomology theories, see Switzer [**258**], Chapter 7. Other examples include cobordism and stable cohomotopy.

LEMMA 2.5.3. *If B' is a closed subspace of a compact space B, then the sequence*

$$\tilde{K}(B/B') \to \tilde{K}(B) \to \tilde{K}(B')$$

is exact.

PROOF. The composite map is clearly zero. Conversely, every element of $\tilde{K}(B)$ is represented by an actual vector bundle. If $E \to B$ is a vector bundle whose restriction to B' represents the zero element of $\tilde{K}(B')$, then after adding a suitable trivial bundle we may assume that the restriction to B' is trivial. A given trivialisation enables us to form a vector bundle over B/B' whose pullback to B is the required bundle. (But be warned that different trivialisations may give rise to different elements of $\tilde{K}(B/B')$, so the first map in this sequence is not necessarily injective).   □

So to make $\tilde{K}(B)$ into part of a generalised cohomology theory, for $n \geq 0$ we set

$$\tilde{K}^{-n}(B) = \tilde{K}(S^n B), \qquad K^{-n}(B) = \tilde{K}^{-n}(B \cup (\text{point})).$$

In a similar way, if we work with real instead of complex vector bundles, we write BO for $\bigcup_{n \geq 1} BO(n)$, and $KO(B)$, $\widetilde{KO}(B)$, $KO^{-n}(B)$, $\widetilde{KO}^{-n}(B)$ for the corresponding functors. Thus for example the analogue of Proposition 2.5.2 is the statement that if B is compact then $KO(B) \cong [B; BO \times \mathbb{Z}]$.

THEOREM 2.5.4 (Bott periodicity theorem [**46**]).
We have natural isomorphisms

$$K^{-n-2}(B) \cong K^{-n}(B), \qquad KO^{-n-8}(B) \cong KO^{-n}(B).$$

PROOF. It is beyond the scope of this book to include a proof of the Bott periodicity theorem. There are now several good references available for the complex case. For example, see Atiyah [**19**], Husemoller [**137**], Milnor [**191**] or Switzer [**258**]. These last two references also contain a proof for the real case. Note that it is an easy exercise using the adjunction between S and Ω (see the exercise in Section 1.2) and Proposition 2.5.2, to show that the Bott periodicity theorem is equivalent to the statements

$$\Omega^2 BU \simeq BU \times \mathbb{Z}, \quad \Omega^8 BO \simeq BO \times \mathbb{Z}.$$

In fact these homotopy equivalences are part of a larger pattern:

$\Omega BU \simeq U$	$\Omega U \simeq BU \times \mathbb{Z}$
$\Omega BO \simeq O$	$\Omega O \simeq O/U \times \mathbb{Z}/2$
$\Omega(O/U) \simeq U/Sp$	$\Omega(U/Sp) \simeq BSp \times \mathbb{Z}$
$\Omega BSp \simeq Sp$	$\Omega Sp \simeq Sp/U$
$\Omega(Sp/U) \simeq U/O$	$\Omega(U/O) \simeq BO \times \mathbb{Z}$

where U, O and Sp are the infinite unitary, orthogonal and symplectic groups, regarded as embedded in each other via the obvious maps, which double dimension where necessary.   □

The Bott periodicity theorem allows us to define the functors K^n and KO^n for all $n \in \mathbb{Z}$. The following table shows the values of these functors on a single point, as can easily be seen from the above homotopy equivalences.

n (mod 2)	0	1
$K^{-n}(\text{point}) = K(S^n)$	$\mathbb{Z}$	0

n (mod 8)	0	1	2	3	4	5	6	7
$KO^{-n}(\text{point}) = KO(S^n)$	$\mathbb{Z}$	$\mathbb{Z}/2$	$\mathbb{Z}/2$	0	$\mathbb{Z}$	0	0	0

The point here is really that to give a generalised cohomology theory h^*, it suffices to define h^n for n (sufficiently) positive, since the isomorphisms $\tilde{h}^n \cong \tilde{h}^{n+1}S$ extend the definition in the negative direction. It is the periodicity which allows us to make the definition of K^n and KO^n for n positive.

We next describe Atiyah's calculation [18] of the K-theory of BG for G a finite group (or more generally, Atiyah and Segal [21], for G a compact Lie group). Suppose M is a finite dimensional (real or complex) representation of G. Then we may form the space

$$EG \times_G M = (EG \times M)/G$$

where G acts on EG on the right and on M on the left. There is an obvious map

$$EG \times_G M \to EG \times_G (\text{point}) = BG$$

with fibre M, and making $EG \times_G M$ into a (real or complex) vector bundle over BG. The idea is that these vector bundles are sufficient to determine the K-theory of BG. In practice, there are a few technical difficulties, which we now describe.

The first problem is that we have only defined $K(B)$ for a compact space B. However, Proposition 2.5.2 gives us a way to extend this definition to all paracompact spaces, and in particular to all CW-complexes. We define

$$K(B) = [B; BU \times \mathbb{Z}].$$

In general, if B^n is the n-skeleton of B, it is not true that $K(B) = \varprojlim K(B^n)$, but we have the Milnor exact sequence (Theorem 1.9.1)

$$0 \to \varprojlim{}^1 K^{-1}(B^n) \to K(B) \to \varprojlim K(B^n) \to 0$$

It turns out that for B the classifying space of a finite (or compact Lie) group, the $\varprojlim^1$ term vanishes, so that an element of $K(BG)$ can be thought of as being given by a consistent family of elements of K-theory of the skeleta. Note also that for B non-compact, the Grothendieck group of vector bundles, as defined in Definition 2.5.1, need not coincide with either $K(B)$ or $\varprojlim K(B^n)$.

But a vector bundle on B does give rise to an element of $K(B)$, so we have by the above construction a natural map

$$\mathcal{R}(\mathbb{C}G) \to K(BG)$$

where $\mathcal{R}(\mathbb{C}G)$ is the Grothendieck ring of finite dimensional complex representations, as in Section 5.2 of Volume I.

THEOREM 2.5.5 (Atiyah Completion Theorem). *Let I_G be the kernel of the augmentation map $\mathcal{R}(\mathbb{C}G) \to \mathbb{Z}$ sending a representation to its dimension, and denote by $\mathcal{R}(\mathbb{C}G)^{\wedge}$ the completion*

$$\mathcal{R}(\mathbb{C}G)^{\wedge} = \varprojlim \mathcal{R}(\mathbb{C}G)/I_G^n.$$

Then the above map $\mathcal{R}(\mathbb{C}G) \to K(BG)$ induces an isomorphism

$$\mathcal{R}(\mathbb{C}G)^{\wedge} \cong K(BG).$$

Moreover, $K^1(BG) = 0$.

PROOF. There are two proofs available for this theorem. Atiyah's original proof [18] for finite groups involves starting with cyclic groups, going up to solvable groups by induction, and then treating the arbitrary finite group using the Brauer induction theorem. Atiyah and Segal [21] produced a more conceptual proof for all compact Lie groups. Their proof starts with the circle group S^1, then with the general torus $(S^1)^n$. The next step, involving the theory of elliptic operators, reduces the unitary group $U(n)$ to its maximal torus. The final step reduces an arbitrary compact Lie group to the unitary group by means of a faithful unitary representation.   □

2.6. Characteristic classes

In this section we introduce cohomology classes associated to vector bundles. Real and complex representations of a group G give rise to vector bundles over BG in a natural way, and so we obtain cohomology classes associated to representations.

We start by describing the cohomology of $BGL_n(\mathbb{C}) = BU(n) = G^n(\mathbb{C}^{\infty})$. If $n = 1$ we have $U(1) = S^1$ and $BU(1) = BS^1 = \mathbb{C}P^{\infty}$. It is not hard to calculate that $H^*(\mathbb{C}P^{\infty}; \mathbb{Z})$ is a polynomial ring $\mathbb{Z}[x]$ on a generator x in degree two. A choice of sign for x corresponds to a choice of orientation for the sphere $S^2 = \mathbb{C}P^1 \subseteq \mathbb{C}P^{\infty}$. We choose the sign to correspond to the natural orientation coming from the complex structure.

For a general n, we have an inclusion $(S^1)^n \subseteq U(n)$ as the diagonal matrices, and hence a map $\phi : (\mathbb{C}P^{\infty})^n \to BU(n)$. This induces a map on cohomology rings

$$\phi^* : H^*(BU(n); \mathbb{Z}) \to H^*((\mathbb{C}P^{\infty})^n; \mathbb{Z}).$$

The right hand side is a polynomial ring in generators $x_1, \ldots, x_n$ of degree two. Since $(S^1)^n$ is contained in the group of monomial matrices in $U(n)$, the image of ϕ^* lies in the invariants of the symmetric group Σ_n permuting the n copies of $\mathbb{C}P^\infty$, namely in the subring $\mathbb{Z}[\sigma_1, \ldots, \sigma_n] \subseteq \mathbb{Z}[x_1, \ldots, x_n]$ generated by the elementary symmetric functions $\sigma_i(x_1, \ldots, x_n)$ of degree $2i$.

It turns out that ϕ^* is injective, and its image exactly equals $\mathbb{Z}[\sigma_1, \ldots, \sigma_n]$. For a proof, see Husemoller [**137**], Section 18.3 or Milnor and Stasheff [**192**], Section 14.3. We define $c_i \in H^{2i}(BU(n); \mathbb{Z})$ to be the preimage of the element $\sigma_i \in H^{2i}((\mathbb{C}P^\infty)^n; \mathbb{Z})$ under ϕ^*, so that

$$H^*(BU(n); \mathbb{Z}) = \mathbb{Z}[c_1, \ldots, c_n].$$

Now if ξ is a complex vector bundle over a paracompact base space B, with classifying map $f : B \to G^n(\mathbb{C}^\infty) = BU(n)$ (see Theorem 2.4.10), we define the **Chern classes** of ξ to be

$$c_i(\xi) = f^*(c_i) \in H^{2i}(B; \mathbb{Z})$$

for $1 \leq i \leq n$, and $c_i(\xi) = 0$ if $i > n$. The **total Chern class** of ξ is defined to be the inhomogeneous element

$$c(\xi) = 1 + c_1(\xi) + \cdots + c_n(\xi) \in H^*(B; \mathbb{Z}).$$

PROPOSITION 2.6.1. *Chern classes of complex vector bundles over paracompact base spaces are characterised by the following properties:*

(i) *(Naturality) If $\rho : B' \to B$ is a map of base spaces and ξ is a vector bundle over B, then $c_i(\rho^*(\xi)) = \rho^*(c_i(\xi)) \in H^{2i}(B'; \mathbb{Z})$.*

(ii) $c(\xi_1 \oplus \xi_2) = c(\xi_1).c(\xi_2)$.

(iii) *If γ is the canonical bundle over $G^1(\mathbb{C}^\infty) = \mathbb{C}P^\infty$ then $c_1(\gamma)$ is the canonical generator (called x above) for $H^2(\mathbb{C}P^\infty; \mathbb{Z})$.*

PROOF. Properties (i) and (iii) are clear from the definition. Property (ii) follows from the observation that $1 + c_1 + \cdots + c_n \in H^*(BU(n); \mathbb{Z})$ has image $\prod_{i=1}^n (1 + x_i)$ in $H^*((\mathbb{C}P^\infty)^n; \mathbb{Z})$ and this expression is clearly multiplicative over sums of bundles.

To show that these properties characterise the Chern classes, we argue as follows. If c_i' are classes satisfying (i), (ii) and (iii), then by property (iii) c_i' agrees with c_i for the canonical bundle on $\mathbb{C}P^\infty$. Hence by properties (i) and (ii), c_i' agrees with c_i for the sum of the canonical bundles on $(\mathbb{C}P^\infty)^n$. Applying property (i) to the map $\phi : (\mathbb{C}P^\infty)^n \to BU(n)$, we see that c_i' agrees with c_i for the canonical bundle on $BU(n)$. Now by applying property (i) to the classifying map, it follows that c_i' agrees with c_i for an arbitrary bundle over a paracompact base space. $\square$

Now suppose G is a topological group having the homotopy type of a CW-complex (for example a discrete group), and suppose M is a finite dimensional complex representation of G. Then $EG \times_G M$ is a complex vector bundle over BG, and we may define Chern classes $c_i(M) \in H^{2i}(G, \mathbb{Z})$ to be the Chern classes of this vector bundle. The representation M is given by a group homomorphism $\rho : G \to U(n)$, and hence gives rise to a map $B\rho : BG \to BU(n)$. The canonical bundle over $BU(n)$ may be thought of as being obtained by the above construction from the canonical n dimensional representation of $U(n)$, and so by naturality we have

$$c_i(M) = (B\rho)^*(c_i) \in H^{2i}(G, \mathbb{Z}).$$

The corresponding theory for real vector bundles and real representations goes as follows. We have $BO(1) = B(\mathbb{Z}/2) = \mathbb{R}P^\infty$, and $H^*(\mathbb{R}P^\infty; \mathbb{Z}/2)$ is a polynomial ring $\mathbb{Z}/2[x]$ on a generator x in degree one. We have an inclusion $(\mathbb{Z}/2)^n = O(1) \subseteq O(n)$ as the diagonal matrices, and hence a map $\phi : (\mathbb{R}P^\infty)^n \to BO(n)$. The induced map on cohomology

$$\phi^* : H^*(BO(n); \mathbb{Z}/2) \to H^*((\mathbb{R}P^\infty)^n; \mathbb{Z}/2) = \mathbb{Z}/2[x_1, \dots, x_n]$$

is injective, and again the image is the subring

$$\mathbb{Z}/2[\sigma_1, \dots, \sigma_n] \subseteq \mathbb{Z}/2[x_1, \dots, x_n]$$

generated by the elementary polynomials $\sigma_i(x_1, \dots, x_n)$ of degree i. See for example Husemoller [**137**], Section 18.5 or Milnor and Stasheff [**192**], Section 7. We define $w_i \in H^i(BO(n); \mathbb{Z}/2)$ to be the preimage of $\sigma_i \in H^i((\mathbb{R}P^\infty)^n; \mathbb{Z}/2)$ under ϕ^*, so that

$$H^*(BO(n); \mathbb{Z}/2) = \mathbb{Z}/2[w_1, \dots, w_n].$$

Now if ξ is a real vector bundle over a paracompact base space B, with classifying map $f : B \to G^n(\mathbb{R}^\infty) = BO(n)$ (see Theorem 2.4.9), we define the **Stiefel–Whitney classes** of ξ to be

$$w_i(\xi) = f^*(w_i) \in H^i(B; \mathbb{Z}/2)$$

for $1 \le i \le n$, and $w_i(\xi) = 0$ if $i > n$. The **total Stiefel–Whitney class** of ξ is defined to be

$$w(\xi) = 1 + w_1(\xi) + \dots + w_n(\xi) \in H^*(B; \mathbb{Z}/2).$$

PROPOSITION 2.6.2. *The Stiefel–Whitney classes of complex vector bundles over paracompact base spaces are characterised by the following properties:*

(i) *(Naturality) If $\rho : B' \to B$ is a map of base spaces and ξ is a vector bundle over B, then $w_i(\rho^*(\xi)) = \rho^*(w_i(\xi)) \in H^i(B'; \mathbb{Z}/2)$.*

(ii) $w(\xi_1 \oplus \xi_2) = w(\xi_1).w(\xi_2)$.

(iii) *If γ is the canonical bundle over $G^1(\mathbb{R}^\infty) = \mathbb{R}P^\infty$ then $w_1(\gamma)$ is the canonical generator (called x above) for $H^1(\mathbb{R}P^\infty; \mathbb{Z}/2)$.*

PROOF. The proof exactly parallels the proof in the complex case.   □

If G is a topological group having the homotopy type of a CW-complex, and M is a finite dimensional real representation of G, then $EG \times_G M$ is a real vector bundle over BG, and we define Stiefel–Whitney classes $w_i(M) \in H^i(G, \mathbb{Z}/2)$ to be the Stiefel–Whitney classes of this vector bundle. If the representation M is given by a group homomorphism $\rho : G \to O(n)$, then just as before we have

$$w_i(M) = (B\rho)^*(w_i) \in H^i(G, \mathbb{Z}/2).$$

2.7. Transfer

In this section, we see that the transfer map $\mathrm{Tr}_{H,G}$ which we investigated in Section 3.6 of Volume I for ordinary cohomology has an analogue for any generalised cohomology theory. The reader should be warned that unlike in ordinary cohomology (Proposition 3.6.17 of Volume I), in general $\mathrm{res}_{G,H}$ followed by $\mathrm{Tr}_{H,G}$ is not multiplication by $|G:H|$.

Generalised cohomology theories are stable under suspension in the sense that $\tilde{h}^r(X) \cong \tilde{h}^{r+1}(SX)$, and in the next section, we interpret the transfer as a "stable map" from BH to BG. We shall discuss (without proof) the Segal conjecture, which is essentially the statement that all stable maps between classifying spaces are generated by the transfers and the group homomorphisms.

For any covering map $p : X \to B$ of CW-complexes with finite fibres of size n, and any generalised cohomology theory h^*, we define a transfer map $p_! : h^*(X) \to h^*(B)$. In case H is a subgroup of a group G of finite index, we saw in Theorem 2.4.12 that there is a covering map $BH \to BG$ with fibres G/H. In this case, and taking h^* to be ordinary cohomology theory H^*, we obtain the transfer map $\mathrm{Tr}_{H,G}$ which we defined in Section 3.6 of Volume I.

We define the map $p_!$ in two stages. First we define a **pretransfer map**

$$PTr : B \to E\Sigma_n \times_{\Sigma_n} X^n$$

as follows. We set

$$\bar{X} = \{(x_1,\dots,x_n) \in X^n \mid \{x_1,\dots,x_n\} \text{ is a fibre of } p\}.$$

There is an obvious covering map $\bar{X} \to B$ with fibres of size $n!$, and a free action of Σ_n, so that $\bar{X} \to B$ is the principal Σ_n-bundle corresponding to $X \to B$. It follows that there is a classifying map $B \to B\Sigma_n$ so that $\bar{X} \to B$ is the pullback

$$
\begin{array}{ccc}
\bar{X} & \longrightarrow & E\Sigma_n \\
\downarrow & & \downarrow \\
B & \longrightarrow & B\Sigma_n.
\end{array}
$$

The maps $\bar{X} \hookrightarrow X^n$ and $\bar{X} \to E\Sigma_n$ give a map $\bar{X} \to E\Sigma_n \times X^n$ and hence, by quotienting out the action of Σ_n on both sides, a map $B \to E\Sigma_n \times_{\Sigma_n} X^n$ which is the required pretransfer map.

The second stage of the construction of $p_!$ is the Dyer–Lashof map. The representability theorem for cohomology theories (see for example Switzer [**258**], Chapter 9) says that for CW-complexes X, $\tilde{h}^r(X) = [X; K_r]$ where K_r are CW-complexes with homotopy equivalences $K_r \simeq \Omega K_{r+1}$. In fact, we shall not need to have recourse to this theorem in this book, since we give explicit spaces K_r every time we introduce a generalised cohomology theory. For example if $\tilde{h}^r(X) = \tilde{H}^r(X;\pi)$ then $K_r = K(\pi,r)$. Note also that conversely, given such spaces K_r, the definition $\tilde{h}^r(X) = [X; K_r]$ defines a reduced generalised cohomology theory. The corresponding unreduced theory is $h^r(X) = [X_+; K_r]$. The Dyer–Lashof map is a certain map

$$DL : E\Sigma_n \times_{\Sigma_n} (K_r)^n \to K_r$$

built using the fact that K_r is homotopic to the k-fold loop space on K_{r+k} for all k. Now if $\alpha \in h^r(X)$ is represented by a map $\hat{\alpha} : X_+ \to K_r$, then the transfer $p_!(\alpha) \in h^r(B)$ is represented by the composite map

$$B_+ \xrightarrow{\text{PTr}} (E\Sigma_n \times_{\Sigma_n} X^n)_+ \xrightarrow{1 \times \hat{\alpha}^n} E\Sigma_n \times_{\Sigma_n} (K_r)^n \xrightarrow{DL} K_r.$$

It remains to define the Dyer–Lashof map DL [**98**]. To define this, we use a specific model for $E\Sigma_n$, called the space of **little cubes**. Boardman and Vogt [**43**] define the space $\mathcal{C}_k(n)$ of little cubes as follows. Let I^k be an k dimensional cube, and identify the sphere S^k with $I^k/\dot{I}^k$. Then $\mathcal{C}_k(n)$ is the space of maps

$$S^k \to \bigvee_{j=1}^{n} S^k$$

having the property that there are n disjoint subcubes I_j^k of the form $[a_1, b_1] \times \cdots \times [a_k, b_k]$ inside the left hand copy of S^k, such that I_j^k is mapped by the linear map

$$(x_1, \ldots, x_k) \mapsto \left(\frac{x_1 - a_1}{b_1 - a_1}, \ldots, \frac{x_k - a_k}{b_k - a_k} \right)$$

to the jth sphere on the right, and points outside these subcubes are sent to the basepoint. We topologise $\mathcal{C}_k(n)$ as a subspace of the space of maps. The group Σ_n acts freely by permuting the n little cubes I_j^k.

LEMMA 2.7.1. $\mathcal{C}_k(n)$ is $(k-2)$-connected.

PROOF. There is an obvious fibration $\mathcal{C}_k(n) \to \mathcal{C}_k(n-1)$ whose fibre is the space of subcubes $I_n^k \subseteq I^k \setminus \bigcup_{j=1}^{n-1} I_j^k$. Replacing such a subcube by its centre is a homotopy equivalence of this space with $I^k \setminus \bigcup_{j=1}^{n-1} I_j^k$, and this space is homotopy equivalent to I^k with $(n-1)$ points removed. So this fibre is $(k-2)$-connected, and the lemma follows by induction on n using the long exact sequence of homotopy groups of a fibration (Theorem 1.6.6).   □

Now there are obvious inclusions $\mathcal{C}_k(n) \hookrightarrow \mathcal{C}_{k+1}(n)$, and we define

$$\mathcal{C}(n) = \bigcup_{k=1}^{\infty} \mathcal{C}_k(n)$$

with the weak topology with respect to the union.

PROPOSITION 2.7.2. $\mathcal{C}(n)$ is a contractible CW-complex on which Σ_n acts freely.

PROOF. Since $\mathcal{C}(n)$ was given the weak topology with respect to the union, it is a CW-complex. By the lemma, we have $\pi_i(\mathcal{C}(n)) = 0$ for all $i \geq 0$, and so by Theorem 1.5.8, $\mathcal{C}(n)$ is contractible. $\square$

It follows that we may use $\mathcal{C}(n)$ as our model for $E\Sigma_n$, and $\mathcal{C}(n)/\Sigma_n = B\Sigma_n$.

Now suppose $K_r \simeq \Omega^k K_{r+k} = \mathrm{Map}_*(S^k, s_0; K_{r+k}, x_0)$ as above. Then composition of maps gives a well defined map

$$\mathcal{C}_k(n) \times_{\Sigma_n} (\Omega^k K_{r+k})^n \to \Omega^k K_{r+k}$$

and hence a map

$$\mathcal{C}_k(n) \times_{\Sigma_n} (K_r)^n \to K_r$$

well defined up to homotopy. Passing to the limit, this gives a map

$$E\Sigma_n \times_{\Sigma_n} (K_r)^n \xrightarrow{DL} K_r$$

well defined up to homotopy, and this is the Dyer–Lashof map.

This completes the description of the transfer for finite coverings. In fact, there is a generalisation due to Becker and Gottlieb [26], for bundles with compact fibre and structure group a compact Lie group. We now describe a version of their construction, modified to look as much like the above discussion as possible.

Let G be a compact Lie group and F a compact smooth manifold on which G acts smoothly and faithfully. Let $\xi = (p : E \to B)$ be a principal G-bundle and $\xi[F] = (p_F : X = E \times_G F \to B)$ the associated bundle with structure group G and fibre F (see Definition 2.3.3). According to Definition 2.4.3 (and Theorem 2.4.6) there is a map $f : B \to BG$ with $f^*(\xi_G) = \xi$, so that we have a pullback square

$$
\begin{array}{ccc}
E & \longrightarrow & EG \\
\downarrow{\scriptstyle p} & & \downarrow \\
B & \xrightarrow{f} & BG.
\end{array}
$$

Also, the obvious map $E \times F \to X$ gives us a map

$$E \to \mathrm{Map}(F, X) = \mathrm{Map}_*(F_+, X),$$

where F_+ denotes F with disjoint basepoint. These maps give us a map $E \to EG \times \mathrm{Map}_*(F_+, X)$, and hence by quotienting out the action of G on both sides, a map

$$P\mathrm{Tr} : B \to EG \times_G \mathrm{Map}_*(F_+, X)$$

which is the appropriate pretransfer map for this situation.

As before, to define the Dyer–Lashof map

$$DL : EG \times_G \mathrm{Map}_*(F_+, K_r) \to K_r$$

we need a specific model for EG. By a theorem of Mostow [**196**], we may choose a fixed smooth embedding $i : F \hookrightarrow V = \mathbb{R}^m$ of F into a finite dimensional real orthogonal representation V of G, so that the G-action of F is compatible with the inclusion. We use as our model for EG the set of embeddings of F in $\mathbb{R}^\infty$ of the form

$$F \xrightarrow{i} V = \mathbb{R}^m \to \mathbb{R}^\infty$$

where this last map is an injective linear map. In other words, we have embedded G in a real orthogonal group $O(m)$ and used the Grassmannian model for $EO(m)$ as our model for EG. In particular it is clear that this EG is a contractible space on which G acts freely.

Now choose $\varepsilon > 0$ small enough so that in an ε-neighbourhood of F in V, every point has a unique closest point in F, so that this neighbourhood is locally a product of F and a small ball. Such an ε exists because F is compact, and the resulting neighbourhood is called a **tubular neighbourhood** of F in V.

Any element $\gamma \in EG$ can be regarded as a map $F \hookrightarrow \mathbb{R}^k \subseteq \mathbb{R}^\infty$ for k large enough, where $\mathbb{R}^k$ is the span of the first k coordinates in $\mathbb{R}^\infty$. We denote by F_ε the corresponding tubular neighbourhood of F in $\mathbb{R}^k$. Denote by D_ε^k an (abstract) open ball of real dimension k and diameter ε. Then there is a map

$$\tilde\gamma : S^k \to (F \times D_\varepsilon^k) \cup \{\infty\} = F_+ \wedge S^k$$

defined as follows. We regard the left hand S^k as $\mathbb{R}^k \cup \{\infty\}$. Then $\tilde\gamma$ sends ∞, together with all points in $\mathbb{R}^k$ not in F_ε, to ∞. It sends a point in F_ε to the pair consisting of the closest point in F and the displacement vector (expressing the displacement from this closest point) in D_ε^k.

Now suppose $K_r \simeq \Omega^k K_{r+k}$. Then we have

$$\mathrm{Map}_*(F_+, K_r) \cong \mathrm{Map}_*(F_+ \wedge S^k, K_{r+k})$$

$$\beta \mapsto \tilde\beta.$$

The map sending a pair of elements $\gamma \in EG$ and $\beta \in \mathrm{Map}_*(F_+, K_r)$ to the composite $\tilde\beta \circ \tilde\gamma \in \mathrm{Map}_*(S^k, K_{r+k}) = K_r$ gives us a map

$$DL : EG \times_G \mathrm{Map}_*(F_+, K_r) \to K_r,$$

and this is the Dyer–Lashof map.

Just as before, if $\alpha \in h^r(X)$ is represented by $\hat\alpha : X_+ \to K_r$, then the transfer $(p_F)_!(\alpha) \in h^r(B)$ (or just $\mathrm{Tr}(\alpha)$) is represented by the composite map

$$B_+ \xrightarrow{\mathrm{PTr}} (EG \times_G \mathrm{Map}_*(F_+, X))_+ \xrightarrow{1 \times \hat\alpha_*} EG \times_G \mathrm{Map}_*(F_+, K_r) \xrightarrow{DL} K_r.$$

Note that if $F \hookrightarrow V'$ is another embedding as in Mostow's theorem, then there is a one parameter family of embeddings $F \hookrightarrow V \oplus V'$, and so the corresponding Dyer–Lashof maps are homotopic.

It is clear from the construction of the transfer that it is natural, in the sense that if

$$
\begin{array}{ccc}
X' & \xrightarrow{\ \mu\ } & X \\
\downarrow & & \downarrow \\
B' & \xrightarrow{\ \lambda\ } & B
\end{array}
$$

is a pullback of fibrations with fibre F and structure group G, then for $\alpha \in h^r(X)$ we have

$$\mathrm{Tr}\mu^*(\alpha) = \lambda^*\mathrm{Tr}(\alpha) \in h^r(B').$$

LEMMA 2.7.3. *Suppose* $\tilde{\gamma} : S^k \to F_+ \wedge S^k$ *represents* $\gamma \in EG$ *as above, and denote by* π *the second projection* $F_+ \wedge S^k \to S^k$. *Then the composite* $\pi \circ \tilde{\gamma} : S^k \to S^k$ *has degree equal to the Euler characteristic* $\chi(F) = \sum_{i\geq 0}(-1)^i \dim H^i(F)$.

PROOF. We first remark that this is clear for the original case where F was a finite set of size $n = \chi(F)$, since each component of the map $S^k \to \bigvee_{j=1}^n S^k$ has degree one and we are just adding.

We shall make the general case look like the finite case by using a little elementary Morse theory. If you read Milnor [**191**] up to the end of p. 36, you will know enough Morse theory to understand this proof.

Choose a Morse function $\phi : F \to \mathbb{R}$; i.e., a smooth function with isolated critical points. Morse theory says that F is homotopy equivalent to a CW-complex with one d-cell for each critical point on index d. Since F is compact, there are only finitely many critical points. By scaling up the Morse function if necessary, we may assume that the neighbourhoods $|\mathrm{grad}\,\phi| < \varepsilon$ (ε as above) of the critical points are disjoint open balls in F.

Now the map $\pi \circ \tilde{\gamma}$ is really just the displacement map $S^k \to D_\varepsilon^k \cup \{\infty\} = S^k$. This displacement map is perpendicular to $\mathrm{grad}\,\phi$ at the closest point on F, and so we can just add these to give a map $\pi \circ \tilde{\gamma} + \mathrm{grad}\,\phi$, which we regard as a map from S^k to $D_\varepsilon^k \cup \{\infty\}$. It is homotopic to $\pi \circ \tilde{\gamma}$ by the homotopy $\pi \circ \tilde{\gamma} + t\,\mathrm{grad}\,\phi$ ($0 \leq t \leq 1$), and therefore has the same degree. But now $\pi \circ \tilde{\gamma} + \mathrm{grad}\,\phi$ sends everything to $\{\infty\}$ except for non-overlapping spherical neighbourhoods of the critical points. The degree of the map on this spherical neighbourhood is $(-1)^d$ if the index is d (Milnor [**191**], bottom of p. 36). So the total degree of the map $S^k \to S^k$ is the sum over the cells in the CW-decomposition of F of (-1) to the power of the dimension of the cell; this is the Euler characteristic $\chi(F)$.   □

PROPOSITION 2.7.4. *The restriction map followed by the transfer*

$$H^*(B) \xrightarrow{\ p_F^*\ } H^*(X) \xrightarrow{\ \mathrm{Tr}\ } H^*(B)$$

is equal to multiplication by $\chi(F)$.

PROOF. By the lemma, if $F \to *$ is a trivial bundle with fibre F over a point, then the transfer of the identity element in $H^0(F)$ is $\chi(F)$ times the identity in $H^0(*)$.

Without loss of generality, we may suppose B is connected. Using naturality of the transfer on the square

$$
\begin{array}{ccc}
F & \longrightarrow & X \\
\downarrow & & \downarrow{\scriptstyle p_F} \\
* & \longrightarrow & B
\end{array}
$$

we see that the transfer of the identity element in $H^0(X)$ is $\chi(F)$ times the identity element in $H^0(B)$. Now using naturality of the transfer on the square

$$
\begin{array}{ccc}
X & \xrightarrow{(p_F,1)} & B \times X \\
\downarrow{\scriptstyle p_F} & & \downarrow{\scriptstyle 1 \times p_F} \\
B & \xrightarrow{\Delta} & B \times B
\end{array}
$$

for $\alpha \in H^*(B)$ we have

$$\operatorname{Tr} p_F^*(\alpha) = \operatorname{Tr}(p_F,1)^*(\alpha \times 1) = \Delta^* \operatorname{Tr}(\alpha \times 1) = \Delta^*(\alpha \times \chi(F)) = \chi(F).\alpha.$$

$\square$

2.8. Stable cohomotopy and the Segal conjecture

In Theorem 2.5.5, we described the K-theory of a classifying space of a finite group as the completion of the representation ring with respect to the augmentation ideal. In this section, we describe another generalised cohomology theory, namely stable cohomotopy π_s^*. Graeme Segal conjectured that this theory evaluated on BG should bear exactly the same relation to the Burnside ring as $K^*(BG)$ does to the representation ring; namely, it should equal the completion of the Burnside ring at the augmentation ideal. This conjecture was proved by G. Carlsson [77].

To explain stable cohomotopy, we first explain the notion of stable maps. If X and Y are finite dimensional pointed CW-complexes, then a map from X to Y gives rise to a map between the suspensions $SX \to SY$ in an obvious way. This gives us a well defined map

$$[X;Y] \to [SX;SY]$$

and we define **stable maps** (i.e., stabilised with respect to suspension) from X to Y to be

$$\{X;Y\} = \varinjlim_m [S^m X; S^m Y].$$

Note that if $\alpha, \beta : S^m X \to S^m Y$ then we can define the sum $S(\alpha) + S(\beta) : S^{m+1}X \to S^{m+1}Y$ by dividing the suspension coordinate into two halves. Thus $\{X;Y\}$ has the structure of an abelian group, and $\{X;X\}$ has the structure of a ring, with multiplication given by composition.

Now recall from Section 1.2 that

$$[S^m X; S^m Y] = [X; \Omega^m S^m Y]$$

so that

$$\{X, Y\} = \varinjlim_m [X; \Omega^m S^m Y] = [X; \varinjlim_m \Omega^m S^m Y]$$

where again the maps $\Omega^m S^m Y = \mathrm{Map}_*(S^m, S^m Y) \to \Omega^{m+1} S^{m+1} Y$ are given by suspension as above. The fact that we may pass the limit sign through the brackets depends on the fact that X is a finite dimensional complex, and for a general CW-complex X the second expression is better behaved and so we define in general

$$\{X; Y\} = [X; \Omega^\infty S^\infty Y]$$

where $\Omega^\infty S^\infty$ denotes $\varinjlim_m \Omega^m S^m$. Composition of maps gives in an obvious way a map

$$\{Y; Z\} \times \{X; Y\} \to \{X; Z\}.$$

We can now define the **stable homotopy** π_*^s and **stable cohomotopy** π_s^* of a space via

$$\pi_r^s(X) = \{S^r; X_+\}, \quad \pi_s^r(X) = \{X_+; S^r\} = [X_+; \varinjlim_m \Omega^m S^{r+m}].$$

where X_+ denotes X with a disjoint basepoint.

Stable cohomotopy is a generalised cohomology theory. Note that the exactness axiom fails if we use ordinary cohomotopy $[X; S^r]$ instead of stabilising with respect to suspension. Also note that the space $\varinjlim_m \Omega^m S^{r+m}$ makes sense for r negative as well as positive. More generally, for any space Y, the functors

$$h_Y^r : X \mapsto \{X; S^r Y\}$$

form a generalised cohomology theory. Note that it is easy to see that

$$\Omega^\infty S^\infty (S^r Y) = \Omega(\Omega^\infty S^\infty (S^{r+1} Y)).$$

Now the transfer map (Section 2.7) gives us some elements of $\pi_s^0(BG)$, for G a finite group. Namely, if H is a subgroup of G, then there is a finite covering map $BH \to BG$. Regarding $\{-; BH_+\}$ as a generalised cohomology theory, we have a transfer map

$$\{BH_+; BH_+\} \to \{BG_+; BH_+\}.$$

The image of the identity element under this map is a stable map from BG_+ to BH_+ (i.e., a map $\mathrm{Tr}_{H,G} : BG_+ \to \Omega^\infty S^\infty BH_+$) which is also called the transfer. Composing with the element of $\{BH_+; S^0\}$ given by identifying BH to a point, we obtain an element of $\pi_s^0(BG)$ corresponding to H. This element only depends on the conjugacy class of H as a subgroup of G, and so this way we obtain a map from the Burnside ring $b(G)$ (see Section 5.4 of Volume I) to stable cohomotopy

$$b(G) \to \pi_s^0(BG).$$

The Segal conjecture says that the above map induces an isomorphism between the completion $b(G)^\wedge$ and $\pi_s^0(BG)$. The completion is taken with respect to the kernel I_G of the augmentation map $b(G) \to \mathbb{Z}$ sending a permutation representation to the number of points being permuted,

$$b(G)^\wedge = \varprojlim_n b(G)/I_G^n.$$

As in the case of K-theory (Theorem 2.5.5), the Segal conjecture also contains information about $\pi_s^n(BG)$ for $n \neq 0$. In case $G = 1$, the groups $\pi_s^n(\text{point}) = \pi_{-n}^s(\text{point})$ are the stable homotopy groups of spheres for n negative (the calculation of which is still an open problem), and zero for n positive. In general, Segal's conjecture includes the statement that $\pi_s^n(BG) = 0$ for n positive. There is no trivial reason why this should be true, as there is (Hurewicz theorem) in case $G = 1$.

THEOREM 2.8.1 (Segal conjecture; Carlsson). *The transfer maps described above induce an isomorphism*

$$b(G)^\wedge \to \pi_s^0(BG).$$

Furthermore, for $n > 0$, $\pi_s^n(BG) = 0$.

PROOF. The proof depends on developing an equivariant version of stable cohomotopy theory, and is beyond the scope of this book. For details see Carlsson [77]. □

It has been pointed out by Adams [2] (this is developed in detail in Lewis, May and McClure [166]) that as a consequence of the Segal conjecture, one knows all the stable maps between BG' and BG for finite groups G' and G. If H' is a subgroup of G' and $\phi : H' \to G$ is a group homomorphism, then the composite of $B\phi \in \{BH'_+; BG_+\}$ and $\text{Tr}_{H',G'} \in \{BG'_+; BH'_+\}$ gives us an element of $\{BG'_+; BG_+\}$ which we write as $\zeta_{H',\phi}$.

If $\zeta_{H',\phi} \in \{BG'_+; BG_+\}$ and $\zeta_{H'',\phi'} \in \{BG''_+; BG'_+\}$ are such maps, then we can calculate the composite $\zeta_{H',\phi} \circ \zeta_{H'',\phi'} \in \{BG''_+; BG_+\}$ by using the Mackey decomposition theorem as follows. Factor $\phi' : H'' \to G'$ as a surjection $H'' \to K'$ followed by an inclusion $i' : K' \hookrightarrow G'$. Then the Mackey formula enables us to express the composite of the stable maps

$$BK' \xrightarrow{B(i')} BG' \xrightarrow{\text{Tr}} BH'$$

as a sum of stable maps

$$BK' \xrightarrow{\text{Tr}} B(K' \cap {}^x H') \xrightarrow{c_{x^{-1}}} B({}^{x^{-1}} K' \cap H') \xrightarrow{B(i')} BH'$$

where x runs over a set of double coset representatives $K'\backslash G'/H'$ and $c_{x^{-1}}$ denotes conjugation by x^{-1}. Then the composite

$$BH'' \xrightarrow{B(\phi')} BK' \xrightarrow{\text{Tr}} B(K' \cap {}^x H')$$

is equal to the composite

$$BH'' \xrightarrow{\text{Tr}} B((\phi')^{-1}(K' \cap {}^x H')) \xrightarrow{B(\phi')} B(K' \cap {}^x H'),$$

and so we have expressed the composite as a sum of basis elements of the given kind

$$\boxed{\zeta_{H',\phi} \circ \zeta_{H'',\phi'} = \sum_{x \in K'\backslash G'/H'} \zeta_{(\phi')^{-1}(K' \cap {}^x H'),\phi \circ i' \circ c_{x^{-1}} \circ \phi'}.}$$

This formula for the composite is illustrated in the following diagram:

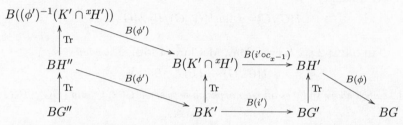

Denote by $b(G', G)$ the free abelian group whose basis elements are labelled $\zeta_{H',\phi}$, one for each conjugacy class of pairs (H', ϕ), where H' is a subgroup of G' and $\phi : H' \to G$. Then the above formula for multiplying basis elements gives a bilinear map

$$b(G', G) \times b(G'', G') \to b(G'', G).$$

This makes $b(G, G)$ into a ring, which may be thought of as a non-commutative analogue of the Burnside ring $b(G)$. This multiplication also makes $b(G', G)$ into a $b(G, G)$-$b(G', G')$-bimodule. We have a map

$$b(G', G) \to \{BG'_+; BG_+\}$$

sending the basis element $\zeta_{H',\phi}$ to the corresponding stable map. We wish to state that this is an isomorphism after completing $b(G', G)$ with respect to a suitable filtration.

Before describing this completion, we give another interpretation of $b(G', G)$ and the above map. Namely, $b(G', G)$ is the abelian group with generators the finite $G' \times G$-sets which are free on restriction to G (i.e., to $1 \times G$); and relations $[X] + [Y] = [X \dot\cup Y]$. The basis element $\zeta_{H',\phi}$ described above corresponds to the transitive permutation representation of $G' \times G$ on the cosets of

$$\Delta_{H',\phi} = \{(h', \phi(h')), h' \in H'\} \subseteq G' \times G.$$

Clearly every finite G-free $G' \times G$-set can be written as a disjoint union of such transitive $G' \times G$-sets in an essentially unique way.

If X is a finite G-free $G' \times G$-set, then the corresponding stable map from BG'_+ to BG_+ can be described as follows. The space $EG' \times_{G'} X$ admits a free G-action, and so the covering map $EG' \times_{G'} X \to EG' \times_{G'} X/G$ is a principal G-bundle. Thus (Theorem 2.4.6) there is a corresponding map $EG' \times_{G'} X/G \to BG$, well defined up to homotopy. The stable map from BG'_+ to BG_+ corresponding to X is the transfer for the finite covering $EG' \times_{G'} X/G \to BG'$ (and send the disjoint basepoint in BG'_+ to the disjoint basepoint in $(EG' \times_{G'} X/G)_+$) followed by the above map $(EG' \times_{G'} X/G)_+ \to BG_+$.

We make $b(G', G)$ into a $b(G')$-module as follows. If X is a finite G-free $G' \times G$-set and Y is a finite G'-set, then $Y \times X$ is a $G' \times G' \times G$-set, and we restrict to the diagonal copy of $G' \times G$ to get a finite G-free $G' \times G$-set. We write $b(G', G)^\wedge$ for the completion of $b(G', G)$ at the augmentation ideal $I_{G'}$ of $b(G')$; namely

$$b(G', G)^\wedge = \varprojlim_n b(G', G)/I_{G'}^n . b(G', G).$$

THEOREM 2.8.2 (Lewis, May, McClure [166]). *The map*

$$b(G', G) \to \{BG'_+; BG_+\}$$

described above induces an isomorphism between $b(G', G)^\wedge$ and $\{BG'_+; BG_+\}$. □

Finally, we point out that as a consequence of this theorem, stable splittings of BG_+ as a wedge sum correspond to decompositions of the identity in $b(G, G)^\wedge$ as a sum of primitive orthogonal idempotents. This subject is pursued further in Martino and Priddy [175] and Benson and Feshbach [40].

2.9. Cohomology of general linear groups

In this section we describe Quillen's calculation [212] of the cohomology of general linear groups over a finite field, and in the next section we describe how this led to Quillen's definition of the algebraic K-groups of a ring (this particular case calculates the algebraic K-groups of a finite field). In fact, it is possible to calculate the cohomology of these general linear groups entirely algebraically (Kroll [158]), but this hides the most striking features of Quillen's calculation.

The idea of Quillen's calculation is to think of the Adams operation ψ^q, where q is a power of a prime p, as a map from BU to itself, take its "homotopy fixed points" $F\psi^q$, and to show that there is a map $BGL_n(\mathbb{F}_q) \to F\psi^q$ giving interesting cohomology classes by pulling back, and in the limit inducing a cohomology equivalence $BGL(\mathbb{F}_q) \to F\psi^q$, where $GL(\mathbb{F}_q)$ is the union of the $GL_n(\mathbb{F}_q)$ embedded in each other using the top left hand corner. The details are as follows.

The first problem is to represent ψ^q as a map from BU to itself. According to Section 2.5, for B compact, ψ^q is a natural transformation on $\tilde{K}(B)$. We would like to apply Yoneda's lemma (2.1.4 of Volume I), and say that because $\tilde{K}(B)$ is representable as $[B; BU]$, the natural transformation ψ^q is representable by a map $BU \to BU$. The problem is that BU is not compact. However, it has a CW decomposition $BU = \bigcup X^m$ with only cells in even dimension (this follows from the row echelon form for matrices—see Milnor and Stasheff [192], Section 6). One can show inductively (using the Bott periodicity theorem) that on a finite CW-complex which only has cells in even dimension, the functor K^{-1} vanishes, and so the Milnor exact sequence (see Theorem 1.9.1)

$$0 \to \varprojlim{}^1 K^{-1}(X^m) \to K(BU) \to \varprojlim K(X^m) \to 0$$

reduces to $K(BU) = \varprojlim K(X^m)$. Thus the operations ψ^q are defined on $K(BU)$ and are hence represented by maps $\psi^q : BU \to BU$.

We now describe the process of taking homotopy fixed points. This is similar to the process of taking the homotopy fibre of a map, as described in Section 1.6. If $\phi : X \to X$ is a self-map of a space, then the fixed points X^ψ of ψ are obtained by taking the pullback

$$
\begin{array}{ccc}
X^\phi & \longrightarrow & X \\
\downarrow & & \downarrow{\scriptstyle \Delta} \\
X & \xrightarrow{(1,\phi)} & X \times X
\end{array}
$$

where Δ is the diagonal map. The **homotopy fixed points** $X^{h\phi}$ are obtained by replacing Δ by a fibration. Namely, if we denote by $\tilde{\Delta} : X^I \to X \times X$ the map taking a path to its endpoints, then $\tilde{\Delta}$ is a fibration and it is homotopy equivalent to Δ (in other words, the inclusion of X into X^I as the constant maps is a homotopy equivalence which when composed with $\tilde{\Delta}$ gives Δ). So we define the homotopy fixed points of ϕ to be the pullback

$$
\begin{array}{ccc}
X^{h\phi} & \longrightarrow & X^I \\
\downarrow & & \downarrow{\scriptstyle \tilde{\Delta}} \\
X & \xrightarrow{(1,\phi)} & X \times X
\end{array}
$$

The vertical maps in this diagram are fibrations with fibre ΩX. Note that there is an obvious map $X^\phi \to X^{h\phi}$.

Now suppose that X has a basepoint x_0 and comes with an additive structure; namely an addition map $X \times X \to X$ which is associative and commutative up to homotopy, and a map $X \to X$ which acts as an additive inverse up to homotopy. BU is an example of such a space, in which the addition is given by the limit of the Whitney sum maps $G^n(\mathbb{C}^\infty) \times G^m(\mathbb{C}^\infty) \to G^{n+m}(\mathbb{C}^\infty)$ and negation is given by the obvious maps $G^n(\mathbb{C}^m) \to G^{m-n}(\mathbb{C}^m)$ (see the remarks in the second paragraph of Section 2.5). Putting these addition and inverse maps together, we get a subtraction map $d : X \to X$.

Denoting by PX the (contractible) path space $\mathrm{Map}_*((I,0),(X,x_0))$, we can extend the above pullback diagram to a diagram

$$
\begin{array}{ccccc}
X^{h\phi} & \longrightarrow & X^I & \longrightarrow & PX \\
\downarrow & & \downarrow{\scriptstyle \tilde{\Delta}} & & \downarrow \\
X & \xrightarrow{(1,\phi)} & X \times X & \xrightarrow{d} & X
\end{array}
$$

Here, the map $X^I \to PX$ sends a path $t \mapsto \omega(t)$ to the path $t \mapsto d(\omega(t), \omega(0))$ obtained by shifting the starting point to the origin. The map $PX \to X$ is the map sending a path ω to its endpoint $\omega(1)$. It follows that $X^{h\phi}$ is exactly the homotopy fibre of the map $1 - \phi = d \circ (1, \phi)$.

We give $X^{h\phi}$ an additive structure as follows. Denoting by $P_{1-\phi}$ the space of ordered pairs (x, ω) consisting of a point x in X and an unbased map $\omega : I \to X$ with $\omega(1) = d(x, \phi(x))$ as in Section 1.6, we know that $P_{1-\phi}$ is homotopy equivalent to X, and $X^{h\phi}$ is the fibre of the map $P_{1-\phi} \to X$ taking (x, ω) to $\omega(1) = d(x, \phi(x))$. The addition map on X gives us maps

$$
\begin{array}{ccc}
P_{1-\phi} \times P_{1-\phi} & \longrightarrow & P_{1-\phi} \\
\downarrow & & \downarrow \\
X \times X & \longrightarrow & X
\end{array}
$$

and hence a map $X^{h\phi} \times X^{h\phi} \to X^{h\phi}$. Similarly the inverse map $X \to X$ gives us an inverse map $X^{h\phi} \to X^{h\phi}$. It is easy to check that these maps give an additive structure on $X^{h\phi}$ in the sense described above.

LEMMA 2.9.1. *If X has an additive structure in the sense described above, and Y is a space with the property that every map $Y \to \Omega X$ is homotopic to the constant map, then $[Y; X^{h\phi}] = [Y; X]^{\phi}$, where the latter denotes the fixed points of composition with ϕ on $[Y; X]$.*

PROOF. A map from Y to $X^{h\phi}$ can be thought of as a map $f : Y \to X$ and a homotopy from f to $\phi \circ f$. So composition with $X^{h\phi} \to X$ gives a surjective map $[Y; X^{h\phi}] \to [Y; X]^{\phi}$. Both X and $X^{h\phi}$ have an additive structure, which induce abelian group structures on these sets of homotopy classes. It is easy to check that the above map is a homomorphism of abelian groups. If $f : Y \to X^{h\phi}$ is in the kernel of this homomorphism, then by the homotopy lifting property for the fibration $X^{h\phi} \to X$, f is homotopic to a map whose image lies in the fibre ΩX of this fibration.

$$
\begin{array}{ccc}
Y \times 0 & \longrightarrow & X^{h\phi} \\
\downarrow & & \downarrow \\
Y \times I & \longrightarrow & X
\end{array}
$$

Since every map $Y \to \Omega X$ is homotopic to the constant map by hypothesis, the lemma is proved. $\qquad\square$

We wish to apply the above lemma in the situation where X is BU and ϕ is ψ^q. Following Quillen, we write $F\psi^q$ for the homotopy fixed point set in this case, namely the homotopy fibre of $1 - \psi^q : BU \to BU$.

By Theorem 2.4.11, ΩBU is homotopy equivalent to the infinite unitary group U, namely the union of the finite unitary groups embedded in each other using the top left hand corner.

If G is a finite group, then we saw in Section 2.5 that there is a natural map from the representation ring to K-theory given by using a representation to form a vector bundle. We compose this with the projection down to reduced K-theory:

$$\mathcal{R}(\mathbb{C}G) \longrightarrow K^0(BG) \quad = [BG; BU \times \mathbb{Z}]$$

$$\downarrow$$

$$\tilde{K}^0(BG) \quad = \quad [BG; BU]$$

This map clearly commutes with the operations ψ^q, which are defined on both sides, see Section 2.5, and Section 5.9 of Volume I, so that we have a well defined map $\mathcal{R}(\mathbb{C}G)^{\psi^q} \to [BG; BU]^{\psi^q}$. The Atiyah Completion Theorem 2.5.5 says that the above map induces an isomorphism $\mathcal{R}(\mathbb{C}G)^\wedge \cong K^0(BG)$, where $\mathcal{R}(\mathbb{C}G)^\wedge$ is the completion of $\mathcal{R}(\mathbb{C}G)$ at the augmentation ideal I_G. So we also have an isomorphism $I_G^\wedge \cong \tilde{K}^0(BG)$. Moreover, this theorem also states that $[BG; \Omega BU] = [BG; U] = K^1(BG) = 0$. So we are able to apply the above lemma to see that

$$[BG; F\psi^q] = [BG; BU^{h\psi^q}] \cong [BG; BU]^{\psi^q}.$$

It follows that there is a well defined map

$$\mathcal{R}(\mathbb{C}G)^{\psi^q} \to [BG; F\psi^q].$$

Now recall from Section 5.9 of Volume I that the Brauer lift gives an isomorphism

$$\mathcal{R}(\mathbb{C}G)^{\psi^q} \cong \mathcal{R}(\mathbb{F}_q G)$$

so that we have a well defined map

$$\mathcal{R}(\mathbb{F}_q G) \to [BG; F\psi^q].$$

So given any finitely generated $\mathbb{F}_q G$-module, we obtain a map, well defined up to homotopy, from BG to $F\psi^q$. Pulling back elements of cohomology of $F\psi^q$ through this map, we obtain elements of the cohomology of G, which can be thought of as **characteristic classes** for modular representations.

We now calculate the homotopy and cohomology of $F\psi^q$. The homotopy is easy, so we begin with that. Regarding $F\psi^q$ as the homotopy fibre of

$1 - \psi^q : BU \to BU$, we obtain a long exact sequence of homotopy groups

$$\cdots \to \pi_j(BU) \xrightarrow{1-\psi^q} \pi_j(BU) \to \pi_{j-1}(F\psi^q)$$
$$\to \pi_{j-1}(BU) \xrightarrow{1-\psi^q} \pi_{j-1}(BU) \to \cdots$$

Now for $r > 0$ we saw in Section 2.5 that $\pi_j(BU) = [S^j; BU] = K(S^j)$ is isomorphic to $\mathbb{Z}$ if j is even and is zero if j is odd. An explicit calculation on S^2 shows that ψ^q acts as multiplication by q on $K(S^2) \cong \mathbb{Z}$ and hence as multiplication by q^j on $K(S^{2j})$ by Bott periodicity. Thus we have

$$\pi_{2j-1}(F\psi^q) = \mathbb{Z}/(q^{2j} - 1), \qquad \pi_{2j}(F\psi^q) = 0.$$

It follows that there is no p-torsion in $\pi_*(F\psi^q)$, and so by the p-local version of the Hurewicz Theorem 1.3.6, $H_i(F\psi^q; \mathbb{F}_p) = 0$ for $i > 0$. Similarly, applying this theorem in characteristic zero, we see that $H_i(F\psi^q; \mathbb{Q}) = 0$ for $i > 0$. Note also that if we examine the above sequence as a long exact sequence of modules for $\pi_1(F\psi^q)$, and we use the fact that $\pi_1(BU) = 0$, we see that $\pi_1(F\psi^q)$ acts trivially on $\pi_n(F\psi^q)$ for all n, so that $F\psi^q$ is a simple space.

We now describe the calculation of the cohomology of $F\psi^q$ with coefficients in $\mathbb{F}_l$, where l is any prime other than p. One applies the Eilenberg–Moore spectral sequence to the pullback square

$$\begin{array}{ccc} F\psi^q & \longrightarrow & BU^I \\ \downarrow & & \downarrow{\tilde{\Delta}} \\ BU & \xrightarrow{(1,\psi^q)} & BU \times BU \end{array}$$

obtained by regarding $F\psi^q$ as a homotopy fixed point set. Look at Section 3.7 for a brief discussion of the Eilenberg–Moore spectral sequence. In this case, the E_2 term is

$$E_2^{**} = \mathrm{Tor}_{H^*(BU \times BU)}^{**}(H^*(BU), H^*(BU^I))$$

and the spectral sequence converges to $H^*(F\psi^q)$ (all cohomology with $\mathbb{F}_l$ coefficients). As a module over

$$H^*(BU \times BU) = \mathbb{F}_l[c_1', c_2', \ldots, c_1'', c_2'', \ldots]$$

via the above maps, we have

$$H^*(BU^I) = H^*(BU \times BU)/(c_1' - c_1'', c_2' - c_2'', \ldots)$$
$$H^*(BU) = H^*(BU \times BU)/(qc_1' - c_1'', q^2 c_2' - c_2'', \ldots)$$

Hence

$$H^*(BU) \otimes_{H^*(BU \times BU)} H^*(BU^I) = \mathbb{F}_l[c_1, c_2, \ldots]/((q-1)c_1, (q^2 - 1)c_2, \ldots)$$
$$= \mathbb{F}_l[c_r, c_{2r}, \ldots]$$

where r is the multiplicative order of q modulo l, and c_{jr} has degree $2jr$.

The standard techniques for calculating Tor over a polynomial ring shows that the E_2 page is a tensor product of a polynomial and an exterior algebra

$$E_2^{**} = \mathbb{F}_l[c_r, c_{2r}, \dots] \otimes \Lambda(e_r, e_{2r}, \dots)$$

where the $c_{jr} \in E_2^{0,2jr}$ and $e_{jr} \in E_2^{-1,2jr}$. Since this is a second quadrant spectral sequence and the differentials take these generators into the first quadrant, we have $E_2 = E_\infty$. It follows that the additive structure of $H^*(F\psi^q; \mathbb{F}_l)$ is the same as E_2^{**}, and the multiplicative structure is the same with the exception that instead of $e_{jr}^2 = 0$ we only know that e_{jr}^2 is in the subring generated by $c_r, c_{2r}, \dots$.

WARNING. You may be inclined to guess from the above calculation that

$$H^*(F\psi^q; \mathbb{Z}) = \mathbb{Z}[c_1, c_2, \dots]/((q-1)c_1, (q^2-1)c_2, \dots).$$

This is easily seen to be false, by noticing that if this were the case, then $H^*(F\psi^q; \mathbb{Z})$ would be concentrated in even degrees, so that the long exact sequence

$$\cdots \to H^{i-1}(F\psi^q; \mathbb{F}_l) \to H^i(F\psi^q; \mathbb{Z}) \xrightarrow{\times l} H^i(F\psi^q; \mathbb{Z}) \to H^i(F\psi^q; \mathbb{F}_l) \to \cdots$$

coming from the short exact sequence of coefficients

$$0 \to \mathbb{Z} \to \mathbb{Z} \to \mathbb{F}_l \to 0$$

would reduce to

$$0 \to H^{2i-1}(F\psi^q; \mathbb{F}_l) \to H^{2i}(F\psi^q; \mathbb{Z}) \xrightarrow{\times l} H^{2i}(F\psi^q; \mathbb{Z}) \to H^{2i}(F\psi^q; \mathbb{F}_l) \to 0.$$

However, we have seen that $H^{2*}(F\psi^q; \mathbb{F}_l)$ contains even products of the e_{jr}'s, which are not in the image of reduction modulo l. For further information on the integral cohomology of $F\psi^q$, see Huebschmann [130] and Jeandupeux [139].

We now complete the determination of the multiplicative structure of the ring $H^*(F\psi^q; \mathbb{F}_l)$ at the same time as calculating $H^*(GL_n(\mathbb{F}_q), \mathbb{F}_l)$, as follows. The canonical modular representation of dimension n of $GL_n(\mathbb{F}_q)$ over $\mathbb{F}_q$ gives, via the Brauer lift as above, a map

$$BGL_n(\mathbb{F}_q) \to F\psi^q$$

well defined up to homotopy. We shall see that the characteristic classes c_{jr}, e_{jr} defined by this map generate $H^*(GL_n(\mathbb{F}_q), \mathbb{F}_l)$. We do this by restricting to a suitable abelian subgroup.

Let C denote a cyclic group of order $q^r - 1$, so that C has an irreducible representation of dimension r over $\mathbb{F}_q$ (via the isomorphism $C \cong \mathbb{F}_{q^r}^\times$). Thus the direct product C^m of m copies of C has a faithful representation of dimension mr over $\mathbb{F}_q$, and hence we have an embedding $C^m \hookrightarrow GL_n(\mathbb{F}_q)$.

LEMMA 2.9.2. *The restriction map*

$$H^*(GL_n(\mathbb{F}_q), \mathbb{F}_l) \to H^*(C^m, \mathbb{F}_l)$$

is injective.

PROOF. We first treat the case where l is odd. In this case, we factorise the embedding of C^m in $GL_n(\mathbb{F}_q)$ as

$$C^m \hookrightarrow \Sigma_m \wr GL_r(\mathbb{F}_q) \hookrightarrow GL_n(\mathbb{F}_q).$$

Using the fact that l is odd, one finds that the index in $GL_n(\mathbb{F}_q)$ of this intermediate group is coprime to l, so that by Corollary 3.6.18 of Volume I, the restriction map in mod l cohomology is injective. Now a theorem of Nakaoka [202] (see also the exercise at the end of Section 4.1) says that the cohomology of a wreath product $\Sigma_m \wr A$ is detected on A^m and $\Sigma_m \times A$. Since the Sylow l-subgroups of Σ_m are iterated wreath products of cyclic groups of order l, it follows that the cohomology of $GL_n(\mathbb{F}_q)$ is detected on l-subgroups of exponent dividing $q^r - 1$. These are all conjugate to subgroups of C^m, and so we are done.

If $l = 2$ (so that $r = 1$ and $m = n$), the index of this wreath product is not odd, so one uses the subgroup $\Sigma_{[n/2]} \wr GL_2(\mathbb{F}_q) \times (\mathbb{F}_q^\times)^{n-2[n/2]}$ of odd index. A separate calculation is needed to check that the mod 2 cohomology of $GL_2(\mathbb{F}_q)$ is detected on abelian 2-subgroups of exponent dividing $q-1$. $\square$

The calculation is now completed as follows. We have maps

$$B(C^m) \to BGL_n(\mathbb{F}_q) \to F\psi^q$$

giving rise to maps in cohomology

$$H^*(F\psi^q; \mathbb{F}_l) \to H^*(GL_n(\mathbb{F}_q), \mathbb{F}_l) \to H^*(C^m, \mathbb{F}_l).$$

Now the normaliser of C in $GL_r(\mathbb{F}_q)$ contains a cyclic group of order r acting on C by sending an element to its qth power. Thus the image of the restriction in cohomology from $GL_n(\mathbb{F}_q)$ to C^m lies in the invariants of $\Sigma_m \wr \mathbb{Z}/r$, and so we have

$$H^*(F\psi^q; \mathbb{F}_l) \to H^*(GL_n(\mathbb{F}_q), \mathbb{F}_l) \to (\bigotimes^m H^*(C, \mathbb{F}_l)^{\mathbb{Z}/r})^{\Sigma_m}.$$

We have seen in Lemma 2.9.2 that the second of these maps is injective. So if we can prove that the composite is surjective, it will follow that the second map is an isomorphism. An explicit calculation with the cohomology of C (cf. Section 3.5 of Volume I) shows that the characteristic classes c_r and e_r of the r dimensional representation of C over $\mathbb{F}_q$ are non-zero. One then needs formulae for the characteristic classes of a direct sum of representations; this amounts to calculating the maps in cohomology determined by the additive structure on $F\psi^q$, which can be calculated from the corresponding formulae for BU. Using these formulae it turns out that the characteristic classes $c_r, c_{2r}, \ldots, c_{mr}$ and $e_r, e_{2r}, \ldots, e_{mr}$ of the n dimensional representation of C^m over $\mathbb{F}_q$ generate the invariants of $\Sigma_m \wr \mathbb{Z}/r$ on the cohomology of C^m, so that the composite map is surjective, and hence the second map is an isomorphism. It also follows that we may evaluate the squares of the e_{jr} with $j \le m$ by restricting to C^m.

Now if l is odd then $H^*(C, \mathbb{F}_l)$ has a polynomial generator in degree two and a generator in degree one squaring to zero (see Section 3.5 of Volume I). If $l = 2$ then $r = 1$ and $m = n$, so that C has order $q - 1$. In this case, if

$q \equiv 1 \mod 4$ then C has order divisible by 4, and $H^*(C, \mathbb{F}_l)$ has the same structure as above. If $q \equiv 3 \mod 4$ then C is a direct product of a cyclic group of order two and a cyclic group of odd order, and so $H^*(C, \mathbb{F}_l)$ is a polynomial ring on a single generator in degree one. Quillen calls the case l odd or $l = 2$ and $q \equiv 1 \mod 4$ the *typical case*, and the case $l = 2$ and $q \equiv 3$ mod 4 the *exceptional case*.

Restriction to C^m shows that the classes e_{jr} square to zero in the typical case, while in the exceptional case we have $e_{jr}^2 = \sum_{a=0}^{jr} c_a c_{2jr-1-a}$.

We summarise all this as follows.

THEOREM 2.9.3 (Quillen [**212**]). *Let q be a prime power, and suppose l is a prime not dividing q. Let r be the multiplicative order of q modulo l. Then $H^*(F\psi^q; \mathbb{F}_l)$ is generated by classes c_{jr} and e_{jr}, $j = 1, 2, \ldots$, with $\deg(c_{jr}) = 2jr$ and $\deg(e_{jr}) = 2jr - 1$, subject to relations*

$$
e_{jr}^2 = \begin{cases} 0 & \text{(typical case)} \\ \displaystyle\sum_{a=0}^{jr} c_a c_{2jr-1-a} & \text{(exceptional case)} \end{cases}
$$

where the typical case is l odd or $l = 2$, $q \equiv 1 \mod 4$, and the exceptional case is $l = 2$, $q \equiv 3 \mod 4$.

Let $n = mr + e$ with $0 \le e < r$. Let C be a cyclic group of order $q^r - 1$. Then we have restriction maps

$$
H^*(F\psi^q; \mathbb{F}_l) \to H^*(GL_n(\mathbb{F}_q), \mathbb{F}_l) \to (\bigotimes^m H^*(C, \mathbb{F}_l)^{\mathbb{Z}/r})^{\Sigma_m}.
$$

The first of these maps is surjective and the second is an isomorphism. The ring $H^*(GL_n(\mathbb{F}_q), \mathbb{F}_l)$ is generated by the classes c_{jr} and e_{jr}, $1 \le j \le m$ subject to the same relations as above. □

Now embed $GL_n(\mathbb{F}_q)$ in $GL_{n+1}(\mathbb{F}_q)$ as the matrices which agree with the identity on the last row and column. Then the natural $(n + 1)$ dimensional module for $GL_{n+1}(\mathbb{F}_q)$ restricts to the direct sum of the natural n dimensional module for $GL_n(\mathbb{F}_q)$ and a one dimensional trivial module. So if we compose the Brauer lift map $BGL_{n+1}(\mathbb{F}_q) \to F\psi^q$ with $BGL_n(\mathbb{F}_q) \to BGL_{n+1}(\mathbb{F}_q)$, we obtain the Brauer lift map $BGL_n(\mathbb{F}_q) \to F\psi^q$. So letting $GL(\mathbb{F}_q)$ be the union of the $GL_n(\mathbb{F}_q)$ with these inclusions, there is a map $BGL(\mathbb{F}_q) \to F\psi^q$ well defined up to homotopy. The above theorem implies that this map is a mod l cohomology isomorphism for all primes l not equal to p. Moreover, Quillen shows in [**212**] Section 11 that the mod p cohomology of $BGL_n(\mathbb{F}_q)$ vanishes in positive degrees at most $d(p - 1)$, where d is defined by $q = p^d$. We also saw that $H^i(F\psi^q; \mathbb{F}_p) = 0 = H^i(F\psi^q; \mathbb{Q})$ for $i > 0$, so the map is a cohomology equivalence at all primes. Since the cohomology groups are finitely generated abelian groups, it is easy to see that this implies that $H^*(F\psi^q; \mathbb{Z}) \to H^*(BGL(\mathbb{F}_q); \mathbb{Z})$ is an isomorphism, and hence also $H_*(BGL(\mathbb{F}_q)) \to H_*(F\psi^q)$ is an isomorphism. However, $BGL(\mathbb{F}_q) \to F\psi^q$ is not a homotopy equivalence, since the spaces have very different fundamental groups. In the next section, we shall describe the Quillen "plus" construction, which changes the fundamental group of a space without changing its homology, and we shall see that the above map induces a homotopy equivalence $BGL(\mathbb{F}_q)^+ \to F\psi^q$.

REMARK. Similar computations may be performed with the other groups of Lie type. For details see Fiedorowicz and Priddy [**110**] and Kleinerman [**152**].

2.10. The plus construction and algebraic K-theory

In the last section we saw that there is a map $BGL(\mathbb{F}_q) \to F\psi^q$ which induces an isomorphism in integral homology. We now introduce the Quillen "plus" construction, which turns homology equivalences into homotopy equivalences, and use this to define the algebraic K-theory of a ring. The homotopy equivalence $BGL(\mathbb{F}_q)^+ \to F\psi^q$ calculates the algebraic K-theory of finite fields.

THEOREM 2.10.1 (Quillen). *Let (X, x_0) be a CW-complex. Let π be a perfect normal subgroup (i.e., equal to its commutator subgroup) of $\pi_1(X, x_0)$. Then there is an inclusion $i : X \hookrightarrow X^+$ with the following properties:*

(i) *(X^+, X) is a relative CW-complex of dimension at most 3.*

(ii) *$i_* : \pi_1(X, x_0) \to \pi_1(X^+, x_0)$ is an epimorphism with kernel π.*

(iii) *Let $\tilde{X}^+$ be a covering space of X^+, and $\tilde{X}$ be the corresponding cover of X with fundamental group $\pi_1(\tilde{X}, x_0) = i_*^{-1}\pi_1(\tilde{X}^+, x_0)$, so that i lifts to a map $\tilde{i} : \tilde{X} \to \tilde{X}^+$. Then $\tilde{i}_* : H_n(\tilde{X}) \to H_n(\tilde{X}^+)$ is an isomorphism for all $n \geq 0$.*

(iv) *If $f : (X, x_0) \to (Z, z_0)$ is a map of spaces with π in the kernel of $f_* : \pi_1(X, x_0) \to \pi_1(Z, z_0)$, then there is a map $f' : (X^+, x_0) \to (Z, z_0)$ with $f' \circ i \simeq f$. If f enjoys properties (ii) and (iii) above of the inclusion i (but f is not necessarily an inclusion), then f' is a homotopy equivalence. In particular, X^+ is uniquely determined up to homotopy.*

PROOF. The idea of the proof is to attach 2-cells to kill π, and then attach 3-cells to make sure the homology is unaffected, which of course by the Hurewicz theorem requires that π is perfect. The details are as follows. Choose a collection of commutators $[y_j, z_j]$ of elements of π, generating π as a normal subgroup of $\pi_1(X, x_0)$. Choose maps $\tilde{y}_j, \tilde{z}_j : (S^1, s_0) \to (X, x_0)$, $j \in J$, representing these generators. Set $\lambda_j = \tilde{y}_j * \tilde{z}_j * \tilde{y}_j^{-1} * \tilde{z}_j^{-1}$, and let Y be the space

$$Y = X \cup_{\lambda_j} \bigcup_{j \in J} \sigma_j^2$$

obtained by attaching 2-cells σ_j^2 to X using the attaching maps λ_j. Thus we have $\pi_1(Y, y_0) \cong \pi_1(X, x_0)/\pi$.

Let $\bar{\sigma}_j^2$ denote the quotient of the cell σ_j^2 obtained by identifying the parts of the boundary corresponding to $\tilde{y}_j$ and $\tilde{y}_j^{-1}$ in opposite directions, so that the characteristic map $\sigma_j^2 \to Y$ factors through $\bar{\sigma}_j^2$, and $\bar{\sigma}_j^2$ is an open cylinder with ends corresponding to $\tilde{z}_j$ and $\tilde{z}_j^{-1}$. Since z_j can be expressed as a product of the commutators generating π, the path $\tilde{z}_j$ is null homotopic in Y, and so the map $\bar{\sigma}_j^2 \to Y$ extends to a map from the 2-sphere (= closed cylinder) obtained by capping off the ends of $\bar{\sigma}_j^2$. This can then be used as the attaching map $\mu_j : (S^2, s_0) \to (Y, x_0)$ for a 3-cell σ_j^3, and we set

$$X^+ = Y \cup_{\mu_j} \bigcup_{j \in J} \sigma_j^3.$$

Note that the attaching maps λ_j and μ_j can be chosen to have their images lying in the 1-skeleton of X, resp. 2-skeleton of Y, so that X^+ is a CW-complex.

We now check properties (i)–(iv) above. Property (i) is clear from the construction. Property (ii) follows from the fact that

$$\pi_2(X^+, Y, x_0) = \pi_1(X^+, Y, x_0) = 0$$

so that $\pi_1(X^+, x_0) = \pi_1(Y, x_0) = \pi_1(X, x_0)/\pi$.

To prove property (iii), suppose $\tilde{X}^+$ is a cover of X^+ corresponding to a subgroup $G/\pi \leq \pi_1(X, x_0)/\pi$, and $\tilde{X}$ is the cover of X corresponding to $G \leq \pi_1(X, x_0)$. Then we have a corresponding cover $\tilde{Y}$ of Y with

$$\tilde{Y} = \tilde{X} \cup_{\phi_g(\lambda_j)} \bigcup_{\substack{j \in J \\ g \in \pi_1(X, x_0)/G}} \sigma_{j,g}^2$$

where the ϕ_g are the covering translations corresponding to coset representatives g of G in $\pi_1(X, x_0)$. Similarly

$$\tilde{X}^+ = \tilde{Y} \cup_{\phi_g(\mu_j)} \bigcup_{\substack{j \in J \\ g \in \pi_1(X, x_0)/G}} \sigma_{j,g}^3.$$

Thus the complex of cellular chains $C_*(\tilde{X}^+, \tilde{X})$ is just

$$\cdots \to 0 \to F \xrightarrow{\cong} F \to 0 \to 0,$$

where F is the free $\mathbb{Z}$-module on the cells $\sigma_{j,g}^3$, respectively $\sigma_{j,g}^2$. The boundary map is the map taking $\sigma_{j,g}^3$ to $\sigma_{j,g}^2$, and so $H_*(\tilde{X}^+, \tilde{X}) = 0$, which implies property (iii).

Finally, if $f : (X, x_0) \to (Z, z_0)$ is a map of spaces with the property that π is in the kernel of $f_* : \pi_1(X, x_0) \to \pi_1(Z, z_0)$, then we form the pushout (Z^+, z_0) of f and i

$$
\begin{array}{ccc}
(X, x_0) & \xrightarrow{\ i\ } & (X^+, x_0) \\
\downarrow{\scriptstyle f} & & \downarrow{\scriptstyle f^+} \\
(Z, z_0) & \xrightarrow{\ i'\ } & (Z^+, z_0).
\end{array}
$$

This is the same as the space obtained from (Z, z_0) by attaching 2-cells and 3-cells by the above recipe using the null homotopic paths $f \circ \tilde{y}_j, f \circ \tilde{z}_j$:

$(S^1, s_0) \to (Z, z_0)$. Thus by (ii), the inclusion i' induces an isomorphism $\pi_1(Z, z_0) \to \pi_1(Z^+, z_0)$ and hence lifts to an inclusion of universal covers $\tilde{i}' : \tilde{Z} \hookrightarrow \tilde{Z}^+$. By (iii), this is a homology isomorphism of spaces with trivial fundamental group, and hence by Whitehead's Theorem 1.4.3, it induces an isomorphism on homotopy groups. So for $n > 1$,

$$\pi_n(Z, z_0) = \pi_n(\tilde{Z}, z_0) = \pi_n(\tilde{Z}^+, z_0) = \pi_n(Z^+, z_0).$$

So we can now apply Whitehead's Theorem 1.5.8 to deduce that the inclusion $i' : (Z, z_0) \hookrightarrow (Z^+, z_0)$ is a homotopy equivalence. Let $\alpha : (Z^+, z_0) \to (Z, z_0)$ be a homotopy inverse for i'. Then we can take for f' the map $\alpha \circ f^+$, so that

$$f' \circ i = \alpha \circ f^+ \circ i = \alpha \circ i' \circ f \simeq f.$$

If f enjoys properties (ii) and (iii) then $f'_* : \pi_1(X^+, x_0) \to \pi_1(Z, z_0)$ is an isomorphism, and so f' lifts to a map $\tilde{f}' : (\tilde{X}^+, x_0) \to (\tilde{Z}, z_0)$ of universal covers, which is a homology isomorphism and hence an isomorphism on homotopy groups by Theorem 1.4.3. So f' is also an isomorphism on homotopy groups, and hence a homotopy equivalence by Theorem 1.5.8. □

If Λ is a ring, let $GL(\Lambda)$ be the union of the groups $GL_n(\Lambda)$ under the obvious inclusions $GL_n(\Lambda) \hookrightarrow GL_{n+1}(\Lambda)$, and regarded as a discrete group. Let $E(\Lambda)$ denote the subgroup generated by the **elementary matrices**, namely the matrices which differ from the identity matrix in a single off-diagonal entry. The identities

$$\begin{pmatrix} ABA^{-1}B^{-1} & 0 \\ 0 & I \end{pmatrix} = \begin{pmatrix} A & 0 \\ 0 & A^{-1} \end{pmatrix} \begin{pmatrix} B & 0 \\ 0 & B^{-1} \end{pmatrix} \begin{pmatrix} (BA)^{-1} & 0 \\ 0 & BA \end{pmatrix}$$

$$\begin{pmatrix} A & 0 \\ 0 & A^{-1} \end{pmatrix} = \begin{pmatrix} I & A \\ 0 & I \end{pmatrix} \begin{pmatrix} I & 0 \\ -A^{-1} & I \end{pmatrix} \begin{pmatrix} I & A \\ 0 & I \end{pmatrix} \begin{pmatrix} 0 & -I \\ I & 0 \end{pmatrix}$$

$$\begin{pmatrix} 0 & -I \\ I & 0 \end{pmatrix} = \begin{pmatrix} I & 0 \\ I & I \end{pmatrix} \begin{pmatrix} I & -I \\ 0 & I \end{pmatrix} \begin{pmatrix} I & 0 \\ I & I \end{pmatrix}$$

show that every commutator in $GL_n(\Lambda)$ is a product of elementary matrices in $GL_{2n}(\Lambda)$, so that $E(\Lambda)$ is the commutator subgroup of $GL(\Lambda)$ and is perfect.

DEFINITION 2.10.2. *The (Quillen) algebraic K groups of a ring Λ are defined as*

$$K_i(\Lambda) = \pi_i(BGL(\Lambda)^+) \quad (i \geq 1).$$

Here, the plus construction is with respect to the commutator subgroup $E(\Lambda)$ of the fundamental group $\pi_1(BGL(\Lambda)) = GL(\Lambda)$. Note that since $GL(\Lambda)$ is a discrete group, $BGL(\Lambda)$ is the same thing as the Eilenberg–Mac Lane space $K(GL(\Lambda), 1)$.

(The group $K_0(\Lambda)$ is defined as the Grothendieck group of finitely generated projective Λ-modules, as usual; see Section 5.1 of Volume I).

THEOREM 2.10.3. *The map $BGL(\mathbb{F}_q) \to F\psi^q$ defined in the last section induces a homotopy equivalence $BGL(\mathbb{F}_q)^+ \simeq F\psi^q$, where the plus construction is applied to the perfect normal subgroup $E(\mathbb{F}_q)$ of elementary matrices. We have*

$$K_{2j-1}(\mathbb{F}_q) = \mathbb{Z}/(q^{2j} - 1), \qquad K_{2j}(\mathbb{F}_q) = 0.$$

PROOF. We saw at the end of the last section that $BGL(\mathbb{F}_q) \to F\psi^q$ is a homology isomorphism. Now $\pi_1(BGL(\mathbb{F}_q)) = GL(\mathbb{F}_q)$, and $\pi_1(F\psi^q) = \mathbb{Z}/(q - 1)$. Since the map is a homology isomorphism, the kernel of the map on π_1 is $E(\mathbb{F}_q)$, the kernel of the determinant map. It follows from Theorem 2.10.1 (iv) that this map extends to a map $BGL(\mathbb{F}_q)^+ \to F\psi^q$ which is a homology isomorphism. Since both $BGL(\mathbb{F}_q)^+$ and $F\psi^q$ are simple spaces, we may apply Whitehead's Theorems 1.4.3 and 1.5.8 to deduce that this map is a homotopy equivalence. Thus $K_i(\mathbb{F}_q) \cong \pi_i(F\psi^q)$, which was calculated in Section 2.9.   □

The groups $K_1(\Lambda)$ and $K_2(\Lambda)$ have explicit algebraic interpretations as follows.

PROPOSITION 2.10.4. (i) $K_1(\Lambda) \cong GL(\Lambda)/E(\Lambda)$
(ii) $K_2(\Lambda) \cong H_2(E(\Lambda), \mathbb{Z})$.

PROOF. (i) This follows directly from the definition

$$K_1(\Lambda) = \pi_1(BGL(\Lambda)^+)$$

since the purpose of the plus construction on $BGL(\Lambda)$ was to kill the perfect subgroup $E(\Lambda)$ of $\pi_1(BGL(\Lambda)) = GL(\Lambda)$.

(ii) Let F be the homotopy fibre (see the exercise at the end of Section 1.6) of the map $BGL(\Lambda) \to BGL(\Lambda)^+$. Thus we have a long exact sequence of homotopy groups (Proposition 1.2.7)

$$\cdots \to \pi_2(BGL(\Lambda)) \to \pi_2(BGL(\Lambda)^+) \qquad GL(\Lambda) \qquad GL(\Lambda)/E(\Lambda)$$

$$\begin{array}{ccccc} & \| & \downarrow & \| & \| \\ & 0 & & & \end{array}$$

$$\pi_1(F) \longrightarrow \pi_1(BGL(\Lambda)) \to \pi_1(BGL(\Lambda)^+) \to 0$$

and so we have a central extension

$$0 \to \pi_2(BGL(\Lambda)^+) \to \pi_1(F) \to E(\Lambda) \to 0.$$

We have a map $F \to B\pi_1(F)$ inducing an isomorphism on π_1. Turn it into a fibration with fibre F'. The five term exact sequence of this fibration (see Proposition 3.2.11 and Section 3.3) is

$$H_2(F) \to H_2(B\pi_1(F)) \to H_1(F') \to H_1(F) \to H_1(B\pi_1(F)) \to 0$$

$$\begin{array}{ccccc} \| & \| & \| & \| \\ 0 & 0 & 0 & 0 \end{array}$$

and hence $H_2(\pi_1(F), \mathbb{Z}) = 0$. The five term exact sequence of the above central extension (see Section 3.5) is

$$H_2(\pi_1(F),\mathbb{Z}) \twoheadrightarrow H_2(E(\Lambda),\mathbb{Z}) \to H_1(\pi_2(BGL(\Lambda)^+),\mathbb{Z}) \twoheadrightarrow H_1(\pi_1(F),\mathbb{Z}) \twoheadrightarrow H_1(E(\Lambda),\mathbb{Z})$$

$$\underset{0}{\|} \qquad\qquad\qquad \underset{\pi_2(BGL(\Lambda)^+)}{\|} \qquad \underset{H_1(F;\mathbb{Z})}{\|} \quad =0$$

and hence $K_2(\Lambda) = \pi_2(BGL(\Lambda)^+) \cong H_2(E(\Lambda), \mathbb{Z})$.   □

REMARK. In Quillen [214], another definition is given for $K_i(\Lambda)$. Namely, one forms a category $Q(_\Lambda\mathbf{Proj})$ whose objects are finitely generated projective Λ-modules, and where an arrow from P to Q is an isomorphism class of diagrams of the form

$$P \xleftarrow{j} P' \xrightarrow{i} Q$$

where P' is another finitely generated projective Λ-module, j is an epimorphism and i is a monomorphism. Two such diagrams are thought of as isomorphic if there is an isomorphism which is the identity on P and Q. Arrows are composed by pulling back

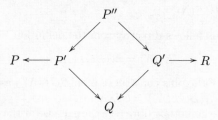

so that P'' is the submodule of $P' \oplus Q'$ consisting of elements (x, y) where x and y have the same image in Q. We now form the classifying space $BQ(_\Lambda\mathbf{Proj})$ of this category (see the exercise in Section 1.8), and set

$$K_i(\Lambda) = \pi_{i+1}(BQ(_\Lambda\mathbf{Proj}), \mathbf{0}), \quad i \geq 0$$

where $\mathbf{0}$ denotes the zero module in $Q(_\Lambda\mathbf{Proj})$, thought of as a basepoint for $BQ(_\Lambda\mathbf{Proj})$.

In Grayson [116], one can find Quillen's proof that there is a homotopy equivalence

$$\Omega BQ(_\Lambda\mathbf{Proj}) \simeq K_0(\Lambda) \times BGL(\Lambda)^+$$

so that

$$\pi_{i+1}(BQ(_\Lambda\mathbf{Proj})) \cong \pi_i(BGL(\Lambda)^+)$$

for $i \geq 1$, and hence this definition agrees with the previous one.

It is proved in Wagoner [275] that if we let $\mu(\Lambda)$ be the ring of infinite matrices (λ_{ij}), $1 \leq i, j < \infty$ with only a finite number of non-zero entries in each row and column, modulo the ideal consisting of those matrices with at most finitely many non-zero entries (this should be thought of as a sort of algebraic suspension of the ring Λ), then

$$\Omega(BGL(\mu(\Lambda))^+) \simeq K_0(\Lambda) \times BGL(\Lambda)^+.$$

It follows that $K_0(\Lambda) \times BGL(\Lambda)^+$ is an n-fold loop space in a canonical way for every value of $n \geq 0$.

2.11. Hochschild homology

Algebraic K-theory of even quite well understood rings is extraordinarily difficult to calculate. For example, one only knows the first few algebraic K-groups of $\mathbb{Z}$. It is therefore useful to have maps to theories which are easier to calculate. In this section, we introduce Hochschild homology of a ring, and define the Dennis trace map from algebraic K-theory to Hochschild homology. In a later section, we introduce a variant of this theory, called cyclic homology, which admits a natural map to Hochschild homology. We shall see that there is a map called the Chern map from algebraic K-theory to cyclic homology, whose composite with the map to Hochschild homology is the Dennis trace map.

Throughout this section, R is a commutative ring of coefficients and Λ is an R-algebra which is projective as an R-module. We first introduce the acyclic Hochschild complex, which is a resolution for the regular representation of Λ as a Λ-bimodule (i.e., as a $\Lambda \otimes_R \Lambda^{\mathrm{op}}$-module), and then use it to calculate Tor and Ext of Λ with a Λ-bimodule A. The case $A = \Lambda$ gives the Hochschild homology and cohomology of the ring Λ.

The acyclic Hochschild complex is defined as follows. We write $S_n(\Lambda)$ for the left $\Lambda \otimes_R \Lambda^{\mathrm{op}}$-module

$$\underbrace{\Lambda \otimes_R \cdots \otimes_R \Lambda}_{n+2 \text{ copies}} \qquad (n \geq -1)$$

where $\Lambda \otimes \Lambda^{\mathrm{op}}$ acts via

$$(\mu \otimes \gamma^*)(\lambda_0 \otimes \cdots \otimes \lambda_{n+1}) = (\mu\lambda_0) \otimes \lambda_1 \otimes \cdots \otimes \lambda_n \otimes (\lambda_{n+1}\gamma).$$

The differential $b'_n : S_n(\Lambda) \to S_{n-1}(\Lambda)$ is defined via

$$b'_n(\lambda_0 \otimes \cdots \otimes \lambda_{n+1}) = \sum_{i=0}^{n} (-1)^i \lambda_0 \otimes \cdots \otimes \lambda_i\lambda_{i+1} \otimes \cdots \otimes \lambda_{n+1}.$$

It is easy to check that $b'_{n-1} \circ b'_n = 0$, so that $S_*(\Lambda)$ is a chain complex. The map $s_n : S_{n-1}(\Lambda) \to S_n(\Lambda)$ given by

$$s_n(\lambda_0 \otimes \cdots \otimes \lambda_n) = \lambda_0 \otimes \cdots \otimes \lambda_n \otimes 1$$

satisfies $b'_{n+1} \circ s_{n+1} - s_n \circ b'_n = \mathrm{id}$ so that $(-1)^n s_n$ is a chain contraction of $S_*(\Lambda)$ (see Section 2.3 of Volume I). It follows that $S_*(\Lambda)$ is exact.

Writing $\tilde{S}_n(\Lambda)$ for

$$\underbrace{\Lambda \otimes_R \cdots \otimes_R \Lambda}_{n \text{ copies}} \qquad (n \geq 0)$$

we have an isomorphism

$$S_n(\Lambda) \cong (\Lambda \otimes \Lambda^{\mathrm{op}}) \otimes \tilde{S}_n(\Lambda)$$

$$\lambda_0 \otimes \cdots \otimes \lambda_{n+1} \longleftrightarrow (\lambda_0 \otimes \lambda_{n+1}^*) \otimes (\lambda_1 \otimes \cdots \otimes \lambda_n)$$

as $\Lambda \otimes \Lambda^{\mathrm{op}}$-modules, so that since $\tilde{S}_n(\Lambda)$ is R-projective, $S_n(\Lambda)$ is a projective $\Lambda \otimes \Lambda^{\mathrm{op}}$-module. It follows that $S_*(\Lambda)$ is a projective resolution of Λ as a $\Lambda \otimes \Lambda^{\mathrm{op}}$-module. It is called the **acyclic Hochschild complex.**

If A is a Λ-bimodule, we regard it as a right $\Lambda \otimes \Lambda^{\mathrm{op}}$-module via $a(\mu \otimes \gamma^*) = \gamma a\mu$. We then define the **Hochschild homology** $H_n(\Lambda, A)$ to be the homology of the **Hochschild complex**

$$(S_n(\Lambda, A), b_n) = (A \otimes_{\Lambda \otimes \Lambda^{\mathrm{op}}} S_n(\Lambda), 1 \otimes b_n').$$

We have

$$S_n(\Lambda, A) = A \otimes_{\Lambda \otimes \Lambda^{\mathrm{op}}} S_n(\Lambda) \cong A \otimes_{\Lambda \otimes \Lambda^{\mathrm{op}}} (\Lambda \otimes \Lambda^{\mathrm{op}}) \otimes_R \tilde{S}_n(\Lambda)$$

$$\cong A \otimes_R \tilde{S}_n(\Lambda)$$

and it is easy to check that the boundary map is given in terms of this by

$$b_n(a \otimes \lambda_1 \otimes \cdots \otimes \lambda_n) = a\lambda_1 \otimes \cdots \otimes \lambda_n +$$

$$\sum_{i=1}^{n-1} (-1)^i a \otimes \lambda_1 \otimes \cdots \otimes \lambda_i\lambda_{i+1} \otimes \cdots \otimes \lambda_n + (-1)^n \lambda_n a \otimes \lambda_1 \otimes \cdots \otimes \lambda_{n-1}.$$

Since $S_*(\Lambda)$ is a projective resolution of Λ as a $\Lambda \otimes \Lambda^{\mathrm{op}}$-module, it follows from Section 1.13 of Volume I that

$$H_n(\Lambda, A) \cong \mathrm{Tor}_n^{\Lambda \otimes \Lambda^{\mathrm{op}}}(A, \Lambda).$$

Similarly we define the **Hochschild cohomology** $H^n(\Lambda, A)$ to be the cohomology of the cochain complex

$$S^n(\Lambda, A) = \mathrm{Hom}_{\Lambda \otimes \Lambda^{\mathrm{op}}}(S_n(\Lambda), A)$$

$$\cong \mathrm{Hom}_{\Lambda \otimes \Lambda^{\mathrm{op}}}(\Lambda \otimes \Lambda^{\mathrm{op}} \otimes \tilde{S}_n(\Lambda), A) \cong \mathrm{Hom}_R(\tilde{S}_n(\Lambda), A)$$

with coboundary

$$(b^n f)(\lambda_1, \ldots, \lambda_{n+1}) = \lambda_1 f(\lambda_2, \ldots, \lambda_n + 1) +$$

$$\sum_{i=1}^{n} (-1)^i f(\lambda_1, \ldots, \lambda_i\lambda_{i+1}, \ldots, \lambda_{n+1}) + (-1)^{n+1} f(\lambda_1, \ldots, \lambda_n)\lambda_{n+1}.$$

Again using the fact that $S_n(\Lambda)$ is a projective resolution of Λ as a $\Lambda \otimes \Lambda^{\mathrm{op}}$-module, we deduce that

$$H^n(\Lambda, A) \cong \mathrm{Ext}_{\Lambda \otimes \Lambda^{\mathrm{op}}}^n(\Lambda, A).$$

We may compare this with our previous definitions of homology and cohomology as follows. If Λ is an augmented algebra over R, and M is a right Λ-module, then we may regard M as a Λ-bimodule by making Λ act trivially on the left via the augmentation. In this case, since $(\Lambda \otimes \Lambda^{\mathrm{op}}) \otimes \tilde{S}_n(\Lambda)$

is a projective resolution of Λ as a left $(\Lambda \otimes \Lambda^{\mathrm{op}})$-module, it follows that $\Lambda \otimes_R \tilde{S}_n(\Lambda)$ is a projective resolution of R as a left Λ-module. So

$$H_n(\Lambda, M) \cong \mathrm{Tor}_n^{\Lambda \otimes \Lambda^{\mathrm{op}}}(M, \Lambda) \cong \mathrm{Tor}_n^{\Lambda}(M, R)$$

agrees with our previous algebraic definition of $H_n(\Lambda, M)$ (Section 2.1, and Section 1.13 of Volume I).

Similarly if M is a left Λ-module regarded as a Λ-bimodule with trivial right action then

$$\mathrm{Hom}_{\Lambda \otimes \Lambda^{\mathrm{op}}}(S_n(\Lambda), M) \cong \mathrm{Hom}_{\Lambda}(\Lambda \otimes \tilde{S}_n(\Lambda), M)$$

and so

$$H^n(\Lambda, M) \cong \mathrm{Ext}_{\Lambda \otimes \Lambda^{\mathrm{op}}}^n(\Lambda, M) \cong \mathrm{Ext}_{\Lambda}^n(R, M)$$

agrees with our previous definition of $H^n(\Lambda, M)$.

The advantage of the Hochschild approach is that it provides homology and cohomology groups for an algebra without augmentation. For example, we could use this definition for a block of a group algebra. The "natural coefficients" to use for the Hochschild homology of a (not necessarily augmented) algebra Λ are the Λ-bimodule Λ itself, giving homology groups $HH_n(\Lambda) = H_n(\Lambda, \Lambda)$. For Hochschild cohomology, we use the Λ-bimodule $\Lambda^* = \mathrm{Hom}_R(\Lambda, R)$ to obtain cohomology groups $HH^n(\Lambda) = H^n(\Lambda, \Lambda)$. It is not hard to see that the Hochschild (co)homology of a direct sum of algebras, with these natural coefficients, is the direct sum of the Hochschild (co)homology of the summands.

THEOREM 2.11.1. *Suppose that* $_\Lambda P_\Gamma$ *and* $_\Gamma Q_\Lambda$ *are bimodules inducing a Morita equivalence between* Λ *and* Γ *(see Section 2.2 of Volume I). Then for any* Λ-*bimodule* M *we have*

$$H_n(\Lambda, M) \cong H_n(Q \otimes_\Lambda M \otimes_\Lambda P), \quad H^n(\Lambda, M) \cong H^n(Q \otimes_\Lambda M \otimes_\Lambda P).$$

In particular, $HH_n(\Lambda) \cong HH_n(\Gamma)$ *and* $HH^n(\Lambda) \cong HH^n(\Gamma)$.

PROOF. We have adjunctions

$$\mathrm{Hom}_\Lambda(P \otimes_\Gamma -, -) \cong \mathrm{Hom}_\Gamma(-, Q \otimes_\Lambda -),$$
$$\mathrm{Hom}_\Lambda(-, P \otimes_\Gamma -) \cong \mathrm{Hom}_\Gamma(Q \otimes_\Lambda -, -)$$

and hence by dimension shifting we have

$$\mathrm{Ext}_\Lambda^n(P \otimes_\Gamma -, -) \cong \mathrm{Ext}_\Gamma^n(-, Q \otimes_\Lambda -),$$
$$\mathrm{Ext}_\Lambda^n(-, P \otimes_\Gamma -) \cong \mathrm{Ext}_\Gamma^n(Q \otimes_\Lambda -, -).$$

Hence we have

$$\begin{aligned} H^n(\Lambda, M) &\cong \mathrm{Ext}_{\Lambda \otimes \Lambda^{\mathrm{op}}}^n(\Lambda, M) = \mathrm{Ext}_{\Lambda \otimes \Lambda^{\mathrm{op}}}^n(P \otimes_\Gamma Q, M) \\ &\cong \mathrm{Ext}_{\Gamma \otimes \Lambda^{\mathrm{op}}}^n(Q, Q \otimes_\Lambda M) \cong \mathrm{Ext}_{\Gamma \otimes \Gamma^{\mathrm{op}}}^n(\Gamma, Q \otimes_\Lambda M \otimes_\Lambda P) \\ &\cong H^n(\Gamma, Q \otimes_\Lambda M \otimes_\Lambda P). \end{aligned}$$

The dual proof works in homology.   □

THEOREM 2.11.2. *The additive structures of the Hochschild homology and cohomology of a group algebra RG are given by*

(i) $HH_n(RG) \cong \bigoplus_{g \in_G G} H_n(C_G(g), R)$

(ii) $HH^n(RG) \cong \prod_{g \in_G G} H^n(C_G(g), R)$.

PROOF. We shall prove (ii); the argument for (i) is dual. We have

$$HH^n(RG) \cong \mathrm{Ext}^n_{RG \otimes RG^{\mathrm{op}}}(RG, RG^*).$$

Now $RG^{\mathrm{op}} \cong RG$ (via $g \mapsto g^{-1}$, Section 3.1 of Volume I) and $RG \otimes RG \cong R(G \times G)$. The $R(G \times G)$-module structure on RG is given by the two-sided action

$$(g_1, g_2) : g \mapsto g_1 g g_2^{-1},$$

and so it is just the permutation module $R_{\Delta(G)} \uparrow^{G \times G}$ on the cosets of the diagonal

$$\Delta(G) = \{(g, g) \mid g \in G\} \subseteq G \times G.$$

Similarly, RG^* is the coinduced module $RG_{\Delta(G)} \Uparrow^{G \times G}$ (cf. Section 2.8 of Volume I). So by the Eckmann–Shapiro Lemma (see Corollary 2.8.4 of Volume I) we have

$$HH^n(RG) \cong \mathrm{Ext}^n_{R(G \times G)}(R_{\Delta(G)} \uparrow^{G \times G}, R_{\Delta(G)} \Uparrow^{G \times G})$$

$$\cong \mathrm{Ext}^n_{R\Delta(G)}(R_{\Delta(G)}, R_{\Delta(G)} \Uparrow^{G \times G} \downarrow_{\Delta(G)}).$$

By the Mackey Decomposition Theorem (3.3.4 of Volume I, or rather the appropriate analogue for not necessarily finite groups),

$$R_{\Delta(G)} \Uparrow^{G \times G} \downarrow_{\Delta(G)} \cong \prod_{g \in_G G} R_{\Delta(C_G(g))} \Uparrow^{\Delta(G)}$$

and so by another application of the Eckmann–Shapiro Lemma we obtain

$$HH^n(RG) \cong \prod_{g \in_G G} \mathrm{Ext}^n_{RG}(R, R_{C_G(g)} \Uparrow^G)$$

$$\cong \prod_{g \in_G G} \mathrm{Ext}^n_{RC_G(g)}(R, R) \cong \prod_{g \in_G G} H^n(C_G(g), R). \qquad \square$$

The multiplicative structure of $HH^n(RG)$ given by Yoneda composition is described in terms of the above additive decomposition in Section 2.2 of Volume I.

We now describe the Dennis trace map from algebraic K-theory to Hochschild homology. It is the composite of the following maps. The Hurewicz map goes from $K_n(\Lambda) = \pi_n(BGL(\Lambda)^+)$ to

$$H_n(BGL(\Lambda)^+) = H_n(BGL(\Lambda)) = H_n(GL(\Lambda), \mathbb{Z}).$$

By Theorem 2.11.2, $H_n(GL(\Lambda), \mathbb{Z})$ sits as a summand of $HH_n(\mathbb{Z}GL(\Lambda))$, corresponding to the centraliser of the identity element. The Dennis trace map is the composite

$$K_n(\Lambda) \xrightarrow{\text{Hurewicz}} H_n(GL(\Lambda), \mathbb{Z}) \to HH_n(\mathbb{Z}GL(\Lambda)) \to HH_n(\Lambda).$$

The map

$$HH_n(\mathbb{Z}GL(\Lambda)) = \varinjlim_r HH_n(\mathbb{Z}GL_r(\Lambda)) \to HH_n(\Lambda)$$

is defined as follows. The obvious maps $\mathbb{Z}GL_r(\Lambda) \to \mathrm{Mat}_r(\Lambda)$ induce a diagram

$$
\begin{array}{ccccc}
HH_n(\mathbb{Z}GL_r(\Lambda)) & \longrightarrow & HH_n(\mathrm{Mat}_r(\Lambda)) & \cong & HH_n(\Lambda) \\
\downarrow & & \downarrow & & \| \\
HH_n(\mathbb{Z}GL_{r+1}(\Lambda)) & \longrightarrow & HH_n(\mathrm{Mat}_{r+1}(\Lambda)) & \cong & HH_n(\Lambda)
\end{array}
$$

and then we pass to the limit. The isomorphism $HH_n(\mathrm{Mat}_r(\Lambda)) \cong HH_n(\Lambda)$ comes from Theorem 2.11.1, since $\mathrm{Mat}_r(\Lambda)$ is Morita equivalent to Λ via the matrix trace (cf. Section 2.2 of Volume I).

The inclusion of $GL_r(\Lambda)$ into $GL_{r+1}(\Lambda)$ is defined by

$$A \mapsto \begin{pmatrix} A & 0 \\ 0 & 1 \end{pmatrix}$$

while the inclusion of $\mathrm{Mat}_r(\Lambda)$ into $\mathrm{Mat}_{r+1}(\Lambda)$ is defined by

$$A \mapsto \begin{pmatrix} A & 0 \\ 0 & 0 \end{pmatrix}.$$

It follows that the above diagram commutes modulo the image of $HH_n(\mathbb{Z}) \to HH_n(\Lambda)$, which is zero for $n > 0$.

2.12. Free loops on BG

In the last section we saw that the Hochschild (co)homology of a group algebra is the direct sum over conjugacy classes of ordinary (co)homology of the centralisers of elements of G. In this section, we interpret this in terms of the free loop space LBG.

For any space X, we define the free loop space LX to be the space of all (unbased) maps $S^1 \to X$, with the usual (compact-open) topology.

LEMMA 2.12.1. *Let the group G act on the set G by conjugation, and on EG in the usual way. Denote by $EG \times_G G$ the quotient space $(EG \times G)/G$. Then there is a natural map*

$$\rho : LBG \to EG \times_G G$$

which is a homotopy equivalence.

PROOF. Let Y denote the space consisting of maps from the unit interval I to EG with the property that the images of the two endpoints 0 and 1 are in the same G-orbit. Thus G acts freely on Y, and the quotient is $Y/G = LBG$.

We have a map $\tilde{\rho} : Y \to EG \times G$ given by taking a map $f : I \to EG$ to $(f(0), g)$, where g is the unique element of G with $f(0)g = f(1)$. This map $\tilde{\rho}$ is a fibration, and since EG is contractible, it is easy to see that the fibres of $\tilde{\rho}$ are contractible. Moreover, $\tilde{\rho}$ commutes with the G-action, and so passes to a map

$$\rho : Y/G = LBG \to EG \times_G G.$$

Since $\tilde{\rho}$ induces a bijection on G-orbits of connected components, ρ induces a bijection on connected components. Examining the square

$$
\begin{array}{ccc}
Y & \xrightarrow{\ \tilde{\rho}\ } & EG \times G \\
\downarrow & & \downarrow \\
Y/G & \xrightarrow{\ \rho\ } & EG \times_G G,
\end{array}
$$

on each G-orbit of connected components of Y, $\tilde{\rho}$ and the vertical maps induce isomorphisms on homotopy groups, and hence so does ρ. So ρ is a homotopy equivalence by Whitehead's Theorem 1.5.8.   □

PROPOSITION 2.12.2. *Hochschild (co)homology of RG is the same as ordinary (co)homology of LBG with coefficients in R.*

PROOF. It follows from the above lemma that

$$H_*(LBG; R) \cong H_*(EG \times_G G; R)$$

$$= H_*\Big(\bigcup_{g \in_G G} BC_G(g); R \Big) = \bigoplus_{g \in_G G} H_*(C_G(g), R)$$

so that by Theorem 2.11.2, this is the same as Hochschild homology of RG. The same argument holds in cohomology.   □

There is another approach, using simplicial sets, which does not go via the decomposition as a sum over centralisers, and which we now describe.

We regard the acyclic Hochschild complex $S_*(RG)$ of RG as the chain complex of the simplicial R-module associated (in the manner described at the end of Section 1.8) to the simplicial set SG whose n-simplices are symbols $g_0 \otimes \cdots \otimes g_{n+1}$ and with faces and degeneracies

$$d_i(g_0 \otimes \cdots \otimes g_{n+1}) = g_0 \otimes \cdots \otimes g_i g_{i+1} \otimes \cdots \otimes g_{n+1}$$

$$s_i(g_0 \otimes \cdots \otimes g_{n+1}) = g_0 \otimes \cdots \otimes g_i \otimes 1 \otimes g_{i+1} \otimes \cdots \otimes g_{n+1}.$$

We let G act by conjugation on SG

$$g(g_0 \otimes g_1 \otimes \cdots \otimes g_{n+1}) = gg_0 \otimes g_1 \otimes \cdots \otimes g_{n+1}g^{-1}$$

and SG/G is a simplicial set whose corresponding chain complex of R-modules is the Hochschild complex for RG with coefficients in RG. We write $g_0 \otimes \cdots \otimes g_n$ for the n-simplex in SG/G corresponding to the G-orbit of $g_0 \otimes \cdots \otimes g_n \otimes 1$ in SG. Thus the face and degeneracy maps for SG/G are given by

$$d_i(g_0 \otimes \cdots \otimes g_n) = g_0 \otimes \cdots \otimes g_i g_{i+1} \otimes \cdots \otimes g_n \qquad 0 \le i \le n-1$$
$$d_n(g_0 \otimes \cdots \otimes g_n) = g_n g_0 \otimes \cdots \otimes g_{n-1}$$
$$s_i(g_0 \otimes \cdots \otimes g_n) = g_0 \otimes \cdots \otimes g_i \otimes 1 \otimes g_{i+1} \otimes \cdots \otimes g_n \qquad 0 \le i \le n.$$

PROPOSITION 2.12.3. *As simplicial sets, we have*

$$\mathrm{Hom}(S[1], NG) = SG/G$$

where NG is the nerve of G regarded as a category with one object (so that $|NG| = BG$), and $S[1]$ is the simplicial circle.

PROOF. Denote by X the simplicial set of maps $f : \Delta[1] \to N\mathcal{C}(G)$, the nerve of the Cayley category of G (see Exercise 3 in Section 2.4) with the property that $f(0)$ and $f(1)$ are in the same G-orbit. Thus an n-simplex in X is a simplicial map $\Delta[1] \times \Delta[n] \to N\mathcal{C}(G)$ with endpoint conditions which we discuss below, and X/G is the simplicial set $\mathrm{Hom}(S[1], NG)$. We shall show that X is isomorphic to SG as a simplicial set with G-action.

Denote by σ and τ the unique non-degenerate simplices of dimensions 1 and n in $\Delta[1]$ and $\Delta[n]$, and denote by $\bar{s}_i \sigma$ the simplex $(s_0)^i (s_1)^{n-i-1} \sigma$ of $\Delta[1]$, which is only non-degenerate in the ith direction. Then $\Delta[1] \times \Delta[n]$ has exactly $n+1$ non-degenerate $(n+1)$-simplices $(\bar{s}_i \sigma, s_i \tau)$, $0 \le i \le n$, and a simplicial map $f : \Delta[1] \times \Delta[n] \to N\mathcal{C}(G)$ is determined by its value on these simplices. Since $d_i(\bar{s}_i \sigma, s_i \tau) = d_i(\bar{s}_{i-1} \sigma, s_{i-1} \tau)$, the images of these simplices are related by

$$d_i f(\bar{s}_i \sigma, s_i \tau) = d_i f(\bar{s}_{i-1} \sigma, s_{i-1} \tau).$$

Moreover, the condition that $f(0)$ and $f(1)$ are in the same G-orbit reduces to the simplicial condition that $d_0 f(\bar{s}_0 \sigma, s_0 \tau)$ and $d_{n+1} f(\bar{s}_n \sigma, s_n \tau)$ are in the same G-orbit.

We choose the notation so that $d_{n+1} f(\bar{s}_n \sigma, s_n \tau)$ is the n-simplex

$$g_0 \xrightarrow{g_1} g_0 g_1 \xrightarrow{g_2} \cdots \xrightarrow{g_n} g_0 \cdots g_n$$

in $N\mathcal{C}(G)$ (in bar notation, Section 3.3 of Volume I, this is $g_0[g_1| \cdots |g_n]$). Thus $f(\bar{s}_n \sigma, s_n \tau)$ is some $(n+1)$-simplex of the form

$$g_0 \xrightarrow{g_1} g_0 g_1 \xrightarrow{g_2} \cdots \xrightarrow{g_n} g_0 \cdots g_n \to \alpha_{n+1}$$

for some value of α_{n+1} in G. Since $d_n f(\bar{s}_n \sigma, s_n \tau) = d_n f(\bar{s}_{n-1} \sigma, s_{n-1} \tau)$, we see that $f(\bar{s}_{n-1} \sigma, s_{n-1} \tau)$ is some simplex of the form

$$g_0 \xrightarrow{g_1} g_0 g_1 \xrightarrow{g_2} \cdots \xrightarrow{g_{n-1}} g_0 \cdots g_{n-1} \to \alpha_n \to \alpha_{n+1}$$

for some value of α_n. Continuing this way, we see that for suitable values of α_i, $1 \le i \le n+1$, $f(\bar{s}_i \sigma, s_i \tau)$ is a simplex of the form

$$g_0 \xrightarrow{g_1} g_0 g_1 \xrightarrow{g_2} \cdots \xrightarrow{g_i} g_0 \cdots g_i \to \alpha_{i+1} \to \cdots \to \alpha_{n+1}.$$

In particular, $d_0 f(\bar{f}_0 \sigma, s_0 \tau)$ is of the form

$$\alpha_1 \to \cdots \to \alpha_{n+1}$$

and is in the G-orbit of

$$g_0 \xrightarrow{g_1} g_0 g_1 \xrightarrow{g_2} \cdots \xrightarrow{g_n} g_0 \cdots g_n.$$

For the sake of symmetry, we choose the notation so that $g_{n+1} = \alpha_{n+1}$ and hence $\alpha_i = g_{n+1}^{-1} g_n^{-1} \cdots g_i^{-1}$. Thus $d_{n+1} f(\bar{s}_n \sigma, s_n \tau) = g_0 \cdots g_{n+1} f(\bar{s}_0 \sigma, s_0 \tau)$.

We have thus shown that every n-simplex in X is a simplicial map $f : \Delta[1] \times \Delta[n] \to NC(G)$ where $f(\bar{s}_i \sigma, s_i \tau)$ is the $(n+1)$-simplex

$$g_0 \xrightarrow{g_1} g_0 g_1 \xrightarrow{g_2} \cdots \xrightarrow{g_i} g_0 \cdots g_i \xrightarrow{g_i^{-1} \cdots g_0^{-1} g_{n+1}^{-1} \cdots g_{i+1}^{-1}} g_{n+1}^{-1} \cdots g_{i+1}^{-1} \xrightarrow{g_{i+1}} \cdots \xrightarrow{g_n} g_{n+1}^{-1},$$

and it is easy to check that a map of this form does indeed define an n-simplex in X. The correspondence which associates the n-simplex $g_0 \otimes \cdots \otimes g_{n+1}$ in SG to the above n-simplex is easily seen to respect the face and degeneracy maps and the G-action, and hence gives an isomorphism of simplicial sets between $\mathrm{Hom}(S[1], NG)$ and SG/G.   □

2.13. Cyclic homology

Cyclic homology was first introduced by Connes [83], and independently by Tsygan [272] who called it "additive K-theory." These authors worked with algebras in characteristic zero, and used the quotient of the Hochschild complex by the actions of cyclic groups. Loday and Quillen [167] noticed that Tsygan's proof of the long exact sequence connecting cyclic homology and Hochschild homology could be expressed in terms of a certain double complex, which provides the appropriate definition of cyclic homology over an arbitrary commutative ring of coefficients. We advise the reader not already familiar with double complexes to read Section 3.4 on the spectral sequence of a double complex before continuing with this section.

Recall from Section 2.11 that in the Hochschild complex for computing the Hochschild homology groups $HH_*(\Lambda) = H_*(\Lambda, \Lambda)$, the elements of degree n are $\Lambda \otimes_R \tilde{S}_n(\Lambda) = \Lambda^{\otimes(n+1)}$, with differential

$$b(\lambda_0 \otimes \cdots \otimes \lambda_n) = \sum_{i=0}^{n-1} (-1)^i \lambda_0 \otimes \cdots \otimes \lambda_i \lambda_{i+1} \otimes \cdots \otimes \lambda_n$$

$$+ (-1)^n \lambda_n \lambda_0 \otimes \cdots \otimes \lambda_{n-1}.$$

We also reindex the acyclic Hochschild complex so that $S_{n-1}(\Lambda) = \Lambda^{\otimes(n+1)}$ sits in degree n (rather than $n-1$), with differential

$$b'(\lambda_0 \otimes \cdots \otimes \lambda_n) = \sum_{i=0}^{n-1} (-1)^i \lambda_0 \otimes \cdots \otimes \lambda_i \lambda_{i+1} \otimes \cdots \otimes \lambda_n.$$

We have an action of $\langle t_n \mid (t_n)^{n+1} = 1 \rangle \cong \mathbb{Z}/(n+1)$ on $\Lambda^{\otimes(n+1)}$ given by

$$t_n(\lambda_0 \otimes \cdots \otimes \lambda_n) = \lambda_n \otimes \lambda_0 \otimes \cdots \otimes \lambda_{n-1}.$$

Let N be the sum of the distinct powers of $(-1)^n t_n$,

$$N = \sum_{j=0}^{n} ((-1)^n t_n)^j.$$

and let $\varepsilon = 1 - (-1)^n t_n$. One can easily check that $N\varepsilon = \varepsilon N = 0$, and N and ε are related to the differentials b and b' via the formulae

$$b\varepsilon = \varepsilon b', \qquad b'N = Nb.$$

From the first of these formulae, one sees that b takes elements in the image of ε to elements in the image of ε, so that b passes down to a differential on the quotient by the action of $(-1)^n t_n$

$$b : \Lambda_\varepsilon^{\otimes(n+1)} \to \Lambda_\varepsilon^{\otimes(n)},$$

where $\Lambda_\varepsilon^{\otimes(n+1)}$ is defined to be $\Lambda^{\otimes(n+1)}/\mathrm{Im}(\varepsilon)$. If the coefficient ring R is a field of characteristic zero, we define the cyclic homology of Λ to be the homology of this chain complex

$$HC_n(\Lambda) = \frac{\mathrm{Ker}(b : \Lambda_\varepsilon^{\otimes(n)} \to \Lambda_\varepsilon^{\otimes(n-1)})}{\mathrm{Im}(b : \Lambda_\varepsilon^{\otimes(n+1)} \to \Lambda_\varepsilon^{\otimes(n)})}.$$

PROPOSITION 2.13.1. *There is a long exact sequence (the* **Connes sequence***)*

$$\cdots \to HH_n(\Lambda) \xrightarrow{I} HC_n(\Lambda) \xrightarrow{S} HC_{n-2}(\Lambda) \xrightarrow{B} HH_{n-1}(\Lambda) \xrightarrow{I} HC_{n-1}(\Lambda) \to \cdots$$

PROOF. We write down the following double complex.

The columns are alternate copies of the Hochschild complex and the acyclic Hochschild complex. The signs on the acyclic Hochschild complex have been negated so that the squares anticommute (by the above formulae relating ε, N, b and b'), in accordance with the definition of a double complex (Section 3.4).

The nth row of the above double complex is the complex described in Section 3.5 of Volume I for calculating $H_*(\mathbb{Z}/(n+1), \Lambda^{\otimes(n+1)})$. Since Λ is an algebra over a field of characteristic zero, this vanishes in degrees greater than zero, and in degree zero it just gives $\Lambda_\varepsilon^{\otimes(n+1)}$. So if we look at the spectral sequence obtained by doing the horizontal differential first in the above double complex (i.e., the spectral sequence of the transpose of this double complex; see Section 3.4), we see that the E^1 term

$$E_{pq}^1 = H_q(\mathbb{Z}/(p+1), \Lambda^{\otimes(p+1)})$$

vanishes for $q > 0$ and is the complex for calculating $HC_*(\Lambda)$ for $q = 0$. So this spectral sequence shows that $HC_*(\Lambda)$ is isomorphic to the homology of the total complex $CC_*(\Lambda)$ of the above double complex.

Now the total complex X_* of the first two columns of the above double complex has the Hochschild complex as a subcomplex, and the acyclic Hochschild complex as the quotient. The long exact sequence of homology for this short exact sequence of complexes shows that the homology of X_* is just $HH_*(\Lambda)$.

The complex X_* is a subcomplex of $CC_*(\Lambda)$, and the quotient is just the same complex with a shift, $CC_*(\Lambda)[-2]$, so we have a short exact sequence of complexes

$$0 \to X_* \to CC_*(\Lambda) \to CC_*(\Lambda)[-2] \to 0.$$

Taking homology, we obtain the required long exact sequence.   □

If Λ is defined over a more general coefficient ring R, then the homology of the cyclic groups fails to vanish in the above proof. So we *define* the **cyclic chains** CC_* to be the total complex of the double complex appearing in the above proof, and the **cyclic homology** $HC_*(\Lambda)$ to be the homology of this chain complex. With this new definition, the rest of the above proof works, and shows that for a general Λ we still have the long exact sequence of the above proposition.

THEOREM 2.13.2. *A Morita equivalence between R-algebras Λ and Γ gives rise to an isomorphism $HC_n(\Lambda) \cong HC_n(\Gamma)$ for all $n \geq 0$.*

PROOF. (McCarthy [182]; see also Kassel [150]) The idea here is to show that the isomorphism in Hochschild homology described in Theorem 2.11.1 comes from a chain equivalence of Hochschild complexes which commutes with the operations t_n and b', so that the result follows from the long exact Connes sequence by induction on n. The details are as follows.

Suppose that $_\Lambda P_\Gamma$ and $_\Gamma Q_\Lambda$ are bimodules and $\phi : P \otimes_\Gamma Q \to \Lambda$ and $\psi : Q \otimes_\Lambda P \to \Gamma$ are surjective bimodule homomorphisms as in Definition 2.2.2 of Volume I, satisfying the identities $x\psi(y \otimes z) = \phi(x \otimes y)z$ and $y\phi(z \otimes w) = \psi(y \otimes z)w$ for x and z in P and y and w in Q. Choose elements p_i and q_i, $1 \leq i \leq s$, with $\phi(\sum_{i=1}^s p_i \otimes q_i) = 1 \in \Lambda$, and elements p'_j and q'_j, $1 \leq j \leq t$ with $\psi(\sum_{j=1}^t q'_j \otimes p'_j) = 1 \in \Gamma$. We define chain maps on the Hochschild complex

$$\alpha_n : \Lambda^{\otimes(n+1)} \to \Gamma^{\otimes(n+1)}$$

$$(\lambda_0, \ldots, \lambda_n) \mapsto \sum_{(i_0, \ldots, i_n)} (\psi(q_{i_0} \otimes \lambda_0 p_{i_1}), \psi(q_{i_1} \otimes \lambda_1 p_{i_2}), \ldots, \psi(q_{i_n} \otimes \lambda_n p_{i_0}))$$

where the sum is over all sequences $(i_0, \ldots, i_n)$ of integers between 0 and s inclusive, and

$$\beta_n : \Gamma^{\otimes(n+1)} \to \Lambda^{\otimes(n+1)}$$

$$(\gamma_0, \ldots, \gamma_n) \mapsto \sum_{(j_0, \ldots, j_n)} (\phi(p'_{j_0} \otimes \gamma_0 q'_{j_1}), \phi(p'_{j_1} \otimes \gamma_1 q'_{j_2}), \ldots, \phi(p'_{j_n} \otimes \gamma_n q'_{j_0}))$$

where the sum is over all sequences $(j_0, \ldots, j_n)$ of integers between 0 and t inclusive.

Using the identities satisfied by ϕ and ψ, we see that

$$\beta_n \circ \alpha_n : (\lambda_0, \ldots, \lambda_n) \mapsto$$

$$\sum_{\substack{(i_0, \ldots, i_n) \\ (j_0, \ldots, j_n)}} (\phi(p'_{j_0} \otimes q_{i_0}) \lambda_0 \phi(p_{i_1} \otimes q'_{j_1}), \ldots, \phi(p'_{j_n} \otimes q_{i_n}) \lambda_n \phi(p_{i_0} \otimes q'_{j_0})).$$

We define $h_n : \Lambda^{\otimes(n+1)} \to \Lambda^{\otimes(n+2)}$ via

$$h_n(\lambda_0, \ldots, \lambda_n) = \sum_{m=0}^{n} (-1)^m \sum_{\substack{(i_0, \ldots, i_n) \\ (j_0, \ldots, j_n)}} (\lambda_0 \phi(p_{i_0} \otimes q'_{j_0}), \phi(p'_{j_0} \otimes q_{i_0}) \lambda_1 \phi(p_{i_1} \otimes q'_{j_1}),$$

$$\ldots, \phi(p'_{j_{m-1}} \otimes q_{i_{m-1}}) \lambda_m \phi(p_{i_m} \otimes q'_{j_m}), \phi(p'_{j_m} \otimes q_{i_m}), \lambda_{m+1}, \ldots, \lambda_n).$$

Since $\sum_{i_r=1}^{s} \sum_{j_r=1}^{t} \phi(p_{i_r} \otimes q'_{j_r}) \phi(p'_{j_r} \otimes q_{i_r}) = 1$, it is easy to verify that

$$\beta_n \circ \alpha_n - 1 = b \circ h_n + h_{n-1} \circ b$$

so that h is a chain homotopy from $\beta \circ \alpha$ to the identity. Reversing the rôles of Λ and Γ, we see that $\alpha \circ \beta$ is also homotopic to the identity.

Since α and β commute with the Hochschild boundary b, they induce an isomorphism on Hochschild homology by Proposition 2.3.5 of Volume I. Moreover, α and β commute with the operations t_n and b', and hence give rise to maps between $CC_*(\Lambda)$ and $CC_*(\Gamma)$. So we have a map of long exact Connes sequences

$$\cdots \to HC_{n-1}(\Lambda) \xrightarrow{B} HH_n(\Lambda) \xrightarrow{I} HC_n(\Lambda) \xrightarrow{S} HC_{n-2}(\Lambda) \xrightarrow{B} HH_{n-1}(\Lambda) \to \cdots$$

$$\alpha_* \downarrow \cong \qquad \alpha_* \downarrow \cong \qquad \alpha_* \downarrow \qquad \alpha_* \downarrow \cong \qquad \alpha_* \downarrow \cong$$

$$\cdots \to HC_{n-1}(\Gamma) \xrightarrow{B} HH_n(\Gamma) \xrightarrow{I} HC_n(\Gamma) \xrightarrow{S} HC_{n-2}(\Gamma) \xrightarrow{B} HH_{n-1}(\Gamma) \to \cdots$$

Now arguing by induction on n, we see that $\alpha_* : HC_n(\Lambda) \to HC_n(\Gamma)$ is an isomorphism for all $n \geq 0$. $\qquad \square$

Since every other column of the double complex defining $CC_*(\Lambda)$ is contractible, with contracting homotopy $(-1)^n s_n$, we can simplify this complex as follows. We set

$$B = (-1)^n \varepsilon s_n N : \Lambda^{\otimes(n+1)} \to \Lambda^{\otimes(n+2)}$$

(which can be seen to correspond to the map B in Proposition 2.13.1), and we write $B_*(\Lambda)$ for the total complex of the following double complex.

$$
\begin{array}{ccccc}
\vdots & & \vdots & & \vdots \\
\downarrow b & & \downarrow b & & \downarrow b \\
\Lambda^{\otimes 3} & \xleftarrow{\ B\ } & \Lambda^{\otimes 2} & \xleftarrow{\ B\ } & \Lambda \\
\downarrow b & & \downarrow b & & \\
\Lambda^{\otimes 2} & \xleftarrow{\ B\ } & \Lambda & & \\
\downarrow b & & & & \\
\Lambda & & & &
\end{array}
$$

There is an injective map of complexes $B_*(\Lambda) \to CC_*(\Lambda)$ given by sending an element x in degree (i, j) to the sum of x in degree $(2i, j - i)$ and $(-1)^n s_n N x$ in degree $(2i, j - i - 1)$. The quotient complex has a filtration in which the filtered quotients are copies of the acyclic Hochschild complex, and is hence acyclic. So the long exact sequence in homology shows that $B_*(\Lambda) \to CC_*(\Lambda)$ is a homology equivalence, and hence the cyclic homology of Λ is equal to the homology of $B_*(\Lambda)$.

Denote by $R[B]$ the algebra over the coefficient ring R with one generator B satisfying $B^2 = 0$. Then the complex

$$
\cdots \to R[B] \xrightarrow{\ B\ } R[B] \xrightarrow{\ B\ } R[B]
$$

is a resolution of R over $R[B]$, and the above double complex can be thought of as the tensor product over $R[B]$ of this complex and the Hochschild complex $S_*(\Lambda, \Lambda)$. We thus have the following.

PROPOSITION 2.13.3. $HC_*(\Lambda) \cong \mathrm{Tor}_*^{R[B]}(R, S_*(\Lambda, \Lambda))$. $\qquad\square$

Finally, we may simplify the complex $B_*(\Lambda)$ further by replacing the Hochschild complex by its normalisation. Namely, we quotient by the subcomplex of degenerate elements (which is contractible) so that we replace $\Lambda^{\otimes(n+1)}$ by $\bar{\Lambda}^{\otimes n} \otimes \Lambda$, where $\bar{\Lambda}$ is the quotient Λ/R of Λ by the multiples of the identity element. Thus $B_*(\Lambda)$ is chain homotopy equivalent to the total complex $\bar{B}(\Lambda)$ of the following double complex.

$$
\begin{array}{ccccc}
\vdots & & \vdots & & \vdots \\
\downarrow b & & \downarrow b & & \downarrow b \\
\bar{\Lambda}^{\otimes 2} \otimes \Lambda & \xleftarrow{\ B\ } & \bar{\Lambda} \otimes \Lambda & \xleftarrow{\ B\ } & \Lambda \\
\downarrow b & & \downarrow b & & \\
\bar{\Lambda} \otimes \Lambda & \xleftarrow{\ B\ } & \Lambda & & \\
\downarrow b & & & & \\
\Lambda & & & &
\end{array}
$$

One has to check that the operator B is well defined on the reduced complex; this is routine.

Cyclic cohomology $HC^*(\Lambda)$ is defined dually. Namely, the Hochschild complex for computing $HH^*(\Lambda) = H^*(\Lambda, \Lambda^*)$ is given by

$$S^n(\Lambda, \Lambda^*) = \operatorname{Hom}_R(\tilde{S}_n(\Lambda), \Lambda^*) \cong \operatorname{Hom}_R(\Lambda^{\otimes(n+1)}, R) = (\Lambda^{\otimes(n+1)})^*.$$

Thus t_n acts on $S^n(\Lambda, \Lambda^*)$, and we can form a double complex

$$
\begin{array}{ccccccc}
\vdots & & \vdots & & \vdots & & \\
\uparrow b & & \uparrow -b' & & \uparrow b & & \\
(\Lambda^{\otimes 3})^* & \xrightarrow{\varepsilon} & (\Lambda^{\otimes 3})^* & \xrightarrow{N} & (\Lambda^{\otimes 3})^* & \xrightarrow{\varepsilon} & \cdots \\
\uparrow b & & \uparrow -b' & & \uparrow b & & \\
(\Lambda^{\otimes 2})^* & \xrightarrow{\varepsilon} & (\Lambda^{\otimes 2})^* & \xrightarrow{N} & (\Lambda^{\otimes 2})^* & \xrightarrow{\varepsilon} & \cdots \\
\uparrow b & & \uparrow -b' & & \uparrow b & & \\
\Lambda^* & \xrightarrow{\varepsilon} & \Lambda^* & \xrightarrow{N} & \Lambda^* & \xrightarrow{\varepsilon} & \cdots
\end{array}
$$

whose total complex is $CC^*(\Lambda) = \operatorname{Hom}_R(CC_*(\Lambda), R)$. We define the **cyclic cohomology** $HC^*(\Lambda)$ to be the cohomology of this complex. Just as with cyclic homology, there is a long exact Connes sequence

$$\cdots \to HC^{n-1}(\Lambda) \xrightarrow{I} HH^{n-1}(\Lambda) \xrightarrow{B} HC^{n-2}(\Lambda) \xrightarrow{S} HC^n(\Lambda) \xrightarrow{I} HH^n(\Lambda) \to \cdots$$

and an isomorphism

$$HC^*(\Lambda) \cong \operatorname{Ext}^*_{R[B]}(R, S^*(\Lambda, \Lambda^*)).$$

2.14. Cyclic sets

In order to calculate the cyclic homology of a group ring, we need to introduce Connes' notion of a cyclic object in a category. This is a variant of the concept of a simplicial object (Section 1.8), and cyclic sets bear the same relation to topological spaces with an action of the circle group S^1 that simplicial sets do to topological spaces (Dwyer, Hopkins and Kan [97]).

We shall show that the cyclic homology of a group ring RG is the same as the ordinary homology of $ES^1 \times_{S^1} LBG$, where S^1 acts on the free loop space LBG by rotating the loops. Just as in the case of Hochschild homology, this admits a decomposition as a sum over conjugacy classes of elements, which we investigate in the next section. Instead of the classifying space of the centraliser of the corresponding element, one uses a sort of extended centraliser obtained by gluing in a copy of the circle or the real line, depending on whether the element has finite or infinite order. Our approach is close to that of Burghelea and Fiedorowicz [56] and Burghelea [54], but see also Goodwillie [115] and Jones [142].

Let $\bigwedge$ be the category whose objects are the finite cyclically ordered sets, and whose arrows are the functions preserving the cyclic ordering. More precisely (Connes [84]), an object in $\bigwedge$ consists of a finite set together with an injection into S^1, and an arrow in $\bigwedge$ from one such object to another consists of a homotopy class of monotonic (but not necessarily strictly monotonic) continuous maps $S^1 \to S^1$ of degree one such that the image of the first finite set is contained in the second. In particular, there are n arrows from a cyclically ordered set of size n to one of size one. A **cyclic object** X in a category $\mathcal{C}$ is a contravariant functor $\bigwedge \to \mathcal{C}$. A **cyclic set** is a cyclic object in the category of sets. If X is a cyclic object, we write X_n for the image of the cyclically ordered set $\{0, \dots, n\}$ (with the cyclic ordering $0 < 1 < \cdots < n < 0$) in $\mathcal{C}$. The face and degeneracy maps $d_i : X_n \to X_{n-1}$ ($0 \le i \le n$) and degeneracy maps $s_i : X_n \to X_{n+1}$ ($0 \le i \le n$) are defined in exactly the same way as for a simplicial object, but we also have a new map $t_n : X_n \to X_n$ which is the image of the map

$$\{0, \dots, n\} \to \{0, \dots, n\}$$
$$j \mapsto j - 1 \quad (\mathrm{mod}\ n + 1).$$

These maps satisfy the relations given in Section 1.8 for the s_j and d_j, together with the following relations between these and the t_n:

$$d_i t_n = t_{n-1} d_{i-1} \qquad\qquad s_i t_n = t_{n+1} s_{i-1} \qquad 1 \le i \le n$$
$$d_0 t_n = d_n \qquad\qquad\qquad s_0 t_n = t_{n+1}^2 s_n$$
$$(t_n)^{n+1} = 1$$

All maps and relations coming from $\bigwedge$ follow from these. So giving a cyclic object X in $\mathcal{C}$ is the same as giving objects X_n in $\mathcal{C}$, $0 \le n < \infty$, and maps $d_i : X_n \to X_{n-1}$, $s_i : X_n \to X_{n+1}$ and $t_n : X_n \to X_n$ satisfying these relations.

A **cyclic map** $f : X \to Y$ between cyclic objects is a natural transformation of functors; this amounts to giving maps $f_n : X_n \to Y_n$ satisfying $f_{n-1} d_i = d_i f_n$, $f_{n+1} s_i = s_i f_n$ and $f_n t_n = t_n f_n$ for all $0 \le i \le n$. Thus cyclic objects in a category $\mathcal{C}$ form a category **Cycl** $\mathcal{C}$.

There is an obvious functor $\triangle \to \bigwedge$ taking a totally ordered finite set to the corresponding cyclically ordered set, and regarding a monotonic function as a function preserving the cyclic ordering. By composing with this functor, a cyclic object gives rise to a simplicial object, and so we have an obvious forgetful functor **Cycl** $\mathcal{C} \to$ **Simp** $\mathcal{C}$.

EXAMPLES. 1. The simplicial set SG/G corresponding to the Hochschild complex (see Section 2.12) has the structure of a cyclic set with

$$t_n(g_0 \otimes g_1 \otimes \cdots \otimes g_n) = g_n \otimes g_0 \otimes \cdots \otimes g_{n-1}.$$

More generally, the Hochschild complex of a ring Λ (see Section 2.11) is a cyclic R-module with

$$t_n(\lambda_0 \otimes \lambda_1 \otimes \cdots \otimes \lambda_n) = \lambda_n \otimes \lambda_0 \otimes \cdots \otimes \lambda_{n-1}.$$

2. Let $S[1]$ be the simplicial circle; namely the simplicial set with one non-degenerate zero-simplex $*$, one non-degenerate 1-simplex σ, and no other

non-degenerate simplices. If X is a simplicial set, then the simplicial set $\mathrm{Hom}(S[1], X)$ is a cyclic set as follows. Writing $(\bar{s}_i\sigma, s_i\tau)$, $0 \leq i \leq n$, for the $n+1$ non-degenerate $(n+1)$-simplices in $S[1] \times \Delta[n]$ as in the proof of Proposition 2.12.3, we let t_n act via

$$t_n f(\bar{s}_i\sigma, s_i\tau) = f(\bar{s}_{i+1}\sigma, s_{i+1}\tau),$$

where the subscript $i+1$ is to be read modulo $n+1$. With these structures of cyclic sets, it is straightforward (but tedious!) to check that the isomorphism given in Proposition 2.12.3 is an isomorphism of cyclic sets.

3. If Y is a topological space with a continuous action of the circle group $S^1 = \{e^{2\pi i\theta}, 0 \leq \theta \leq 1\}$, we give the singular simplicial set on Y the structure of a cyclic set as follows. If $f : \Delta^n \to Y$ is a singular n-simplex, then we set

$$t_n f(x_0, \ldots, x_n) = e^{2\pi i x_0} f(x_n, x_0, \ldots, x_{n-1}).$$

It is easy to check that this definition satisfies the appropriate relations. Thus $\mathrm{Sing}(-)$ is a functor from spaces with S^1-action to cyclic sets.

4. The forgetful functor $\mathbf{Cycl}\ \mathcal{C} \to \mathbf{Simp}\ \mathcal{C}$ described above has a left adjoint, which assigns to each simplicial object X the free cyclic object $S[1] \bowtie X$ whose n-simplices are pairs (t_n^j, x), $0 \leq j \leq n$, $x \in X_n$ with cyclic structure given by

$$d_i(t_n^j, x) = (t_{n-1}^j, d_{i-j}x) \qquad\qquad i - j \geq 0$$
$$d_i(t_n^j, x) = (t_{n-1}^{j-1}, d_{i-j+n+1}x) \qquad i - j < 0$$
$$d_n(t_n^n, x) = (t_{n-1}^0, d_0 x)$$
$$s_i(t_n^j, x) = (t_{n+1}^j, s_{i-j}x) \qquad\qquad i - j \geq 0$$
$$s_i(t_n^j, x) = (t_{n+1}^{j-1}, s_{i-j+n+1}x) \qquad i - j < 0$$
$$t_n^i(t_n^j, x) = (t_n^{i+j}, x) \qquad\qquad i + j \leq n$$
$$t_n^i(t_n^j, x) = (t_n^{i+j-n-1}, x) \qquad\quad i + j > n.$$

In particular, $S[1] = S[1] \bowtie *$ is a cyclic set.

The underlying simplicial set of $S[1] \bowtie X$ is not in general isomorphic to $S[1] \times X$, but their topological realisations are homeomorphic via the map

$$|S[1] \bowtie X| \to |S[1] \times X|$$
$$((t_n^0, x), (x_0, \ldots, x_n)) \mapsto ((*, x), (x_0, \ldots, x_n))$$
$$((t_n^j, x), (x_0, \ldots, x_n)) \mapsto ((\bar{s}_{j-1}\sigma, x), (x_{n-j+1}, \ldots, x_n, x_0, \ldots, x_{n-j}))$$
$$1 \leq j \leq n.$$

Thus for example $|S[1] \bowtie S[1]|$ is the triangulation in which the torus is cut into two triangles using the trailing diagonal instead of the leading diagonal in $|S[1] \times S[1]|$.

If X is a cyclic set, then there is an evaluation map

$$S[1] \bowtie X \to X$$

$$(t_n^j, x) \mapsto t_n^j x$$

which gives rise to a map of topological realisations $S^1 \times |X| \to |X|$. In case $X = S[1]$, this map $S^1 \times S^1 \to S^1$ is the multiplication map which makes S^1 a group. Note that with the normal triangulation of $S^1 \times S^1$ coming from the simplicial product, multiplication does not correspond to a simplicial map.

Now there is an obvious identification between $(S[1] \bowtie S[1]) \bowtie X$ and $S[1] \bowtie (S[1] \bowtie X)$ whose topological realisation shows that the above map $S^1 \times |X| \to |X|$ gives an action of the group S^1 on $|X|$. It follows that topological realisation is a functor from cyclic sets to spaces with S^1-action.

It is not hard to check that the functors we have just described are adjoint functors between topological spaces with S^1-action and cyclic sets

$$\mathrm{Map}_{S^1}(|X|, Y) \overset{\mathrm{nat}}{\cong} \mathrm{Hom}_{\mathbf{CyclSet}}(X, \mathrm{Sing}(Y)).$$

WARNING. Let X be a simplicial set. While the map

$$|\mathrm{Hom}(S[1], X)| \to \mathrm{Map}(S^1, X) = LX$$

is a homotopy equivalence and commutes with the S^1-action, it does not necessarily have a homotopy inverse as a map of spaces with S^1-action. For a discussion related to this point, see Section 6.4.

We now define the cyclic homology $HC_*(X; R)$ of a cyclic set X. We write down the following double complex:

$$
\begin{array}{ccccccc}
\vdots & & \vdots & & \vdots & & \\
\downarrow{\scriptstyle b} & & \downarrow{\scriptstyle -b'} & & \downarrow{\scriptstyle b} & & \\
RX_2 & \xleftarrow{\;\varepsilon\;} & RX_2 & \xleftarrow{\;N\;} & RX_2 & \xleftarrow{\;\varepsilon\;} & \cdots \\
\downarrow{\scriptstyle b} & & \downarrow{\scriptstyle -b'} & & \downarrow{\scriptstyle b} & & \\
RX_1 & \xleftarrow{\;\varepsilon\;} & RX_1 & \xleftarrow{\;N\;} & RX_1 & \xleftarrow{\;\varepsilon\;} & \cdots \\
\downarrow{\scriptstyle b} & & \downarrow{\scriptstyle -b'} & & \downarrow{\scriptstyle b} & & \\
RX_0 & \xleftarrow{\;\varepsilon\;} & RX_0 & \xleftarrow{\;N\;} & RX_0 & \xleftarrow{\;\varepsilon\;} & \cdots
\end{array}
$$

Here, RX_n denotes the free R-module on X_n, with

$$\varepsilon = 1 - (-1)^n t_n, \quad N = \sum_{j=0}^{n}((-1)^n t_n)^j, \quad b = \sum_{i=0}^{n}(-1)^i d_i, \quad b' = \sum_{i=0}^{n-1}(-1)^i d_i.$$

Just as in Section 2.13, we have $b\varepsilon = \varepsilon b'$ and $b'N = Nb$, so that this is a double complex. We write $CC_*(X; R)$ for the total complex of the above double complex, and $HC_*(X; R)$ for its homology. Since $H_*(X; R)$ is the homology of the first column, and the second column is contractible with contracting homotopy $(-1)^n s_n$, we obtain by the same argument as in Section 2.13 a long exact sequence

$$\cdots \to H_n(X;R) \xrightarrow{I} HC_n(X;R) \xrightarrow{S} HC_{n-2}(X;R) \xrightarrow{B} H_{n-1}(X;R) \xrightarrow{I} \cdots$$

Again we set

$$B = (-1)^n \varepsilon s_n N : \Lambda^{\otimes(n+1)} \to \Lambda^{\otimes(n+2)}$$

and we write $B_*(X;R)$ for the total complex of the following double complex

$$
\begin{array}{ccccc}
\vdots & & \vdots & & \vdots \\
\downarrow b & & \downarrow b & & \downarrow b \\
RX_2 & \xleftarrow{B} & RX_1 & \xleftarrow{B} & RX_0 \\
\downarrow b & & \downarrow b & & \\
RX_1 & \xleftarrow{B} & RX_0 & & \\
\downarrow b & & & & \\
RX_0 & & & &
\end{array}
$$

so that we have a homology equivalence $B_*(X;R) \to CC_*(X;R)$. Again, we can replace $B_*(X;R)$ by the reduced version $\bar{B}(X;R)$ if we wish.

Similarly, starting with the dual complex

$$CC^*(X;R) = \mathrm{Hom}_R(CC_*(X;R),R)$$

we obtain the cyclic cohomology $HC^*(X;R)$.

PROPOSITION 2.14.1. *If X is a cyclic set, then there are natural isomorphisms*

(i) $HC_*(X;R) \cong \mathrm{Tor}_*^{R[B]}(R, C_*(X;R)) \cong H_*(ES^1 \times_{S^1} |X|;R).$
(ii) $HC^*(X;R) \cong \mathrm{Ext}^*_{R[B]}(R, C^*(X;R)) \cong H^*(ES^1 \times_{S^1} |X|;R).$

PROOF. We only prove (i), since (ii) is dual. The first isomorphism follows just as in Section 2.13 from the fact that the double complex defining $\bar{B}_*(X;R)$ can be thought of as the tensor product over $R[B]$ of the complex

$$\cdots \to R[B] \xrightarrow{B} R[B] \xrightarrow{B} R[B]$$

and the reduced Hochschild complex $C_*(X;R)$.

For the second isomorphism, we argue as follows (cf. Section 3.11). Given any topological group G whose underlying topological space is a CW-complex, EG is also a CW-complex. If G acts on a CW-complex Y in such a way that $G \times Y \to Y$ is a cellular map, then $C_*(Y;R)$ is a module over the algebra $C_*(G;R)$ and $C_*(EG \times Y;R)$ is a free resolution of $C_*(Y;R)$. We have

$$C_*(EG \times_G Y;R) = R \otimes_{C_*(G;R)} C_*(EG \times Y;R)$$

and so

$$H_*(EG \times_G Y;R) \cong \mathrm{Tor}_*^{C_*(G;R)}(R, C_*(Y;R)).$$

We use this in case $G = S^1$ to deduce that

$$H_*(ES^1 \times_{S^1} |X|; R) \cong \operatorname{Tor}_*^{C_*(S[1];R)}(R, C_*(X; R)).$$

So it remains to identify the action of $C_*(S[1]; R)$ on $C_*(X; R)$ with the action of $R[B]$. Namely, using the Eilenberg–Zilber map (see Section 1.8) we have

$$C_*(S[1]; R) \otimes C_*(X; R) \xrightarrow{\cong} C_*(S[1] \times X; R) \quad \to \quad C_*(X; R)$$

$$\sigma \otimes x \mapsto \sum_{i=0}^{n}(-1)^{in}(\bar{s}_i\sigma, s_i x) \mapsto \sum_{i=0}^{n}(-1)^{in}t_n^i(s_i x)$$

Since we are using the reduced Hochschild complex, it is easy to check that the latter is the formula for $B(x)$ in $C_*(X; R)$. $\qquad\square$

COROLLARY 2.14.2. (i) $HC_*(RG) \cong H_*(ES^1 \times_{S^1} LBG; R)$.
(ii) $HC^*(RG) \cong H^*(ES^1 \times_{S^1} LBG; R)$.

PROOF. Again we only prove (i), since (ii) is dual. We remarked in the discussion of Example 2 above that the isomorphism $\operatorname{Hom}(S[1], NG) = SG/G$ of Proposition 2.12.3 is an isomorphism of cyclic sets. Thus the cyclic homology of RG, which is the cyclic homology of the cyclic set SG/G, is equal to the cyclic homology of $\operatorname{Hom}(S[1], NG)$. By the above proposition, this is equal to

$$H_*(ES^1 \times_{S^1} |\operatorname{Hom}(S[1], NG)|; R) = H_*(ES^1 \times_{S^1} LBG). \qquad\square$$

2.15. Extended centralisers

In the last section, we saw that

$$HC_*(RG) \cong H_*(ES^1 \times_{S^1} LBG; R).$$

In this section, we decompose the latter as a direct sum over conjugacy classes of ordinary group cohomology of extended centralisers. Dually in cohomology we obtain a direct product decomposition.

DEFINITION 2.15.1. *Suppose g is an element of a (discrete) group G. The* **extended centraliser** $\hat{C}_G(g)$ *is the topological group defined by the pushout diagram*

$$
\begin{array}{ccc}
1 & \mapsto & g \\[4pt]
\mathbb{Z} & \longrightarrow & C_G(g) \\
\downarrow & & \downarrow \\
\mathbb{R} & \longrightarrow & \hat{C}_G(g).
\end{array}
$$

In other words, $\hat{C}_G(g) = (\mathbb{R} \times C_G(g))/\sim$ where $\sim$ is the equivalence relation given by $(\lambda, gx) \sim (\lambda + 1, x)$.

Thus $\hat{C}_G(g)/C_G(g) \cong \mathbb{R}/\mathbb{Z} = S^1$, and so by Theorem 2.4.12 (ii) there is a fibration $BC_G(g) \to B\hat{C}_G(g)$ with fibre S^1.

THEOREM 2.15.2. $ES^1 \times_{S^1} LBG = \bigcup_{g \in_G G} B\hat{C}_G(g).$

PROOF. As in the proof of Lemma 2.12.1, we let Y denote the space of maps $I \to EG$ such that the images of the two endpoints 0 and 1 are in the same G-orbit. Recall that G acts freely on Y with quotient $Y/G = LBG$.

We can write Y as a disjoint union over elements $g \in G$ of spaces Y_g, where Y_g is the subspace consisting of maps $f : I \to EG$ with $f(0)g = f(1)$. The action of an element $h \in G$ takes Y_g to $Y_{h^{-1}gh}$, and so $LBG = Y/G$ is a disjoint union over conjugacy classes of $g \in G$ of the spaces $Y_g/C_G(g)$. Each Y_g is contractible, and $C_G(g)$ acts freely on it.

Given a map $f : I \to EG$ in Y_g, we can extend to a map $\tilde{f} : \mathbb{R} \to EG$ as follows. If $\lambda \in \mathbb{R}$ is equal to $n + t$ with $n \in \mathbb{Z}$ and $0 \le t \le 1$, we set $\tilde{f}(\lambda) = f(t)g^n$. It is easily seen that this definition "matches up" at the integers to give a continuous map.

The action of S^1 on LBG preserves the connected components $Y_g/C_G(g)$, and lifts to an action of $\mathbb{R}$ on Y_g, given by $(\lambda.\tilde{f})(\mu) = \tilde{f}(\lambda + \mu)$ for $\lambda, \mu \in \mathbb{R}$. Thus $\mathbb{R} \times C_G(g)$ acts on $ES^1 \times Y_g$, by letting $\mathbb{R}$ act on ES^1 via the map $\mathbb{R} \to \mathbb{R}/\mathbb{Z} = S^1$ and on Y_g as above, and $C_G(g)$ act trivially on ES^1 and freely on Y_g as above. Now $1 \in \mathbb{R}$ acts in the same way as $g \in C_G(g)$, so that this passes down to an action of $\hat{C}_G(g)$ on $ES^1 \times Y_g$, which can be seen to be free. Thus we can use $ES^1 \times Y_g$ as our model for $E\hat{C}_G(g)$, so that the quotient is

$$ES^1 \times_{S^1} (Y_g/C_G(g)) = (ES^1 \times Y_g)/\hat{C}_G(g) = B\hat{C}_G(g). \qquad \square$$

COROLLARY 2.15.3. (i) $HC_*(RG) = \bigoplus_{g \in_G G} H_*(B\hat{C}_G(g); R).$

(ii) $HC^*(RG) = \prod_{g \in_G G} H^*(B\hat{C}_G(g); R).$

PROOF. This follows from Corollary 2.14.2 and the above theorem. $\square$

We can now define the **Chern map** $\mathrm{Ch}_{n,r} : K_n(\Lambda) \to HC_{n+2r}(\Lambda)$ $(r \ge 0)$ of Connes and Karoubi [147, 148, 149]. Recall that we defined the Dennis trace map at the end of Section 2.11 as a composite of maps, one of which was the map

$$H_n(GL(\Lambda), \mathbb{Z}) \to HH_n(\mathbb{Z}GL(\Lambda))$$

which embeds the left hand side as the summand of the right hand side corresponding to the identity element in the decomposition of Theorem 2.11.2. Now according to the above discussion, the summand of $HC_n(\mathbb{Z}GL(\Lambda))$ corresponding to the centraliser of the identity element is

$$H_n(S^1 \times GL(\Lambda), \mathbb{Z}) = H_n(GL(\Lambda), \mathbb{Z}) \oplus H_{n-2}(GL(\Lambda), \mathbb{Z}) \oplus \cdots$$

(Recall that $H_n(S^1, \mathbb{Z}) = H_n(\mathbb{C}P^\infty; \mathbb{Z})$ is isomorphic to $\mathbb{Z}$ for n even and is zero for n odd). Thus there is an obvious map

$$H_n(GL(\Lambda), \mathbb{Z}) \to HC_{n+2r}(\mathbb{Z}GL(\Lambda))$$

for each value of $r \geq 0$ (related by composition with the S operator in the long exact Connes sequence). The Chern map $\mathrm{Ch}_{n,r}$ is defined to be the composite

$$K_n(\Lambda) \xrightarrow{\mathrm{Hurewicz}} H_n(GL(\Lambda), \mathbb{Z}) \to HC_{n+2r}(\mathbb{Z}GL(\Lambda)) \to HC_{n+2r}(\Lambda).$$

The map $HC_{n+2r}(\mathbb{Z}GL(\Lambda)) \to HC_{n+2r}(\Lambda)$ is defined in a manner analogous to the map in Hochschild homology described at the end of Section 2.11.

CHAPTER 3

Spectral sequences

3.1. Introduction to spectral sequences

A *spectral sequence* is a fairly complicated algebraic gadget for making calculations in homological algebra. In this chapter, we shall introduce the machinery of spectral sequences, and show how to apply them in a number of different situations. The most important spectral sequence for us will be the Lyndon–Hochschild–Serre spectral sequence associated to a group extension. The reader who wishes to avoid topology can read the sections on the spectral sequence of a filtered complex, a double complex and a group extension for an account of this spectral sequence. As always, we shall only give a sketchy account of the topological aspects of the material anyway, and make the algebraic aspects as self-contained as possible.

This introduction is designed for the orientation of the reader, and is not necessary for the logical structure of what follows.

Here is a guide to the interconnections between the various topics introduced in this chapter.

$$
\left.\begin{array}{l}
\left.\begin{array}{l}
\text{Hurewicz fibration} \\
\text{fibre bundle}
\end{array}\right\} \quad \rightarrow \text{Serre fibration} \\[2em]
\left.\begin{array}{l}
\text{group extension} \\
\text{composite functor} \\
\text{pullback of fibrations} \\
\text{(Eilenberg–Moore)}
\end{array}\right\} \quad \rightarrow \text{double complex}
\end{array}\right\} \rightarrow
\begin{array}{c}
\text{filtered complex} \\
\downarrow \\
\text{exact couple} \\
\downarrow \\
\text{spectral sequence}
\end{array}
$$

Thus we shall be interested in the following spectral sequences. R will always denote a commutative ring of coefficients.

(i) If $F \to E \to B$ is a Serre fibration then there are spectral sequences in cohomology and homology,

$$
H^p(B; H^q(F; R)) \Rightarrow H^{p+q}(E; R), \quad H_p(B; H_q(F; R)) \Rightarrow H_{p+q}(E; R).
$$

This notation is meant to suggest that the left-hand side is the initial data for the calculation and the right-hand side is what we wish to calculate. There are certain "differentials" and "extension problems" involved in going from the left to the right, and this makes up the inner workings of the spectral sequence, of which more later.

In fact there is a technical difficulty here. The group $\pi_1(B)$ acts in a natural way on $H^*(F; R)$, and if the action is not trivial, we should really regard $H^*(F; R)$ as a *local system of coefficients* $\mathcal{H}^*(F; R)$ on B, and the spectral sequence takes the form

$$H^p(B; \mathcal{H}^q(F; R)) \Rightarrow H^{p+q}(E; R)$$

(and similarly in homology). We shall simplify the exposition by making the assumption that $\pi_1(B)$ acts trivially on $H^*(F; R)$.

If one wishes to develop the spectral sequence of a group extension topologically, one has to treat these more general local systems, unless the normal subgroup is central. Thus we shall be content to set up the spectral sequence of a group extension algebraically. This is our next topic.

(ii) If N is a normal subgroup of a (not necessarily finite) group G then the fibration $BN \to BG \to B(G/N)$ gives rise to spectral sequences

$$H^p(G/N, H^q(N, R)) \Rightarrow H^{p+q}(G, R), \quad H_p(G/N, H_q(N, R)) \Rightarrow H_{p+q}(G, R).$$

More generally, if M is an RG-module (not necessarily trivial!) then there are spectral sequences

$$H^p(G/N, H^q(N, M)) \Rightarrow H^{p+q}(G, M), \quad H_p(G/N, H_q(N, M)) \Rightarrow H_{p+q}(G, M).$$

The latter is known as the Lyndon–Hochschild–Serre spectral sequence of the group extension. We set up these spectral sequences algebraically, using the theory of double complexes, and concentrate on the case of cohomology. This is because in cohomology the spectral sequence $H^*(G/N, H^*(N, R))$ has a ring structure over which the spectral sequence $H^*(G/N, H^*(N, M))$ is a module.

(iii) We shall give a brief sketch of the Eilenberg–Moore spectral sequence

$$\mathrm{Tor}^{**}_{H^*(B;R)}(H^*(B'; R), H^*(E; R)) \Rightarrow H^*(E'; R)$$

associated to a pullback of fibrations

$$\begin{array}{ccccc}
F & \longrightarrow & E' & \longrightarrow & B' \\
\| & & \downarrow & & \downarrow \\
F & \longrightarrow & E & \longrightarrow & B
\end{array}$$

in which $\pi_1(B)$ acts trivially on $H^*(F; R)$.

In the case where B' consists of just the basepoint of B, we have $E' = F$, and so the spectral sequence takes the form

$$\mathrm{Tor}^{**}_{H^*(B;R)}(R, H^*(E; R)) \Rightarrow H^*(F; R)$$

so that it computes the cohomology of the fibre from that of the base space and total space.

One situation of interest in which we can apply this spectral sequence is the case of the fibration

$$K(G, 1) \to K(G/Z, 1) \to K(Z, 2)$$

of Eilenberg–Mac Lane spaces, where Z is a central subgroup of G. Dave Rusin [**226**] has used this to calculate the mod 2 cohomology of all groups of order 32 (there are 51 of them!). The point is that there are fewer non-zero differentials (and very often none at all) than in the Lyndon–Hochschild–Serre spectral sequence associated to the same group extension.

(iv) Using the Bott periodicity theorem in K-theory, one can obtain a spectral sequence

$$H^*(G, \mathbb{Z}) \Rightarrow R(G)^\wedge$$

due to Atiyah, going from the integral cohomology ring to the completion of the ordinary character ring at the augmentation ideal. Again, we shall only sketch the construction of this spectral sequence.

We now explain in a little more detail how spectral sequences arise.

(1) Given a CW-complex X and a sequence of subcomplexes

$$X_0 \subseteq X_1 \subseteq \cdots \subseteq X = \bigcup X_i$$

we get a long exact sequence for each adjacent pair

$$\cdots \to H^n(X_{p-1}; R) \to H^n(X_p; R) \to H^n(X_p, X_{p-1}; R)$$
$$\to H^{n+1}(X_{p-1}; R) \to \cdots$$

$$\cdots \to H_n(X_{p-1}; R) \to H_n(X_p; R) \to H_n(X_p, X_{p-1}; R)$$
$$\to H_{n-1}(X_{p-1}; R) \to \cdots$$

which interlock in a way we shall examine more closely later. What we wish to do is to assemble all this data in a sensible fashion and use it to calculate $H^*(X; R)$.

(2) We did not really need a CW-complex in order to carry this out. All we needed was a (co)chain complex $\mathbf{X}$ (which might be the (co)chains on a CW-complex, or might arise from projective resolutions of modules, or anything else), and a sequence of subcomplexes

$$\mathbf{X} = F^0\mathbf{X} \supseteq F^1\mathbf{X} \supseteq \cdots \supseteq \bigcap F^i\mathbf{X} = \{0\}.$$
$$F_0\mathbf{X} \subseteq F_1\mathbf{X} \subseteq \cdots \subseteq \mathbf{X} = \bigcup F_i\mathbf{X}.$$

(3) If $p : E \to B$ is a Serre fibration of CW-complexes, we take as our subcomplexes of E the preimages of the skeletal filtration of B.

$$p^{-1}(B^0) \subseteq p^{-1}(B^1) \subseteq \cdots \subseteq E = \bigcup p^{-1}(B^i).$$

By taking (co)chains we have filtrations as above:

$$F^i\mathbf{X} = \mathrm{Ker}(C^*(E) \to C^*(p^{-1}(B^{i-1}))), \quad F_i\mathbf{X} = C_*(p^{-1}(B^i)).$$

In this case it turns out that under suitable hypotheses, the initial data for the spectral sequence amounts to knowing $H^*(B; H^*(F; R))$, and so we can use this to calculate $H^*(E; R)$.

Before we set up spectral sequences in detail, we now try to give some of the flavour of what a spectral sequence looks like. You are not expected to understand yet where all the information is coming from.

The initial data usually comes as a doubly indexed set of abelian groups (or R-modules). This is written as E_2^{pq} in cohomology or E_{pq}^2 in homology (the reason for the 2 will only become apparent when we come to study the spectral sequence of a double complex). For example for a group extension we have

$$E_2^{pq} = H^p(G/N, H^q(N, R)), \quad E_{pq}^2 = H_p(G/N, H_q(N, R)).$$

These are examples of *first quadrant* spectral sequences, where E_2^{pq} (resp. E_{pq}^2) is zero whenever $p < 0$ or $q < 0$.

The next piece of data is that there is a naturally defined *differential* $d_2 : E_2^{pq} \to E_2^{p+2,q-1}$ satisfying $d_2 \circ d_2 = 0$. This information is usually depicted diagrammatically as follows.

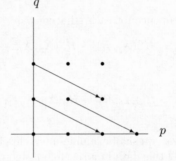

Thus we may form the homology of this complex

$$E_3^{pq} = H(E_2^{pq}, d_2) = \frac{\mathrm{Ker}(d_2 : E_2^{pq} \to E_2^{p+2,q-1})}{\mathrm{Im}(d_2 : E_2^{p-2,q+1} \to E_2^{pq})}.$$

On E_3^{pq} there is a naturally defined differential $d_3 : E_3^{pq} \to E_3^{p+3,q-2}$ satisfying $d_3 \circ d_3 = 0$ as in the following diagram, and we define $E_4^{pq} = H(E_3^{pq}, d_3)$, and continue in this way.

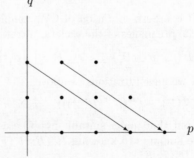

Now for a given p and q, for n large enough (e.g. $n > \max(p, q+1)$) we have

$$d_n : E_n^{pq} \to E_n^{p+n,q-n+1} = 0, \quad d_n : 0 = E_n^{p-n,q+n-1} \to E_n^{pq}$$

and so $E_{n+1}^{pq} = E_n^{pq}$. Thus we may define E_∞^{pq} to be the common value of E_n^{pq} for n large. If the spectral sequence is not first quadrant, there are slightly more delicate convergence problems, but the reader should not worry about that at this stage.

The way E_∞^{pq} is related to what we are trying to calculate is as follows. There is a filtration

$$H^{p+q}(G, R) = F^0 H^{p+q}(G, R) \supseteq F^1 H^{p+q}(G, R) \supseteq$$
$$\cdots \supseteq F^{p+q+1} H^{p+q}(G, R) = 0$$

such that

$$\boxed{F^p H^{p+q}(G, R)/F^{p+1} H^{p+q}(G, R) \cong E_\infty^{pq}.}$$

In other words, the group $H^n(G, R)$ has a filtration in which the quotients are the groups going down a trailing diagonal $p + q = n$ of the E_∞ term of the spectral sequence.

Similarly in homology we have

$$d^n : E_{pq}^n \to E_{p-n,q+n-1}^n, \quad d^n \circ d^n = 0,$$

$$E_{pq}^{n+1} = \frac{\mathrm{Ker}(d^n : E_{pq}^n \to E_{p-n,q+n-1}^n)}{\mathrm{Im}(d^n : E_{p+n,q-n+1}^n \to E_{pq}^n)}$$

and a filtration

$$0 = F_{-1} H_{p+q}(G, R) \subseteq F_0 H_{p+q}(G, R) \subseteq \cdots \subseteq F_{p+q} H_{p+q}(G, R) = H_{p+q}(G, R)$$

such that

$$F_p H_{p+q}(G, R)/F_{p-1} H_{p+q}(G, R) \cong E_{pq}^\infty.$$

Here is a picture of E_{pq}^2.

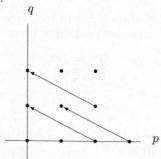

We refer to E_n^{pq} (resp. E_{pq}^n) as the nth *page* of the spectral sequence.

3.2. The spectral sequence of a filtered chain complex

We shall work in this section with the cohomology spectral sequence, and then indicate what changes are necessary to obtain the homology spectral sequence. The ideas are the same in both cases, but the indexing is different. In some sense, homology is indexed as though it were negative degree cohomology.

Suppose

$$\mathbf{X} = F^0\mathbf{X} \supseteq F^1\mathbf{X} \supseteq \cdots \supseteq \bigcap F^i\mathbf{X} = \{0\}.$$

is a filtration of a cochain complex (i.e., a decreasing sequence of subcomplexes). Then the long exact sequences

$$\cdots \to H^n(F^{p+1}\mathbf{X}) \to H^n(F^p\mathbf{X}) \to H^n(F^p\mathbf{X}, F^{p+1}\mathbf{X})$$
$$\to H^{n+1}(F^{p+1}\mathbf{X}) \to \cdots$$

can be fitted together as in the following diagram, where the sequences go alternately down one and to the right two places.

$$\xrightarrow{k_1} H^{n-1}(F^{p+1}\mathbf{X}) \to H^{n-1}(F^{p+1}\mathbf{X}, F^{p+2}\mathbf{X}) \to H^n(F^{p+2}\mathbf{X}) \to H^n(F^{p+2}\mathbf{X}, F^{p+3}\mathbf{X}) \to$$

$$\downarrow{i_1}$$

$$\to H^{n-1}(F^p\mathbf{X}) \xrightarrow{j_1} H^{n-1}(F^p\mathbf{X}, F^{p+1}\mathbf{X}) \xrightarrow{k_1} H^n(F^{p+1}\mathbf{X}) \to H^n(F^{p+1}\mathbf{X}, F^{p+2}\mathbf{X}) \to$$

$$\downarrow{i_1}$$

$$\to H^{n-1}(F^{p-1}\mathbf{X}) \to H^{n-1}(F^{p-1}\mathbf{X}, F^p\mathbf{X}) \to H^n(F^p\mathbf{X}) \xrightarrow{j_1} H^n(F^p\mathbf{X}, F^{p+1}\mathbf{X}) \xrightarrow{k_1}$$

To keep control of all the data here, we set

$$E_1^{pq} = H^{p+q}(F^p\mathbf{X}, F^{p+1}\mathbf{X}), \quad D_1^{pq} = H^{p+q}(F^p\mathbf{X}).$$

as doubly indexed sets of abelian groups (or R-modules, if $\mathbf{X}$ is a cochain complex of R-modules). The exact sequences become

$$D_1^{**} \xrightarrow{i_1} D_1^{**}$$
$$k_1 \nwarrow \quad \swarrow j_1$$
$$E_1^{**}$$

where

$$\deg(i_1) = (-1, 1) \qquad \qquad \mathrm{Ker}(i_1) = \mathrm{Im}(k_1)$$
$$\deg(j_1) = (0, 0) \qquad \qquad \mathrm{Ker}(j_1) = \mathrm{Im}(i_1)$$
$$\deg(k_1) = (1, 0) \qquad \qquad \mathrm{Ker}(k_1) = \mathrm{Im}(j_1).$$

DEFINITION 3.2.1 (Massey). *An* **exact couple** *is an exact triangle of the form*

$$D \xrightarrow{i} D \qquad \operatorname{Ker}(i) = \operatorname{Im}(k)$$
$$k \nwarrow \swarrow j \qquad \operatorname{Ker}(j) = \operatorname{Im}(i)$$
$$E \qquad \operatorname{Ker}(k) = \operatorname{Im}(j).$$

Every time we have an exact couple as above, we obtain a *spectral sequence* as follows. Since $k \circ j = 0$, we have $(j \circ k)^2 = 0$. Thus setting $d = j \circ k$, we may take homology of E with respect to d,

$$H(E, d) = \operatorname{Ker}(d)/\operatorname{Im}(d).$$

This gives rise to the notion of a derived couple.

DEFINITION 3.2.2. *If*

$$D \xrightarrow{i} D$$
$$k \nwarrow \swarrow j$$
$$E$$

is an exact couple, then the **derived couple**

$$D' \xrightarrow{i'} D'$$
$$k' \nwarrow \swarrow j'$$
$$E'$$

is given by setting

$$D' = \operatorname{Im}(i) \subseteq D$$
$$E' = H(E, d) \text{ as a subquotient of } E$$
$$i' = i|_{D'}$$
$$j'(i(x)) = j(x) + \operatorname{Im}(d) \in \operatorname{Ker}(d)/\operatorname{Im}(d)$$
$$k'(z + \operatorname{Im}(d)) = k(z)$$

It is an easy diagram chase to check that the above maps are well defined, and that the derived couple of an exact couple is again an exact couple.

DEFINITION 3.2.3. *We define the exact couple*

$$D_n \xrightarrow{i_n} D_n$$
$$k_n \nwarrow \swarrow j_n$$
$$E_n$$

to be the $(n-1)$*st derived couple of the exact couple*

$$D_1 \xrightarrow{i_1} D_1$$
$$k_1 \nwarrow \swarrow j_1 .$$
$$E_1$$

The sequence (E_1, d_1), (E_2, d_2), ... *is called the* **spectral sequence** *of the original exact couple.*

We now keep track of the double grading. We have

$$\deg(i_n) = \deg(i_{n-1})$$
$$\deg(j_n) = \deg(j_{n-1}) - \deg(i_{n-1})$$
$$\deg(k_n) = \deg(k_{n-1})$$
$$\deg(d_n) = \deg(j_n) + \deg(k_n)$$

and hence

$$\deg(i_n) = (-1, 1)$$
$$\deg(j_n) = (n - 1, -n + 1)$$
$$\deg(k_n) = (1, 0)$$
$$\deg(d_n) = (n, -n + 1).$$

Thus for example (E_2, d_2) and (E_3, d_3) appear as in the pictures on page 96.

For most of the filtered complexes we are interested in, it will turn out that $E_2^{pq} = 0$ for $q < 0$ (and also of course for $p < 0$) so that we shall proceed for the rest of this section under this assumption. This will enable us to avoid a detailed discussion of convergence problems.

Each D_n^{pq} is contained inside D_1^{pq}, and we write D_∞^{pq} for $\bigcap_n D_n^{pq}$. Since each E_n^{pq} is a subquotient of E_{n-1}^{pq}, we may find subgroups $Z_2^{pq} = \mathrm{Ker}(d_1)$ and $B_2^{pq} = \mathrm{Im}(d_1)$ of E_1^{pq} such that $E_2^{pq} = Z_2^{pq}/B_2^{pq}$. We then have subgroups Z_3^{pq} and B_3^{pq} of E_1^{pq} with $Z_3^{pq}/B_2^{pq} = \mathrm{Ker}(d_2)$, $B_3^{pq}/B_2^{pq} = \mathrm{Im}(d_2)$ and $E_3^{pq} = (Z_3^{pq}/B_2^{pq})/(B_3^{pq}/B_2^{pq}) \cong Z_3^{pq}/B_3^{pq}$. Continuing this way and setting $Z_1^{pq} = E_1^{pq}$ and $B_1^{pq} = 0$, we have subgroups

$$E_1^{pq} = Z_1^{pq} \supseteq Z_2^{pq} \supseteq Z_3^{pq} \supseteq \cdots \supseteq B_3^{pq} \supseteq B_2^{pq} \supseteq B_1^{pq} = 0$$

such that $E_n^{pq} \cong Z_n^{pq}/B_n^{pq}$. We now set

$$Z_\infty^{pq} = \bigcap_n Z_n^{pq}, \quad B_\infty^{pq} = \bigcup_n B_n^{pq}, \quad E_\infty^{pq} = Z_\infty^{pq}/B_\infty^{pq}.$$

Now we have a canonical filtration of the cohomology of $\mathbf{X}$:

$$F^p H^{p+q}(\mathbf{X}) = \mathrm{Im}(H^{p+q}(F^p \mathbf{X}) \to H^{p+q}(\mathbf{X})).$$

Our next goal is to prove that

$$F^p H^{p+q}(\mathbf{X})/F^{p+1} H^{p+q}(\mathbf{X}) \cong E_\infty^{pq}.$$

PROPOSITION 3.2.4.

(i) $Z_n^{pq} = \mathrm{Im}(H^{p+q}(F^p \mathbf{X}, F^{p+n}\mathbf{X}) \to H^{p+q}(F^p \mathbf{X}, F^{p+1}\mathbf{X}))$

$\qquad = \mathrm{Ker}(H^{p+q}(F^p \mathbf{X}, F^{p+1}\mathbf{X}) \xrightarrow{\partial_*} H^{p+q+1}(F^{p+1}\mathbf{X}, F^{p+n}\mathbf{X}))$

(ii) $B_n^{pq} = \mathrm{Im}(H^{p+q-1}(F^{p-n+1}\mathbf{X}, F^p \mathbf{X}) \xrightarrow{\partial_*} H^{p+q}(F^p \mathbf{X}, F^{p+1}\mathbf{X}))$

$\qquad = \mathrm{Ker}(H^{p+q}(F^p \mathbf{X}, F^{p+1}\mathbf{X}) \to H^{p+q}(F^{p-n+1}\mathbf{X}, F^{p+1}\mathbf{X}))$

PROOF. (i) The following diagram represents a part of the diagram on page 98.

$$
\begin{array}{c}
H^{p+q+1}(F^{p+n}\mathbf{X}) \\
\downarrow i_1 \\
H^{p+q+1}(F^{p+n-1}\mathbf{X}) \xrightarrow{j_1} H^{p+q+1}(F^{p+n-1}\mathbf{X}, F^{p+n}\mathbf{X}) \\
\downarrow i_1 \\
\vdots \\
\downarrow i_1 \\
H^{p+q+1}(F^{p+2}\mathbf{X}) \\
\downarrow i_1 \\
H^{p+q}(F^p\mathbf{X}, F^{p+1}\mathbf{X}) \xrightarrow{k_1} H^{p+q+1}(F^{p+1}\mathbf{X}) \\
\searrow^{\partial_*} \qquad\qquad \downarrow \\
H^{p+q+1}(F^{p+1}\mathbf{X}, F^{p+n}\mathbf{X})
\end{array}
$$

Referring to this diagram, we have $x \in Z_n^{pq}$ if and only if $d_1(x) = 0$, $d_2(x) = 0, \ldots, d_{n-1}(x) = 0$. This happens if and only if $k_1(x) \in \mathrm{Im}((i_1)^{n-1})$, namely if and only if

$$k_1(x) \in \mathrm{Ker}(H^{p+q+1}(F^{p+1}\mathbf{X}) \to H^{p+q+1}(F^{p+1}\mathbf{X}, F^{p+n}\mathbf{X}))$$

This in turn is true if and only if

$$
\begin{aligned}
x &\in \mathrm{Ker}(H^{p+q}(F^p\mathbf{X}, F^{p+1}\mathbf{X}) \to H^{p+q+1}(F^{p+1}\mathbf{X}, F^{p+n}\mathbf{X})) \\
&= \mathrm{Im}(H^{p+q}(F^p\mathbf{X}, F^{p+n}\mathbf{X}) \to H^{p+q}(F^p\mathbf{X}, F^{p+1}\mathbf{X})).
\end{aligned}
$$

(ii) Referring to the following diagram

$$
\begin{array}{c}
H^{p+q-1}(F^{p-n+1}\mathbf{X}, F^p\mathbf{X}) \\
\downarrow \qquad \searrow^{\partial_*} \\
H^{p+q}(F^p\mathbf{X}) \xrightarrow{j_1} H^{p+q}(F^p\mathbf{X}, F^{p+1}\mathbf{X}) \\
\downarrow i_1 \\
\vdots \\
\downarrow i_1 \\
H^{p+q}(F^{p-n+3}\mathbf{X}) \\
\downarrow i_1 \\
H^{p+q-1}(F^{p-n+1}\mathbf{X}, F^{p-n+2}\mathbf{X}) \xrightarrow{k_1} H^{p+q}(F^{p-n+2}\mathbf{X}) \\
\downarrow i_1 \\
H^{p+q}(F^{p-n+1}\mathbf{X})
\end{array}
$$

we have $x \in B_n^{pq}$ if and only if $x = j_1(y)$ with $(i_1)^{n-2}(y) \in \mathrm{Im}(k_1)$, i.e., with $(i_1)^{n-1}(y) = 0$, or equivalently with

$$y \in \mathrm{Im}(H^{p+q-1}(F^{p-n+1}\mathbf{X}, F^p\mathbf{X}) \to H^{p+q}(F^p\mathbf{X})).$$

This happens if and only if $x \in \mathrm{Im}(\partial_*)$. $\qquad\square$

LEMMA 3.2.5. *If we have a commutative diagram of abelian groups*

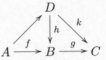

with the bottom row exact, then g induces an isomorphism

$$g : \mathrm{Im}(h)/\mathrm{Im}(f) \xrightarrow{\cong} \mathrm{Im}(k).$$

PROOF. By the first isomorphism theorem

$$\frac{\mathrm{Im}(h)}{\mathrm{Im}(f)} = \frac{\mathrm{Im}(h)}{\mathrm{Ker}(g)} \xrightarrow{g} \mathrm{Im}(gh) = \mathrm{Im}(k) \qquad\square$$

THEOREM 3.2.6. E_n^{pq} *is isomorphic to the image of*

$$H^{p+q}(F^p\mathbf{X}, F^{p+n}\mathbf{X}) \to H^{p+q}(F^{p-n+1}\mathbf{X}, F^{p+1}\mathbf{X}).$$

PROOF. We apply Lemma 3.2.5 to the diagram

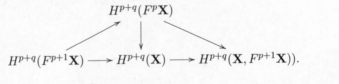

$\qquad\qquad\qquad\qquad\qquad\qquad\qquad\qquad\qquad\qquad\qquad\qquad\square$

COROLLARY 3.2.7. $E_\infty^{pq} = \mathrm{Im}(H^{p+q}(F^p\mathbf{X}) \to H^{p+q}(\mathbf{X}, F^{p+1}\mathbf{X}))$. $\quad\square$

DEFINITION 3.2.8. *We set* $F^p H^{p+q}(\mathbf{X}) = \mathrm{Im}(H^{p+q}(F^p\mathbf{X}) \to H^{p+q}(\mathbf{X}))$.

THEOREM 3.2.9. $F^p H^{p+q}(\mathbf{X})/F^{p+1}H^{p+q}(\mathbf{X}) \cong E_\infty^{pq}$.

PROOF. We apply Lemma 3.2.5 to the diagram

$$H^{p+q}(F^p\mathbf{X})$$

$$H^{p+q}(F^{p+1}\mathbf{X}) \longrightarrow H^{p+q}(\mathbf{X}) \longrightarrow H^{p+q}(\mathbf{X}, F^{p+1}\mathbf{X})).$$

$\qquad\qquad\qquad\qquad\qquad\qquad\qquad\qquad\qquad\qquad\qquad\qquad\square$

We may regard this theorem as saying that whenever we set up a spectral sequence using a filtered complex, we know what it converges to; namely the cohomology of the original complex. The problem is to understand the initial data given by the E_2 term. In the next section we shall give a sketchy account of the E_2 term in the case of the spectral sequence of a fibration. In the following section, we shall see that the problem of understanding the E_2 term is substantially simplified if we start off with a double complex rather than a filtered complex.

We now discuss edge homomorphisms and transgressions. Setting $q = 0$, we see that each E_r^{p0} is a quotient of E_2^{p0}, and E_∞^{p0} is contained in $H^p(\mathbf{X})$. Thus we have homomorphisms

$$E_2^{p0} \twoheadrightarrow E_3^{p0} \twoheadrightarrow \cdots \twoheadrightarrow E_\infty^{p0} \hookrightarrow H^p(\mathbf{X}).$$

The composite homomorphism $E_2^{p0} \xrightarrow{e} H^p(\mathbf{X})$ is called the **horizontal edge homomorphism**.

The **vertical edge homomorphism** $H^q(\mathbf{X}) \xrightarrow{e'} E_2^{0q}$ is the composite map

$$H^q(\mathbf{X}) \twoheadrightarrow E_\infty^{0q} \hookrightarrow \cdots \hookrightarrow E_3^{0q} \hookrightarrow E_2^{0q}.$$

Following through the isomorphisms used to prove Theorem 3.2.9, it is easy to see that this is just the composite of the obvious maps

$$H^q(\mathbf{X}) \to H^q(\mathbf{X}, F^2\mathbf{X}) \to \mathrm{Im}(H^q(\mathbf{X}, F^2\mathbf{X}) \to H^q(\mathbf{X}, F^1\mathbf{X})) = E_2^{0q}.$$

The **transgression** is the differential $d_n : E_n^{0,n-1} \to E_n^{n0}$. In case $n = 2$, this gives rise to the **five term sequence** involving the transgression and edge homomorphisms

$$0 \to E_2^{10} \xrightarrow{e} H^1(\mathbf{X}) \xrightarrow{e'} E_2^{01} \xrightarrow{d_2} E_2^{20} \xrightarrow{e} H^2(\mathbf{X}).$$

This is obtained by splicing together the sequences

$$0 \to E_2^{10} = E_\infty^{10} \to H^1(\mathbf{X}) \to E_\infty^{01} \to 0$$

$$0 \to E_\infty^{01} \to E_2^{01} \xrightarrow{d_2} E_2^{20} \to E_\infty^{20} \to 0$$

and

$$0 \to E_\infty^{20} \to H^2(\mathbf{X}).$$

Thus we have the following.

PROPOSITION 3.2.10. *In the spectral sequence of a filtered cochain complex, there is a five term exact sequence*

$$0 \to E_2^{10} \xrightarrow{e} H^1(\mathbf{X}) \xrightarrow{e'} E_2^{01} \xrightarrow{d_2} E_2^{20} \xrightarrow{e} H^2(\mathbf{X}). \qquad \square$$

We now discuss the dual setup in homology. If we start off with a filtered chain complex

$$F_0\mathbf{X} \subseteq F_1\mathbf{X} \subseteq \cdots \subseteq \mathbf{X} = \bigcup F_i\mathbf{X},$$

then we have an exact couple and spectral sequence

$$E^1_{pq} = H_{p+q}(F_p\mathbf{X}, F_{p-1}\mathbf{X})$$
$$D^1_{pq} = H_{p+q}(F_p\mathbf{X})$$
$$E^n_{pq} = H(E^{n-1}_{pq}, d^n) \qquad \deg(d^n) = (-n, n-1)$$
$$E^\infty_{pq} \cong F_p H_{p+q}(\mathbf{X})/F_{p-1}H_{p+q}(\mathbf{X})$$

where

$$F_p H_{p+q}(\mathbf{X}) = \mathrm{Im}(H_{p+q}(F_p\mathbf{X}) \to H_{p+q}(\mathbf{X})).$$

The edge homomorphisms are $H_p(\mathbf{X}) \xrightarrow{e} E^2_{p0}$ and $E^2_{0q} \xrightarrow{e'} H_q(\mathbf{X})$, the transgression is $d^n : E^n_{0,n-1} \to E^n_{n0}$. The five term sequence is as follows.

PROPOSITION 3.2.11. *In the spectral sequence of a filtered chain complex, there is a five term exact sequence*

$$H_2(\mathbf{X}) \xrightarrow{e} E^2_{20} \xrightarrow{d^2} E^2_{01} \xrightarrow{e'} H_1(\mathbf{X}) \xrightarrow{e} E^2_{10} \to 0. \qquad \square$$

3.3. The spectral sequence of a fibration

In this section we give a sketchy derivation of the spectral sequence of a fibration. Again, we shall work in cohomology rather than homology, and leave the reader to dualise.

Suppose $p : E \to B$ is a Serre fibration with B a CW-complex, and fibre $F = p^{-1}(b_0)$ where b_0 is the basepoint of B. For the purpose of exposition, we shall assume that $\pi_1(B)$ acts trivially on $H^*(F)$ (see Proposition 1.6.7), and then make comments on what to do in the general situation.

Let

$$B^0 \subseteq B^1 \subseteq \cdots \subseteq B$$

be the skeletal filtration of B, and let $E^i = p^{-1}(B^i)$ be the corresponding filtration of E. We examine the spectral sequence arising from this filtration.

According to the definitions, we have $E^{pq}_1 = H^{p+q}(E^p, E^{p-1})$. Now for each p-cell σ^p_α of B, we choose a simplex Δ_α contained in the interior of σ^p_α, and by excision we have

$$H^{p+q}(E^p, E^{p-1}) \cong \bigoplus_\alpha H^{p+q}(p^{-1}(\Delta_\alpha), p^{-1}(\dot\Delta_\alpha)).$$

We choose a path $\omega : I \to B$ with $\omega(0) = b_0$ and $\omega(1) \in \Delta_\alpha$, and let $\Delta_\alpha \vee I$ be the space obtained by identifying $1 \in I$ with $\omega(1) \in \Delta_\alpha$. Then the diagram

$$
\begin{array}{ccc}
0 \times F & \longrightarrow & E \\
\downarrow & \nearrow & \downarrow \\
(\Delta_\alpha \vee I) \times F & \longrightarrow & B
\end{array}
$$

and Proposition 1.6.7 show that we may find a map $\Delta_\alpha \times F \to E$ which is a homotopy equivalence on each fibre over Δ_α, and hence

$$H^{p+q}(p^{-1}(\Delta_\alpha), p^{-1}(\dot{\Delta}_\alpha)) \cong H^{p+q}(\Delta_\alpha \times F, \dot{\Delta}_\alpha \times F) \cong H^q(F)$$

(by the Künneth theorem). Under the assumption that $\pi_1(B)$ acts trivially on $H^*(F)$, this isomorphism is independent of choice of ω, and so we have a canonical isomorphism

$$E_1^{pq} = H^{p+q}(E^p, E^{p-1}) \cong \operatorname{Hom}(H_p(B^p, B^{p-1}), H^q(F)) \cong C^p(B; H^q(F))$$

where $C^p(B; A)$ denotes the cellular cochains on B with coefficients in an abelian group A, namely $\operatorname{Hom}(C_p(B), A)$.

REMARK. If $\pi_1(B)$ has non-trivial action on $H^*(F)$, then we must regard the cohomologies of the fibres as a "local system" $\mathcal{H}^*(F)$ of coefficients on B (cf. the discussion of local coefficient systems investigated in Chapter 13) and we obtain

$$H^{p+q}(E^p, E^{p-1}) \cong C^p(B; \mathcal{H}^q(F)),$$

cellular cochains on B with coefficients in this local system.

Now to compute E_2^{pq} we must see what d_1 does. We follow it through the above isomorphisms as in the following diagram:

$$
\begin{array}{ccc}
E_1^{pq} & \xrightarrow{\ \ d_1\ \ } & E_1^{p+1,q} \\
\cong\downarrow & & \downarrow\cong \\
H^{p+q}(E^p, E^{p-1}) & \xrightarrow{\ \ \delta\ \ } & H^{p+q+1}(E^{p+1}, E^p) \\
\cong\downarrow & & \downarrow\cong \\
\operatorname{Hom}(H_p(B^p, B^{p-1}), H^q(F)) & \xrightarrow{\ \partial^*\ } & \operatorname{Hom}(H_{p+1}(B^{p+1}, B^p), H^q(F)) \\
\cong\downarrow & & \downarrow\cong \\
C^p(B; H^q(F)) & \xrightarrow{\ \ \delta\ \ } & C^{p+1}(B; H^q(F))
\end{array}
$$

Thus we have

$$\boxed{E_2^{pq} \cong H^p(B; H^q(F)).}$$

REMARKS. (i) Introducing a coefficient ring R does not alter the arguments at all, and so we have a spectral sequence

$$H^p(B; H^q(F; R)) \Rightarrow H^{p+q}(E; R).$$

(ii) All the arguments work dually in homology to give a spectral sequence

$$H_p(B; H_q(F; R)) \Rightarrow H_{p+q}(E; R).$$

(iii) If $\pi_1(B)$ acts non-trivially on $H^*(F; R)$ and $H_*(F; R)$ we have to write the above spectral sequences as

$$H^p(B; \mathcal{H}^q(F; R)) \Rightarrow H^{p+q}(E; R), \quad H_p(B; \mathcal{H}_q(F; R)) \Rightarrow H_{p+q}(E; R).$$

(iv) We saw in Theorem 2.4.12 that a normal subgroup N of a group G gives rise to a fibration $BG \to B(G/N)$ with fibre BN, and hence we have spectral sequences

$$H^p(G/N, H^q(N, R)) \Rightarrow H^{p+q}(G, R), \quad H_p(G/N, H_q(N, R)) \Rightarrow H_{p+q}(G, R).$$

We shall develop this spectral sequence algebraically in Section 3.5.

3.4. The spectral sequence of a double complex

In this section we set up the spectral sequence of a double complex. We shall work in cohomology rather than homology, and again the theory works equally well in either. We shall see that the E_2 term is easier to understand in this case than in the case of a filtered complex.

DEFINITION 3.4.1. *A **double complex** is a collection of abelian groups (or modules, etc.) and maps arranged as in the following diagram:*

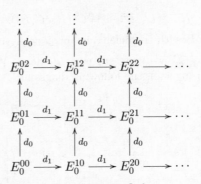

such that the following conditions are satisfied:
(i) *Each row satisfies $d_1 \circ d_1 = 0$.*
(ii) *Each column satisfies $d_0 \circ d_0 = 0$.*
(iii) *$d_0 \circ d_1 + d_1 \circ d_0 = 0$.*
*The **total complex** of a double complex, $\mathbf{X}^n = \mathrm{Tot}(\mathbf{E})$ is given by*

$$\mathbf{X}^n = \bigoplus_{i+j=n} E_0^{ij}$$

with differential

$$d = d_0 + d_1 : \mathbf{X}^n \to \mathbf{X}^{n+1}.$$

Note that the above conditions imply that $d \circ d = 0$.

Given a double complex $\mathbf{E}$, its total complex is *filtered* as follows.

$$D_0^{pq} = F^p\mathbf{X}^{p+q} = \bigoplus_{\substack{i+j=p+q \\ i \geq p}} E_0^{ij}$$

Thus

$$F^p\mathbf{X}^{p+q}/F^{p+1}\mathbf{X}^{p+q} \cong E_0^{pq}$$

and so each layer in the filtration should be thought of as a single *column* of the double complex. The differential on this quotient is just $d_0 : E_0^{pq} \to E_0^{p,q+1}$, since d_1 maps into a lower layer. So we have

$$E_1^{pq} = H(E_0^{pq}, d_0), \quad D_1^{pq} = H(E_0^{pq} \oplus E_0^{p+1,q-1} \oplus \cdots, d_0 + d_1)$$

Now we have two things called d_1, and so we had better check that their meanings are related in some obvious way. Until we have done so, we shall refer to the horizontal differential in the double complex as d_1, and to $j_1 \circ k_1$ as "d_1".

If $x \in E_0^{pq}$ with $d_0(x) = 0$, so that x represents a class $[x] \in E_1^{pq}$, then we compute $k_1[x]$ as follows. Recall that k_1 is defined as the boundary homomorphism associated to the short exact sequence of chain complexes

$$0 \to F^{p+1}\mathbf{X} \to F^p\mathbf{X} \to F^p\mathbf{X}/F^{p+1}\mathbf{X} \to 0$$

i.e.,

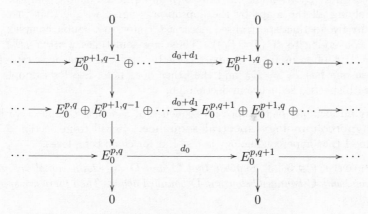

Since $d_0(x) = 0$ we have $(d_0 + d_1)(x, 0, \dots) = (0, d_1(x), 0, \dots)$, and so by the definition of the boundary homomorphism (switchback map) we have $k_1[x] = [(d_1(x), 0, \dots)]$. Hence

$$\text{"}d_1\text{"}[x] = j_1 k_1[x] = [d_1(x)],$$

and so from now on we omit the inverted commas. Thus we have

$$E_2^{pq} = H^p(H^q(\mathbf{E}_0, d_0), d_1)$$

where the d_1 is the map induced by the d_1 of the double complex.

We summarise this in the following theorem:

THEOREM 3.4.2. *Given a double complex (E_0^{pq}, d_0, d_1) there is a spectral sequence with*

$$E_1^{pq} = H(E_0^{pq}, d_0)$$
$$E_2^{pq} = H^p(H^q(\mathbf{E}_0, d_0), d_1)$$
$$E_\infty^{pq} = F^p H^{p+q}(\text{Tot}(\mathbf{E}_0))/F^{p+1} H^{p+q}(\text{Tot}(\mathbf{E}_0)).$$

The shorthand for this theorem is

$$H^p(H^q(\mathbf{E}_0, d_0), d_1) \Rightarrow H^{p+q}(\mathrm{Tot}(\mathbf{E}_0), d_0 + d_1).$$

It may now be seen that (E_0, d_0) and (E_1, d_1) are natural precursors of the remaining pages (E_n, d_n), $n \geq 2$, and d_0 and d_1 have the "right" degrees.

The vertical edge homomorphism for the spectral sequence of a double complex may be described as follows. There is a restriction map

$$H^q(\mathrm{Tot}(\mathbf{E}_0)) \to H^q(E_0^{0*}, d_0)$$

whose image lies in the kernel of d_1, and hence this defines a map

$$H^q(\mathrm{Tot}(\mathbf{E}_0)) \xrightarrow{e'} E_2^{0q},$$

which is just the edge homomorphism.

REMARK. By reversing the rôles of p and q in the double complex (and multiplying all the maps by the appropriate signs) we see that there are actually two distinct spectral sequences associated to a double complex, and both converging to $H^*(\mathrm{Tot}(\mathbf{E}_0))$. These are sometimes written $_I E_n^{pq}$ and $_{II} E_n^{pq}$. It is often useful to compare these two spectral sequences, especially in case one has $E_2 = E_\infty$ and the other does not. See for example the Künneth spectral sequence in Section 3.6.

As an example of the spectral sequence of a double complex, we have the **hypercohomology spectral sequence**. Recall from Section 2.7 of Volume I that hypercohomology is just Ext for chain complexes.

PROPOSITION 3.4.3. *Suppose that* $\mathbf{C}$ *and* $\mathbf{D}$ *are chain complexes of* Λ-*modules, with* $\mathbf{C}$ *bounded below and* $\mathbf{D}$ *bounded above. Then there are spectral sequences*

$$\mathrm{Ext}_\Lambda^p(H_q(\mathbf{C}), \mathbf{D}) \Rightarrow \mathrm{Ext}_\Lambda^{p+q}(\mathbf{C}, \mathbf{D}), \quad \mathrm{Ext}_\Lambda^p(\mathbf{C}, H_{-q}(\mathbf{D})) \Rightarrow \mathrm{Ext}_\Lambda^{p+q}(\mathbf{C}, \mathbf{D}).$$

PROOF. If $\mathbf{P}$ is a projective resolution of $\mathbf{C}$, and $\mathbf{I}$ is an injective resolution of $\mathbf{D}$, then these are the spectral sequences coming from the double complexes $\mathbf{Hom}_\Lambda(\mathbf{C}, \mathbf{I})$ and $\mathbf{Hom}_\Lambda(\mathbf{P}, \mathbf{D})$. $\square$

As an example of the homology spectral sequence of a double complex, we prove the following, which should be compared with the Künneth Theorem (Section 2.7 of Volume I).

PROPOSITION 3.4.4. *Suppose* $\mathbf{C}$ *and* $\mathbf{D}$ *are chain complexes of right, resp. left* Λ-*modules, concentrated in non-negative degrees, such that* $\mathbf{C}$ *consists of projective* Λ-*modules and* $\mathbf{D}$ *is exact in positive degrees. Then* $H_*(\mathbf{C} \otimes_\Lambda \mathbf{D}) \cong H_*(\mathbf{C}) \otimes_\Lambda H_0(\mathbf{D})$. *In particular, if* $\mathbf{C}$ *is also exact in positive degrees, then so is* $\mathbf{C} \otimes_\Lambda \mathbf{D}$, *and* $H_0(\mathbf{C} \otimes_\Lambda \mathbf{D}) \cong H_0(\mathbf{C}) \otimes_\Lambda H_0(\mathbf{D})$.

PROOF. Consider the spectral sequence of the double complex $\mathbf{C} \otimes_\Lambda \mathbf{D}$. Since $\mathbf{C}$ consists of projective modules and $\mathbf{D}$ is exact in positive degrees, the columns of $\mathbf{C} \otimes_\Lambda \mathbf{D}$ are exact. So $E_{pq}^1 = 0$ for $q > 0$, and hence $E_{pq}^1 \cong E_{pq}^\infty$. So we have

$$H_p(\mathbf{C} \otimes_\Lambda \mathbf{D}) \cong E_{p0}^\infty \cong E_{p0}^1 \cong H_p(\mathbf{C}) \otimes_\Lambda H_0(\mathbf{D}). \qquad \square$$

3.5. The spectral sequence of a group extension

Suppose N is a normal subgroup of G and M is an RG-module. Our goal in this section is to construct the Lyndon–Hochschild–Serre spectral sequences

$$H_p(G/N, H_q(N, M)) \Rightarrow H_{p+q}(G, M), \ H^p(G/N, H^q(N, M)) \Rightarrow H^{p+q}(G, M)$$

and we shall concentrate on the latter. The former is treated dually.

We take a projective resolution of R as an $R(G/N)$-module

$$\cdots \to P_1 \xrightarrow{\partial_1} P_0 \xrightarrow{\partial_0} R \to 0$$

and as an RG-module

$$\cdots \to Q_1 \xrightarrow{\partial_1'} Q_0 \xrightarrow{\partial_0'} R \to 0.$$

We form the double complex

$$E_0^{pq} = \mathrm{Hom}_{R(G/N)}(P_p, \mathrm{Hom}_{RN}(Q_q, M)).$$

Note that $\mathrm{Hom}_{RN}(Q_q, M)$ admits a G-action whose restriction to N is trivial, so that we may regard it as an $R(G/N)$-module. The differentials in this double complex are given by taking d_0 to be $(-1)^p$ times the differential induced by ∂_q' and d_1 to be the differential induced by ∂_p. The signs ensure that $d_0 \circ d_1 = -d_1 \circ d_0$.

By Theorem 3.4.2 we have a spectral sequence with

$$E_1^{pq} = \mathrm{Hom}_{R(G/N)}(P_p, H^q(N, M))$$
$$E_2^{pq} = H^p(G/N, H^q(N, M)).$$

REMARK. The above double complex is the algebraic analogue of the construction of the fibration $BG \to B(G/N)$ with fibre BN, given in Theorem 2.4.12, using the space $E(G/N) \times EG$.

It remains to identify E_∞^{pq}.

LEMMA 3.5.1.

$$\mathrm{Hom}_{R(G/N)}(P_p, \mathrm{Hom}_{RN}(Q_q, M)) \cong \mathrm{Hom}_{RG}(P_p \otimes_R Q_q, M)$$

(In the right hand side of this formula, P_p is regarded as an RG-module with N acting trivially).

PROOF. We define maps

$$\phi : \mathrm{Hom}_{R(G/N)}(P_p, \mathrm{Hom}_{RN}(Q_q, M)) \to \mathrm{Hom}_{RG}(P_p \otimes Q_q, M)$$
$$\phi(f)(x \otimes y) = f(x)(y)$$
$$\psi : \mathrm{Hom}_{RG}(P_p \otimes Q_q, M) \to \mathrm{Hom}_{R(G/N)}(P_p, \mathrm{Hom}_{RN}(Q_q, M))$$
$$\psi(\alpha)(x)(y) = \alpha(x \otimes y).$$

It is easy to check that these maps are well defined and inverses to each other. $\qquad\square$

Now by Corollary 2.7.3 of Volume I, the total complex of $\mathbf{P} \otimes \mathbf{Q}$ is exact, and is hence a projective resolution of R by RG-modules. So by the above lemma we have

$$H^*(\mathrm{Tot}(E_0), d_0 + d_1) \cong H^*(\mathrm{Hom}_{RG}(\mathrm{Tot}(\mathbf{P} \otimes \mathbf{Q}), M), \partial + \partial') \cong H^*(G, M).$$

Thus the spectral sequence of Theorem 3.4.2 becomes

$$H^p(G/N, H^q(N, M)) \Rightarrow H^{p+q}(G, M).$$

In exactly the same way, if M_1 is an $R(G/N)$-module (which we also regard as an RG-module with N acting trivially) and M_2 and M_3 are RG-modules, we let $\mathbf{P}$ be a resolution of M_1 as an $R(G/N)$-module and $\mathbf{Q}$ be a resolution of M_2 as an RG-module. We obtain a spectral sequence

$$\mathrm{Ext}^p_{R(G/N)}(M_1, \mathrm{Ext}^q_{RN}(M_2, M_3)) \Rightarrow \mathrm{Ext}^{p+q}_{RG}(M_1 \otimes_R M_2, M_3).$$

EXERCISE. Show that in the spectral sequence

$$H^*(G/N, H^*(N, M)) \Rightarrow H^*(G, M)$$

the horizontal edge homomorphism

$$H^*(G/N, M^N) \xrightarrow{e} H^*(G, M)$$

is the composite of the inflation

$$\inf_{G,N} : H^*(G/N, M^N) \to H^*(G, M^N)$$

with the map $H^*(G, M^N) \to H^*(G, M)$ induced by the inclusion. This composite is also called the inflation map.

Show that the vertical edge homomorphism

$$H^*(G, M) \xrightarrow{e'} H^*(N, M)^{G/N}$$

is the restriction map $\mathrm{res}_{G,N}$.

Thus the five term sequence associated to this spectral sequence (Proposition 3.2.10) is

$$0 \to H^1(G/N, M^N) \xrightarrow{\inf} H^1(G, M) \xrightarrow{\mathrm{res}} H^1(N, M)^{G/N}$$

$$\xrightarrow{d_2} H^2(G/N, M^N) \xrightarrow{\inf} H^2(G, M).$$

This is called the **inflation-restriction sequence**.

Dually, we have a homology spectral sequence

$$H_p(G/N, H_q(N, M)) \Rightarrow H_{p+q}(G, M)$$

and a five term sequence (Proposition 3.2.11) is

$$H_2(G, M) \to H_2(G/N, M_N) \to H_1(N, M)_{G/N}$$

$$\to H_1(G, M) \to H_1(G/N, M_N) \to 0.$$

Here, M_G denotes the quotient of M given by identifying m with gm for all $m \in M$ and $g \in G$.

3.6. The Künneth spectral sequence

DEFINITION 3.6.1. *A double complex P_{pq} is called a* **proper projective resolution** *of the chain complex* $\mathbf{D}$ *if for each value of q the sequences*

$$\cdots \to P_{2,q} \to P_{1,q} \to P_{0,q} \to D_q \to 0$$
$$\cdots \to Z_q(P_{2,*}) \to Z_q(P_{1,*}) \to Z_q(P_{0,*}) \to Z_q(D_*) \to 0$$
$$\cdots \to H_q(P_{2,*}) \to H_q(P_{1,*}) \to H_q(P_{0,*}) \to H_q(D_*) \to 0$$

are projective resolutions.

LEMMA 3.6.2. *Every chain complex has a proper projective resolution.*

PROOF. Choose projective resolutions of $B_q(D_*)$ and $H_q(D_*)$ and apply the horseshoe lemma to the short exact sequences

$$0 \to B_q(D_*) \to Z_q(D_*) \to H_q(D_*) \to 0$$
$$0 \to Z_q(D_*) \to D_q \to B_{q+1}(D_*) \to 0$$

to obtain appropriate projective resolutions P_{pq} of the D_q and maps between them. As usual, it is necessary to negate the maps on every other column to make P_{**} a double complex.   □

THEOREM 3.6.3. *Suppose $\mathbf{C}$ and $\mathbf{D}$ are chain complexes of right, resp. left Λ-modules and suppose the C_n are flat. Then there is a spectral sequence*

$$E^2_{pq} = \bigoplus_{s+t=q} \mathrm{Tor}^\Lambda_p(H_s(\mathbf{C}), H_t(\mathbf{D})) \Rightarrow H_{p+q}(\mathbf{C} \otimes_\Lambda \mathbf{D}).$$

If $\mathbf{C}$ and $\mathbf{D}$ are cochain complexes then there is a spectral sequence with

$$E_2^{pq} = \bigoplus_{s+t=q} \mathrm{Tor}^\Lambda_{-p}(H^s(\mathbf{C}), H^t(\mathbf{D})).$$

This is a "second quadrant spectral sequence" (i.e. $p \leq 0$ and $q \geq 0$) so there are convergence problems, but when the spectral sequence does converge, it does so to $H^{p+q}(\mathbf{C} \otimes_\Lambda \mathbf{D})$.

PROOF. Form a proper projective resolution P_{pq} of $\mathbf{D}$, and tensor with $\mathbf{C}$ in the vertical direction:

$$(\mathbf{C} \otimes_\Lambda \mathbf{P})_{pq} = \bigoplus_{s+t=q} C_s \otimes_\Lambda P_{pt}.$$

Examine the two spectral sequences associated to this double complex. If we do the vertical differential first, each column is $C_* \otimes_\Lambda P_{p*}$. Since $Z(P_{p*})$ and $H(P_{p*})$ are projective (since P_{pq} is a proper projective resolution of $\mathbf{D}$), Corollary 2.7.2 to the Künneth theorem (Volume I) implies that

$$E^1_{pq} = H_q(C_* \otimes P_{p*}) = \bigoplus_{s+t=q} H_s(C_*) \otimes H_t(P_{p*}).$$

Since

$$\cdots \to H_t(P_{2*}) \to H_t(P_{1*}) \to H_t(P_{0*}) \to H_t(D_*) \to 0$$

is a projective resolution, the homology of this complex with respect to the horizontal differential d^1 is

$$E^2_{pq} = \bigoplus_{s+t=q} \mathrm{Tor}_p(H_s(C_*), H_t(D_*)).$$

On the other hand, if we do the horizontal differential first, we obtain a spectral sequence in which E^1 consists of just the zeroth column, since the C_n are flat so that the rows remain exact. The zeroth column is $\mathrm{Tot}(\mathbf{C} \otimes \mathbf{D})$ and so $E^2 = E^\infty = H_*(\mathrm{Tot}(\mathbf{C} \otimes \mathbf{D}))$. Thus the first spectral sequence also converges to the same answer.

The case of cochain complexes is similar, with the stated proviso. □

3.7. The Eilenberg–Moore spectral sequence

In this section, we give an extremely condensed account of the Eilenberg–Moore spectral sequence. For more detailed accounts, see for example Smith [**238, 239**] or McCleary [**183**]. We shall not make much use of this spectral sequence in the rest of this book, but as we already pointed out in the introductory section of this chapter, it can be used as an effective computational device for computing the cohomology of central extensions.

Suppose we are given a pullback diagram of fibrations of CW-complexes

$$\begin{array}{ccc} E' & \longrightarrow & E \\ \downarrow & & \downarrow \\ B' & \longrightarrow & B \end{array}$$

and assume for simplicity that $\pi_1(B) = 0$ and that we are working over a field of coefficients. The complex of singular cochains $\Delta^*(B)$ has a structure of a graded algebra (multiplication corresponding to cup product, which is associative but not commutative) with a differential d satisfying $d(ab) = d(a)b + (-1)^{\deg(a)} a d(b)$. Such an object is called a **differential graded algebra** or DGA. There is also an obvious notion of differential graded modules over a DGA, and we have a diagram of DG modules over $\Delta^*(B)$ as follows.

$$\begin{array}{ccc} \Delta^*(E') & \longleftarrow & \Delta^*(E) \\ \uparrow & & \uparrow \\ \Delta^*(B') & \longleftarrow & \Delta^*(B) \end{array}$$

As in Definition 3.6.1, DG modules admit **proper projective resolutions**, and it turns out that if $Q^{**} \to \Delta^*(B')$ is a proper projective resolution, there is a natural map

$$\mathrm{Tot}(Q^{**}) \otimes_{\Delta^*(B)} \Delta^*(E) \to \Delta^*(E')$$

inducing an isomorphism in cohomology

$$\mathrm{Tot}(\mathrm{Tor}^{**}_{\Delta^*(B)}(\Delta^*(B'), \Delta^*(E))) \cong H^*(E').$$

Here, Tor^{**} is defined as follows. If M^* and N^* are right, resp. left DG modules over a DG algebra Γ, and $P^{**} \to M^*$ is a proper projective resolution, we define

$$\mathrm{Tor}^{**}_\Gamma(M^*, N^*) = H^*(P^{**} \otimes_\Gamma M^*)$$

with respect to the total differential $(d_0 + d_1) \otimes 1 + (-1)^m 1 \otimes d_M$. The two gradings on Tor are as follows. The first grading is the homological grading, and the second is the internal grading coming from the grading on the modules.

Regarding $P^{**} \otimes_\Gamma M^*$ as a double complex

$$(P^{**} \otimes_\Gamma M^*)^{pq} = \bigoplus_{s+t=q} P^{ps} \otimes_\Gamma M^t$$

we obtain a spectral sequence

$$E_2^{**} = \mathrm{Tor}^{**}_{H(\Gamma)}(H(M^*), H(N^*)) \Rightarrow \mathrm{Tot}(\mathrm{Tor}^{**}_\Gamma(M^*, N^*)).$$

Applying this to the above isomorphism, we obtain the Eilenberg–Moore spectral sequence

$$\mathrm{Tor}^{**}_{H^*(B)}(H^*(B'), H^*(E)) \Rightarrow H^*(E').$$

As in the case of the Künneth spectral sequence for cochain complexes, this is a second quadrant spectral sequence. In fact, L. Smith has shown that the Eilenberg–Moore spectral sequence can be viewed as a kind of Künneth spectral sequence.

Given a central extension $1 \to Z \to G \to G/Z \to 1$, the corresponding element of $H^2(G/Z, Z)$ (see Section 3.7 of Volume I) corresponds to a map $K(G/Z, 1) \to K(Z, 2)$ (see Section 2.2). If we turn this into a fibration (end of Section 1.6), then the long exact sequence in homotopy (Theorem 1.6.6) shows that the fibre is a $K(G, 1)$, so we have a fibration

$$K(G, 1) \to K(G/Z, 1) \to K(Z, 2),$$

and hence an Eilenberg–Moore spectral sequence

$$\mathrm{Tor}^{**}_{H^*(K(Z,2))}(H^*(\mathrm{point}), H^*(G/Z)) \Rightarrow H^*(G).$$

It is most practical to compute this in case $Z = \mathbb{Z}/2$. Now the spectral sequence (Section 3.3) of the path fibration (end of Section 1.6)

$$K(Z, 1) = \Omega K(Z, 2) \to PK(Z, 2) \to K(Z, 2)$$

converges to zero in degrees other than zero. Following this through, we see that

$$H^*(K(\mathbb{Z}/2, 2); \mathbb{F}_2) = \mathbb{F}_2[u_0, u_1, \dots],$$

a polynomial ring in infinitely many variables u_i of degree $2^i + 1$. To calculate Tor, one uses a Koszul complex (see for example Serre [**233**], Section

IV A, or MacLane [**170**], Section VII.2). See also Section 4.8 for the rôle of the Steenrod operations, and Rusin [**226**] for explicit calculations in this situation.

3.8. The Atiyah spectral sequence

The Atiyah spectral sequence [**18**] is a spectral sequence going from the cohomology of a finite group to its representation theory. It was obtained as a particular example of the Atiyah–Hirzebruch spectral sequence, by using the Bott Periodicity Theorem 2.5.4 and the Atiyah Completion Theorem 2.5.5.

The Atiyah–Hirzebruch spectral sequence is a method of computing the value of an (unreduced) generalised cohomology theory h^* on a space, starting with the ordinary cohomology of the space and the values of h^* on a point. More generally, given any Serre fibration $p : E \to B$ with B a CW-complex, and fibre $F = p^{-1}(b_0)$, we have an Atiyah–Hirzebruch spectral sequence

$$H^p(B; h^q(F)) \Rightarrow h^{p+q}(E).$$

The special case $E = B$ gives us

$$H^p(B; h^q(\text{point})) \Rightarrow h^{p+q}(B).$$

Just as in the spectral sequence of a fibration (Section 3.3), we should either stipulate that $\pi_1(B)$ acts trivially on $h^*(F)$, which certainly happens in case F is a point, or regard $h^*(F)$ as a local system of coefficients on B.

The derivation of the Atiyah–Hirzebruch spectral sequence is identical to the derivation of the spectral sequence of a fibration. Thus for example $E_1^{pq} = C^p(B; h^q(F))$, and d_1 is described in exactly the same way as before.

The Atiyah spectral sequence is the particular case of the Atiyah–Hirzebruch spectral sequence in which $B = E = BG$ and h^* is K-theory. By Theorem 2.5.5, $K^n(BG) = a(\mathbb{C}G)^\wedge$ for n even and zero for n odd. The Bott Periodicity Theorem 2.5.4 shows that $K^*(\text{point}) = \mathbb{Z}$ for n even and zero for n odd. So the spectral sequence

$$H^*(BG; K^*(\text{point})) \Rightarrow K^*(BG)$$

is a first and fourth quadrant spectral sequence with periodicity two in the vertical direction, and with alternate rows $H^*(G, \mathbb{Z})$ and zero. Since every other row is zero, the even differentials d_{2n} are also zero. It is customary to get rid of the vertical repetition, and to consider the spectral sequence to sit along a single line

$$H^*(G, \mathbb{Z}) \Rightarrow a(\mathbb{C}G)^\wedge$$

with differentials d_3, d_5, etc. increasing cohomological degree by 3, 5, etc.

EXERCISE. Investigate the Atiyah spectral sequence for some small finite groups; for example, cyclic groups, the symmetric group of degree three, the dihedral and quaternion groups of order eight.

3.9. Products in spectral sequences

We shall show in this section how product structures on double complexes or filtered complexes give rise to products in the entire spectral sequence in a useful way. For an alternative approach to products using exact couples, see Massey [**178**]. The theory of products in exact couples is not quite as straightforward as one might hope.

DEFINITION 3.9.1. A **pairing** of double complexes $E_0 \times {'E_0} \to {''E_0}$ consists of maps

$$E_0^{pq} \otimes {'E_0^{p'q'}} \to {''E_0^{p+p',q+q'}}$$
$$x \otimes y \mapsto xy$$

satisfying the identities

$$d_0(xy) = d_0(x)y + (-1)^{p+q}xd_0(y)$$
$$d_1(xy) = d_1(x)y + (-1)^{p+q}xd_1(y).$$

A **double complex with ring structure** is a double complex E_0 together with an associative pairing $E_0 \times E_0 \to E_0$. A **module** over a double complex with ring structure E_0 is a double complex ${'E_0}$ together with a pairing $E_0 \times {'E_0} \to {'E_0}$ satisfying the usual associative law for the action of a ring on a module.

For example, in the case of the spectral sequence of a group extension, suppose M_1, M_1' are $R(G/N)$-modules with projective resolutions $\mathbf{P}$, $\mathbf{P}'$, suppose M_2, M_2' are RG-modules with projective resolutions $\mathbf{Q}$, $\mathbf{Q}'$, and suppose M_3, M_3' are also RG-modules. Setting

$$E_0^{pq} = \mathrm{Hom}_{R(G/N)}(P_p, \mathrm{Hom}_{RN}(Q_q, M_3))$$
$${'E_0^{pq}} = \mathrm{Hom}_{R(G/N)}(P_p', \mathrm{Hom}_{RN}(Q_q', M_3'))$$
$${''E_0^{pq}} = \mathrm{Hom}_{R(G/N)}((\mathbf{P} \otimes \mathbf{P}')_p, \mathrm{Hom}_{RN}((\mathbf{Q} \otimes \mathbf{Q}')_q, M_3 \otimes M_3'))$$

there are cup product maps, as in Section 3.2 of Volume I, giving a pairing $E_0 \times {'E_0} \to {''E_0}$.

If $M_1 = M_2 = M_3 = R$ then E_0 is a double complex with ring structure. If $M_1' = M_2' = R$ and $M_3' = M$ is arbitrary then ${'E_0}$ is a module over E_0. We shall see that this gives the entire spectral sequence

$$H^p(G/N, H^q(N, R)) \Rightarrow H^{p+q}(G, R)$$

a ring structure over which the spectral sequence

$$H^p(G/N, H^q(N, M)) \Rightarrow H^{p+q}(G, M)$$

is a module.

We now pass from the double complex to the associated filtered complex. Recall that

$$F^p\mathrm{Tot}(E_0)^{p+q} = \bigoplus_{\substack{i+j=p+q \\ i \geq p}} E_0^{ij}.$$

Thus a pairing $E_0 \times {}'E_0 \to {}''E_0$ gives rise to a pairing

$$\mathrm{Tot}(E_0)^{p+q} \otimes \mathrm{Tot}({}'E_0)^{p'+q'} \to \mathrm{Tot}({}''E_0)^{p+p'+q+q'}$$

satisfying the identity

$$d(xy) = d(x)y + (-1)^{\deg(x)}x d(y)$$

and with the property that the image under this map of the tensor product of subcomplexes $F^p\mathrm{Tot}(E_0)^{p+q} \otimes F^{p'}\mathrm{Tot}({}'E_0)^{p'+q'}$ lies in $F^{p+p'}\mathrm{Tot}({}''E_0)^{p+p'+q+q'}$.

DEFINITION 3.9.2. *A* **pairing** *of filtered cochain complexes* $\mathbf{X} \times {}'\mathbf{X} \to {}''\mathbf{X}$ *consists of maps*

$$X^m \otimes {}'X^n \to {}''X^{m+n}$$

$$x \otimes y \mapsto xy$$

satisfying the identity

$$d(xy) = d(x)y + (-1)^{\deg(x)}x d(y)$$

and with the property that the image under this map of $F^p\mathbf{X} \otimes F^{p'}\,{}'\mathbf{X}$ *lies in* $F^{p+p'}\,{}''\mathbf{X}$.

Such a pairing induces maps

$$H^{p+q}(F^p\mathbf{X}) \times H^{p'+q'}(F^{p'}\,{}'\mathbf{X}) \to H^{p+p'+q+q'}(F^{p+p'}\,{}''\mathbf{X})$$

$$H^{p+q}(F^p\mathbf{X}, F^{p+n}\mathbf{X}) \times H^{p'+q'}(F^{p'}\,{}'\mathbf{X}, F^{p'+n}\,{}'\mathbf{X})$$
$$\to H^{p+p'+q+q'}(F^{p+p'}\,{}''\mathbf{X}, F^{p+p'+n}\,{}''\mathbf{X})$$

and hence in particular with $n = 1$,

$$D_1^{pq} \times {}'D_1^{p'q'} \to {}''D_1^{p+p',q+q'}$$

$$E_1^{pq} \times {}'E_1^{p'q'} \to {}''E_1^{p+p',q+q'}.$$

The maps

$$H^{p+q}(F^p\mathbf{X}, F^{p+n}\mathbf{X}) \to H^{p+q}(F^p\mathbf{X}, F^{p+1}\mathbf{X}) \to H^{p+q}(F^{p-n+1}\mathbf{X}, F^{p+1}\mathbf{X})$$

take products to products, and so by Proposition 3.2.4, the image of $Z_n \times {}'Z_n$ under this product map lies in $''Z_n$, and the images of $Z_n \times {}'B_n$ and $B_n \times {}'Z_n$ lie in $''B_n$. It follows that there are products

$$E_n^{pq} \times {}'E_n^{p'q'} \to {}''E_n^{p+p',q+q'}$$

satisfying the identity

$$d_n(xy) = d_n(x)y + (-1)^{p+q}x d_n(y),$$

and with the property that the product on E_{n+1} is induced by the product on E_n according to the formula

$$[x][y] = [xy].$$

This then induces a product on E_∞ in the obvious way. It also follows from the above that the isomorphism given in Theorem 3.2.6 preserves products.

Since the relevant maps in Theorem 3.2.9 take products to products, it also follows that the isomorphism given in this theorem preserves products. We summarise the situation in the following theorem.

THEOREM 3.9.3. *A pairing $E_0 \times {}'E_0 \to {}''E_0$ of double complexes gives rise to a pairing $\mathbf{X} \times {}'\mathbf{X} \to {}''\mathbf{X}$ of filtered cochain complexes. Such a pairing in turn gives rise to pairings*

$$E_n^{pq} \times {}'E_n^{p'q'} \to {}''E_n^{p+p',q+q'}$$

satisfying $d_n(xy) = d_n(x)y + (-1)^{p+q} x d_n(y)$ and $[x][y] = [xy]$ in E_{n+1}. In the limit this induces a pairing

$$E_\infty^{pq} \times {}'E_\infty^{p'q'} \to {}''E_\infty^{p+p',q+q'}$$

in such a way that the isomorphisms

$$E_\infty^{pq} \cong F^p H^{p+q}(\mathbf{X}) / F^{p+1} H^{p+q}(\mathbf{X})$$

given in Theorem 3.2.9 preserve products.   □

Notice that we may only deduce the product structure on $H^*(\mathbf{X})$ modulo terms lower down in the filtration. It may be quite hard to deduce the exact product structure on $H^*(\mathbf{X})$ from this information.

3.10. Equivariant cohomology and finite generation

In this book, we give two different proofs of finite generation of cohomology of a finite group. The first proof, due to Venkov [**273, 274**] (see also Quillen [**209**]) is topological, and occupies this section. The second proof, due to Evens [**101**], is algebraic, and is given in Section 4.2. In each case, we prove the theorem natural to the theory. In the case of Venkov's proof, the setting is equivariant cohomology.

Given any topological group G of the homotopy type of a CW complex, and a G-space X, we may define the **equivariant cohomology** of G with coefficients in X to be

$$H_G^*(X; R) = H^*(EG \times_G X; R).$$

If $\phi : G \to G'$ is a homomorphism and $f : X \to X'$ is a compatible map (i.e., $f(gx) = \phi(g)f(x)$) we have a square which commutes up to homotopy

$$\begin{array}{ccc}
(EG \times EG') \times_G X & \xrightarrow{\ \simeq\ } & EG' \times_G X \\
\Big\downarrow{\scriptstyle \simeq} & & \Big\downarrow \\
EG \times_G X & \dashrightarrow & EG' \times_{G'} X'
\end{array}$$

The fibres of the left-hand and top map are contractible by Lemma 2.4.7, so by Theorem 1.5.8 they are homotopy equivalences. This allows us to fill in the dotted arrow at the bottom uniquely up to homotopy.

In particular, if we take $G = G'$ and $X = X'$, this shows that the dotted arrow is a homotopy equivalence, so that $H_G^*(X; R)$ is independent of the choice of EG.

In fact this is the topological analogue of hypercohomology, introduced in Section 2.7 of Volume I. Namely, if G is discrete, let $\Delta_*(X;R)$ be the complex of singular chains on X. Then $\Delta_*(EG \times X; R)$ is a projective resolution of $\Delta_*(X;R)$ as a complex of RG-modules, and so

$$
\begin{aligned}
H^*(EG \times_G X; R) &= H^*(\operatorname{Hom}_R(\Delta_*(EG \times_G X; R), R)) \\
&= H^*(\operatorname{Hom}_{RG}(\Delta_*(EG \times X; R), R)) \\
&= \operatorname{Ext}^*_{RG}(\Delta_*(X; R), R).
\end{aligned}
$$

Since there is a fibration $EG \times_G X \to EG \times_G (\text{point}) = BG$ with fibre X, we have a spectral sequence

$$
H^p(BG; H^q(X)) \Rightarrow H^{p+q}_G(X)
$$

The space $EG \times_G X$ is known as the **Borel construction** on X, and so this spectral sequence is called the spectral sequence of the Borel construction, or the spectral sequence of equivariant cohomology.

THEOREM 3.10.1. *Suppose G is a finite group (or more generally a compact Lie group), R is a commutative Noetherian ring of coefficients, and $G \hookrightarrow U(n)$ is an embedding of G into a complex unitary group. If X is a G-space with the property that $H^*(X; R)$ is a finitely generated R-module, then $H^*_G(X; R)$ is a finitely generated module over $H^*(BU; R)$.*

PROOF. It follows from the hypothesis that $H^n(X; R) = 0$ for n large. We have a fibration $U(n) \times_G X \to U(n)/G$ with fibre X, which gives rise to a spectral sequence

$$
H^*(U(n)/G; H^*(X; R)) \Rightarrow H^*(U(n) \times_G X; R).
$$

Now $U(n)/G$ is a finite CW-complex, and so the E_2 term of the above spectral sequence is a finitely generated R-module. Since E_∞ is a subquotient of E_2, this is also a finitely generated R-module. It follows that $H^*(U(n) \times_G X; R)$ has a finite filtration by finitely generated R-modules, and is hence finitely generated.

Since $EU(n) \times_{U(n)} (U(n) \times_G X) = EU(n) \times_G X$ and $EU(n)$ is a contractible space on which G acts freely, we have

$$
H^*_{U(n)}(U(n) \times_G X; R) \cong H^*_G(X; R).
$$

We thus have a map of spectral sequences

$$
\begin{array}{ccc}
H^*(BU(n); H^*(\text{point}; R)) & \Longrightarrow & H^*(BU(n); R) \\
\downarrow & & \downarrow \\
H^*(BU(n); H^*(U(n) \times_G X; R)) & \Longrightarrow & H^*_G(X; R)
\end{array}
$$

which makes the lower spectral sequence into a spectral sequence of modules over the ring $H^*(BU(n); R)$. Now we described the latter in Section 2.6. We have

$$
H^*(BU(n); R) = R[c_1, \ldots, c_n]
$$

with $\deg(c_i) = 2i$. The space $BU(n)$ is simply connected, and so

$$H^*(BU(n); H^*(U(n) \times_G X; R)) = H^*(U(n) \times_G X; R)[c_1, \dots, c_n]$$

is a finitely generated module over the Noetherian ring $H^*(BU(n); R)$ by the Hilbert basis theorem, and so the E_2 page of the lower spectral sequence in the above diagram is a finitely generated $H^*(BU(n); R)$-module. It follows that the E_∞ page is also a finitely generated module, and hence $H^*_G(X; R)$ has a finite filtration by finitely generated $H^*(BU(n); R)$-module and is hence finitely generated.   □

COROLLARY 3.10.2. *If G is finite (or compact Lie) and R is Noetherian, the ring $H^*(G; R)$ is finitely generated as a module over the subring generated by the Chern classes of any faithful complex representation. In particular, it is a finitely generated graded commutative R-algebra.*

PROOF. This is the case where X is a point in the above theorem.   □

CHAPTER 4

The Evens norm map and the Steenrod algebra

4.1. The Evens norm map

We now define the Evens norm map

$$\mathrm{norm}_{H,G} : \mathrm{Ext}^r_{RH}(M_1, M_2) \to \mathrm{Ext}^{nr}_{RG}(M_1 \mathring{\circledast}^G, M_2 \mathring{\circledast}^G)$$

(with a slight twist in case r is odd, to be explained later), where $H \le G$ and $n = |G : H|$. This bears the same relation to transfer that tensor induction does to induction.

Recall from Section 3.15 of Volume I that if M is an RH-module then $M^{\otimes n}$ is an $R(\Sigma_n \wr H)$-module, which restricts via $i : G \hookrightarrow \Sigma_n \wr H$ to the tensor induced module $M \mathring{\circledast}^G$. Similarly if $\mathbf{C}$ is a chain complex of RH-modules then $\mathbf{C} \otimes_R \mathbf{C}$ is a chain complex of $R(H \times H)$-modules, and so on. Since tensor product is graded commutative, we must be careful about signs in order to make $\mathbf{C}^{\otimes n}$ into a chain complex of $R(\Sigma_n \wr H)$-modules. The action of $\Sigma_n \wr H$ on $\mathbf{C}^{\otimes n}$ is given as follows:

$$(\pi; h_1, \ldots, h_n)(x_1 \otimes \cdots \otimes x_n) = (-1)^\nu h_{\pi^{-1}(1)} x_{\pi^{-1}(1)} \otimes \cdots \otimes h_{\pi^{-1}(n)} x_{\pi^{-1}(n)}$$

where

$$\nu = \sum_{\substack{j<k \\ \pi(j)>\pi(k)}} \deg(x_j)\deg(x_k).$$

In other words, we write π as a product of standard transpositions $(j, j+1)$, and for each one we multiply by a sign of $(-1)^{\deg(x_j)\deg(x_{j+1})}$.

By restricting to G via $i : G \hookrightarrow \Sigma_n \wr H$, we obtain a chain complex $\mathbf{C} \mathring{\circledast}^G = i^*(\mathbf{C}^{\otimes n})$ of RG-modules. If $\mathbf{C}$ is exact and the C_r are projective as R-modules then Proposition 3.4.4 shows that $\mathbf{C}^{\otimes n}$ and $\mathbf{C} \mathring{\circledast}^G$ are also exact.

Now if $\alpha : C_r \to M$ is a map of RH-modules, then in order to make $\alpha^{\otimes n} : (C_r)^{\otimes n} \to M^{\otimes n}$ a map of $R(\Sigma_n \wr H)$-modules, we need to be a little careful with the signs. If r is even, there is no problem, but if r is odd we must make odd permutations in Σ_n act with a minus sign. We define $R^{(r)}$ to be the $R(\Sigma_n \wr H)$-module R on which $H \times \cdots \times H$ acts trivially, and Σ_n acts trivially if r is even, and via the sign representation if r is odd. Thus $\alpha^{\otimes n}$ is really a map from $(C_r)^{\otimes n}$ to $M^{\otimes n} \otimes R^{(r)}$. We define $1 \wr \alpha$ to be the composite map

$$1 \wr \alpha : (\mathbf{C}^{\otimes n})_{nr} \to (C_r)^{\otimes n} \to M^{\otimes n} \otimes R^{(r)}$$

121

as a map of $R(\Sigma_n \wr H)$-modules, and hence a map of RG-modules

$$1 \wr \alpha : (\mathbf{C}_{\circledast}^{\otimes G})_{nr} \to M_{\circledast}^{\otimes G} \otimes R^{(r)}.$$

We have

$$\delta(1 \wr \alpha) = \sum_{j=1}^{n} (-1)^{(j-1)r} \alpha^{\otimes(j-1)} \otimes \delta\alpha \otimes \alpha^{\otimes(n-j)}$$

so that if $\delta\alpha = 0$ then $\delta(1 \wr \alpha) = 0$.

If $\mathbf{Q}$ is a projective resolution of M_1 over RH then $\mathbf{Q}^{\otimes n}$ is an exact sequence of not necessarily projective modules for $R(\Sigma_n \wr H)$, resolving $M_1^{\otimes n}$. If $\mathbf{P}$ is a projective resolution of $M_1^{\otimes n}$ (for example we could take $\mathbf{P}$ to be the tensor product of a projective resolution of R for $R\Sigma_n$ with $\mathbf{Q}^{\otimes n}$) then by the remark after Theorem 2.4.2 of Volume I, there is a map of chain complexes $\phi : \mathbf{P} \to \mathbf{Q}^{\otimes n}$ lifting the identity map on $M_1^{\otimes n}$, and any two such maps are chain homotopic. Thus if $\zeta \in \mathrm{Ext}_{RH}^r(M_1, M_2)$ is represented by a cocycle $\hat\zeta : Q_r \to M_2$ then $(1 \wr \hat\zeta) \circ \phi$ is also a cocycle. We must check that the cohomology class it defines is independent of the choice of $\mathbf{Q}$ and of the representative cocycle $\hat\zeta$.

LEMMA 4.1.1. *Suppose* $\mathbf{C}$ *is a chain complex of* RH-*modules, and* f, $g : \mathbf{Q} \to \mathbf{C}$ *are homotopic chain maps. Then*

$$f^{\otimes n} \circ \phi, \; g^{\otimes n} \circ \phi : \mathbf{P} \to \mathbf{C}^{\otimes n}$$

are homotopic chain maps of $R(\Sigma_n \wr H)$-*modules.*

PROOF. Denote by $\mathbf{I}$ the chain complex of R-modules with $I_0 = R \oplus R$ with basis a and b, $I_1 = R$ with basis c and $\partial c = a - b$, and $I_n = 0$ for $n \neq 0, 1$ (in other words, $\mathbf{I}$ is the cellular chain complex of the unit interval). Then a homotopy from f to g is really just a map of chain complexes of RH-modules $\mathbf{h} : \mathbf{I} \otimes \mathbf{Q} \to \mathbf{C}$ with $\mathbf{h}(a \otimes q) = f(q)$ and $\mathbf{h}(b \otimes q) = g(q)$ (so that $\mathbf{h}(c \otimes q) = h(q)$ satisfies $f - g = \partial \circ h + h \circ \partial$). If we can construct a chain map of $R(\Sigma_n \wr H)$-modules

$$\psi : \mathbf{I} \otimes \mathbf{P} \to \mathbf{I}^{\otimes n} \otimes \mathbf{Q}^{\otimes n}$$

with $\psi(a \otimes p) = a^{\otimes n} \otimes \phi(p)$ and $\psi(b \otimes p) = b^{\otimes n} \otimes \phi(p)$, then the composite

$$\mathbf{I} \otimes \mathbf{P} \xrightarrow{\psi} \mathbf{I}^{\otimes n} \otimes \mathbf{Q}^{\otimes n} \cong (\mathbf{I} \otimes \mathbf{Q})^{\otimes n} \xrightarrow{\mathbf{h}^{\otimes n}} \mathbf{C}^{\otimes n}$$

will be a homotopy from $f^{\otimes n} \circ \phi$ to $g^{\otimes n} \circ \phi$.

We construct the map ψ by induction on degree. Note that ψ is already defined on $I_0 \otimes \mathbf{P}$ by the given conditions. If ψ has been constructed up to degree r, then we construct ψ_{r+1} by filling in the left-hand end of the following diagram with a vertical arrow, using the fact that $\mathbf{P}$ consists of projective modules and $\partial \circ \psi_r \circ \partial = 0$.

$$
\begin{array}{ccccc}
I_1 \otimes P_r & \xrightarrow{\ \partial\ } & (\mathbf{I} \otimes \mathbf{P})_r & \xrightarrow{\ \partial\ } & (\mathbf{I} \otimes \mathbf{P})_{(r-1)} \\
\Big\downarrow & & \Big\downarrow{\psi_r} & & \Big\downarrow{\psi_{r-1}} \\
(\mathbf{I}^{\otimes n} \otimes \mathbf{Q}^{\otimes n})_{r+1} & \xrightarrow{\ \partial\ } & (\mathbf{I}^{\otimes n} \otimes \mathbf{Q}^{\otimes n})_r & \xrightarrow{\ \partial\ } & (\mathbf{I}^{\otimes n} \otimes \mathbf{Q}^{\otimes n})_{r-1}
\end{array}
$$

The resulting map ψ is clearly a chain map with the desired properties. $\qquad\square$

By taking $\mathbf{C}$ to be the module M_2 concentrated in degree r, we see that $(1 \wr \hat{\zeta}) \circ \phi$ is independent of the choice of cocycle $\hat{\zeta}$ representing ζ. By taking $\mathbf{C} = \mathbf{Q}'$ to be another resolution of M_1, we see that the definition is independent of the choice of resolution. So we obtain a well defined cohomology class

$$
1 \wr \zeta \in \mathrm{Ext}^{nr}_{R(\Sigma_n \wr H)}(M_1^{\otimes n}, M_2^{\otimes n} \otimes R^{(r)}).
$$

We define the **Evens norm map** via

$$
\mathrm{norm}_{H,G}(\zeta) = i^*(1 \wr \zeta) \in \mathrm{Ext}^{nr}_{RG}(M_1 \mathring{\otimes}^G, M_2 \mathring{\otimes}^G \otimes R^{(r)}).
$$

Note that if r is odd, the exact sign of this norm depends on the order chosen for the coset representatives.

If we tensor induce the trivial module R we obtain $R \mathring{\otimes}^G = R$ and so this also defines a map

$$
\mathrm{norm}_{H,G} : H^r(H, M) \to H^{nr}(G, M \mathring{\otimes}^G \otimes R^{(r)}).
$$

In particular for r even we have a map

$$
\mathrm{norm}_{H,G} : H^r(H, R) \to H^{nr}(G, R).
$$

The following properties of the Evens norm map are direct consequences of the corresponding properties of tensor induction given in Proposition 3.15.1 of Volume I, with suitable attention paid to the signs.

PROPOSITION 4.1.2. *Suppose $H < G$ with $|G : H| = n$.*
(i) *If $\zeta \in \mathrm{Ext}^r_{RH}(M_1, M_2)$ and $\zeta' \in \mathrm{Ext}^s_{RH}(M_3, M_4)$ then*

$$
\mathrm{norm}_{H,G}(\zeta.\zeta') = (-1)^{\frac{n(n-1)}{2}rs} \mathrm{norm}_{H,G}(\zeta).\mathrm{norm}_{H,G}(\zeta')
$$

$$
\in \mathrm{Ext}^{n(r+s)}_{RG}((M_1 \otimes M_3) \mathring{\otimes}^G, (M_2 \otimes M_4) \mathring{\otimes}^G \otimes R^{(r+s)}).
$$

(ii) *If $H' \leq H$ and $\zeta \in \mathrm{Ext}^r_{RH'}(M_1, M_2)$ then*

$$
\mathrm{norm}_{H,G}\mathrm{norm}_{H',H}(\zeta) = \mathrm{norm}_{H',G}(\zeta)
$$

(where the coset representatives of H' in G have been chosen in the order coming from the choices of coset representatives of H' in H and H in G).
(iii) *If $\zeta, \zeta' \in \mathrm{Ext}^r_{RH}(M_1, M_2)$ then*

$$
\mathrm{norm}_{H,G}(\zeta + \zeta') = \mathrm{norm}_{H,G}(\zeta) + \mathrm{norm}_{H,G}(\zeta')
$$

+ *a sum of transfers from proper subgroups K containing the intersection of the conjugates of H.*

(iv) *If $H' \leq H$ and $\zeta \in \mathrm{Ext}^r_{RH'}(M_1, M_2)$ then $\mathrm{norm}_{H,G}\mathrm{tr}_{H',H}(\zeta)$ is a sum of transfers from subgroups K containing the intersection of the conjugates of H', and with $K \cap H \leq H'$.*

(v) *(Mackey formula) If K, $H \leq G$ and $\zeta \in \mathrm{Ext}^r_{RK}(M_1, M_2)$ then*

$$\mathrm{res}_{G,H}\mathrm{norm}_{K,G}(\zeta) = \prod_{H\backslash G/K} \mathrm{norm}_{H\cap {}^gK, H}\mathrm{res}_{{}^gK, H\cap {}^gK}(g\zeta).$$

(vi) *If N is a normal subgroup of G contained in H, and M_1 and M_2 are RH-modules on which N acts trivially, then for $\zeta \in \mathrm{Ext}^r_{R(H/N)}(M_1, M_2)$ we have*

$$\mathrm{inf}_{G/N,G}\mathrm{norm}_{H/N,G/N}(\zeta) = \mathrm{norm}_{H,G}\mathrm{inf}_{H/N,H}(\zeta). \qquad \square$$

REMARK. Graeme Segal [**230**] has constructed a generalised cohomology theory whose degree zero part (additively) is the multiplicative group of formal series $\{a_{2i}\}$, $a_{2i} \in H^{2i}(X; R)$ and a_0 invertible. The transfer map (see Section 2.7) in this theory gives the Evens norm map

$$\mathrm{norm}_{H,G} : H^r(BH; R) \to H^{nr}(BG; R)$$

for r even. Fulton and MacPherson [**113**] give a direct description of the norm map for finite coverings of topological spaces, and describe the Chern classes of an induced representation (or more generally a direct image bundle) in terms of this construction.

THEOREM 4.1.3. *If p divides $|G|$ and p is not invertible in R then for some $r > 0$, $H^r(G, R) \neq 0$. Moreover, if $g \neq 1$ is an element of order p in G, such an element can be chosen to restrict to a non-zero element of $H^r(\langle g \rangle, R)$.*

PROOF. Let $g \neq 1$ be an element of G of order p, and let $H = N_G\langle g \rangle$. By Corollary 3.5.4 of Volume I, there is a non-nilpotent element $\sigma \in H^2(\langle g \rangle, R)$. For $x \in H$, conjugation by x sends σ to $\lambda\sigma$ for some λ with $\lambda^{p-1} = 1$ (since the action of conjugation by x on $\langle g \rangle$ has order dividing $p - 1$). Thus, letting $\alpha = \sigma^{p-1}$, α is an H-invariant non-nilpotent element of $H^{2(p-1)}(\langle g \rangle, R)$. Let $|N_G(\langle g \rangle) : \langle g \rangle| = p^a h$ with p not dividing h, and set

$$z = \mathrm{norm}_{\langle g \rangle, G}(1 + \alpha).$$

Then the Mackey formula 4.1.2 (v) shows that

$$\mathrm{res}_{G,\langle g \rangle}(z) = (1 + \alpha)^{p^a h} = (1 + \alpha^{p^a})^h$$

$$= 1 + h\alpha^{p^a} + \text{terms of higher degree}.$$

Thus z has a non-zero homogeneous part in degree $2(p-1)p^a$. $\qquad \square$

EXAMPLE. Let $G = \mathbb{Z}/p \times \mathbb{Z}/p$ with p odd, and $H = 1 \times \mathbb{Z}/p \subseteq G$. Then by Section 3.5 of Volume I, $H^*(G, \mathbb{F}_p)$ has generators y_1 and y_2 in degree one with $y_1^2 = y_2^2 = 0$, and generators x_1 and x_2 in degree two. $H^*(H, \mathbb{F}_p)$ is generated by y_2 and x_2, which are the restrictions of the elements of the same name in $H^*(G, \mathbb{F}_p)$, while y_1 and x_1 restrict to zero. We shall calculate $\mathrm{norm}_{H,G}(y_2)$ and $\mathrm{norm}_{H,G}(x_2)$. We first deal with the latter.

Since

$$\mathrm{res}_{G,H}\,\mathrm{norm}_{H,G}(x_2) = x_2^p$$

and $\mathrm{norm}_{H,G}(x_2)$ is invariant under automorphisms of the first copy of $\mathbb{Z}/p$, we have

$$\mathrm{norm}_{H,G}(x_2) = x_2^p + \lambda x_1^{p-1} x_2 + \mu x_1^{p-2} y_1 x_2 y_2$$

for some values of the scalars λ and μ. The norm of x_2 is also invariant under the automorphism of $\mathbb{Z}/p \times \mathbb{Z}/p$ which fixes the second generator and sends the first to its product with the second. This automorphism sends y_2 to $y_1 + y_2$, x_2 to $x_1 + x_2$, and fixes y_1 and x_1. Thus we have

$$\mathrm{norm}_{H,G}(x_2) = (x_1 + x_2)^p + \lambda x_1^{p-1}(x_1 + x_2) + \mu x_1^{p-2} y_1 (x_1 + x_2)(y_1 + y_2).$$

Equating this with the previous expression we find that $\lambda = -1$ and $\mu = 0$, so that

$$\mathrm{norm}_{H,G}(x_2) = x_2^p - x_1^{p-1} x_2.$$

Thus by multiplicativity of the norm we have

$$\mathrm{norm}_{H,G}(x_2^r) = (x_2^p - x_1^{p-1} x_2)^r.$$

Applying the same reasoning to y_2, we see that $\mathrm{norm}_{H,G}(y_2)$ is of the form

$$\lambda x_1^{(p-1)/2} y_2 + \mu x_1^{(p-3)/2} y_1 x_2.$$

Note that since y_2 has odd degree, the action of the automorphisms of $\mathbb{Z}/p$ on cohomology has to be tensored with the sign representation. The invariants of this action are discussed in Section 3.6 of Volume I.

Again applying the automorphism fixing the first generator and sending the second to its product with the first, we see that $\lambda = -\mu$. However, this reasoning does not allow us to evaluate λ explicitly. Of course, the sign of λ depends on which ordering is chosen for the cosets of H in G, as is always true in odd degree. We shall see in Section 4.5 that for a suitable choice of this order we have $\lambda = 1/(\frac{p-1}{2})!$, so that

$$\mathrm{norm}_{H,G}(y_2) = x_1^{(p-3)/2}(x_1 y_2 - y_1 x_2)/(\tfrac{p-1}{2})!.$$

We summarise the results of this calculation for future reference. The corresponding result for $p = 2$ is easy to check by the same method.

PROPOSITION 4.1.4. *Let $G = \mathbb{Z}/p \times \mathbb{Z}/p$ with p odd, and $H = 1 \times \mathbb{Z}/p \le G$. Then using the above notation for elements of $H^*(H, \mathbb{F}_p)$ and $H^*(G, \mathbb{F}_p)$ we have*

(i) $\mathrm{norm}_{H,G}(x_2) = x_2^p - x_1^{p-1} x_2$

(ii) $\mathrm{norm}_{H,G}(y_2) = \lambda x_1^{(p-3)/2}(x_1 y_2 - y_1 x_2),$

where λ is a multiplicative constant which will be seen in Section 4.5 to be $1/(\frac{p-1}{2})!$.

If $G = \mathbb{Z}/2 \times \mathbb{Z}/2$ and $H = 1 \times \mathbb{Z}/2 \leq G$ then $H^*(G, \mathbb{F}_2)$ is generated by elements y_1 and y_2 of degree one. If H is the subgroup corresponding to the element y_1, then writing y_2 for the restriction of y_2 to H we have

$$\text{norm}_{H,G}(y_2) = y_2^2 + y_1 y_2. \qquad \square$$

EXERCISE. (Nakaoka's Theorem [202], see also Quillen [213] §3) Suppose that $R = k$ is a field, and $G \subseteq \Sigma_n$ and H are finite groups with wreath product $G \wr H$. If $\mathbf{P}$ is a minimal resolution of k as a kG-module and $\mathbf{Q}$ is a minimal resolution of k as a kH-module, show that $\mathbf{Q}^{\otimes n}$ is a minimal projective resolution of k as a $k(H^n)$-module. Show that if we use $\mathbf{P} \otimes \mathbf{Q}^{\otimes n}$ as a resolution for $G \wr H$, then the differential on

$$\text{Hom}_{k(G \wr H)}(\mathbf{P} \otimes \mathbf{Q}^{\otimes n}, k) \cong \text{Hom}_{kG}(\mathbf{P}, \text{Hom}_{kH}(\mathbf{Q}, k)^{\otimes n})$$

comes entirely from the differential on $\mathbf{P}$. Deduce that the cohomology of $G \wr H$ is given by

$$H^*(G \wr H, k) = H^*(G, H^*(H^n, k)).$$

In other words, in the Lyndon–Hochschild–Serre spectral sequence for the normal subgroup H^n, converging to $H^*(G \wr H, k)$, all the differentials from the E_2 page onwards are zero, and the filtered graded ring $H^*(G \wr H, k)$ is isomorphic to its associated double graded ring E_∞^{**}. Use this to give another proof of Lemma 4.1.1 in case $R = k$.

4.2. Finite generation of cohomology

In this section we present Evens' proof [101] that the cohomology ring of a finite group over a Noetherian ring is finitely generated. For Venkov's topological proof see Section 3.10.

THEOREM 4.2.1 (Evens). *If G is a finite group and M is an RG-module which is Noetherian as a module for R then $H^*(G, M)$ is Noetherian as a module for $H^*(G, R)$.*

COROLLARY 4.2.2. *If R is a Noetherian ring then $H^*(G, R)$ is finitely generated by homogeneous elements, as a ring over R.*

WARNING. If R is not Noetherian then $H^*(G, R)$ is not necessarily finitely generated over R. For example if $G = \mathbb{Z}/p$ and $R = \mathbb{Z} \oplus (\mathbb{Z}/p)^\infty$ (with multiplication given by $(a, b).(c, d) = (ac, cb + ad)$) then there are infinitely many new generators in degree one.

The corollary follows from the theorem (with $M = R$) by applying the following lemma.

LEMMA 4.2.3. *If $H = \bigoplus_{n \geq 0} H^n$ is a graded ring and H is Noetherian (as a ring), then H^0 is Noetherian and H is finitely generated over H^0 by homogeneous elements.*

PROOF. Since there is a surjective map $H \twoheadrightarrow H^0$, H^0 is Noetherian. Let $H^+ = \bigoplus_{n>0} H^n$. This is an ideal in H, and is hence finitely generated. Each generator is a finite sum of homogeneous elements, and so H^+ is generated as an ideal by homogeneous elements $a_1, \ldots, a_n$, say. Then H is generated as a ring over H^0 by $1, a_1, \ldots, a_n$. $\qquad\square$

To prove the theorem, we first reduce to the Sylow p-subgroups of G, and then we proceed by induction on the order of the p-group, using the spectral sequence coming from a central element of order p. The crux of the proof is the fact that this spectral sequence stops at some finite page, and this involves the use of the Evens norm map.

REDUCTION TO SYLOW p-SUBGROUPS. By Proposition 3.7.10 of Volume I, for any α of positive degree in $H^*(G, M)$ we have $|G|.\alpha = 0$. Thus if we set

$$H^+(G, M) = \bigoplus_{n>0} H^n(G, M)$$

then we have

$$H^+(G, M) = \bigoplus_{p \mid |G|} H^+(G, M)_{(p)}$$

where $H^+(G, M)_{(p)}$ denotes the elements in $H^+(G, M)$ annihilated by $|G|_p$. Again using Proposition 3.7.10 of Volume I, we see that if P is a Sylow p-subgroup of G then

$$\mathrm{res}_{G,P} : H^*(G, M)_{(p)} \to H^*(P, M)$$

is injective, and

$$H^*(P, M) = H^*(G, M)_{(p)} \oplus T(M)$$

where

$$T(M) = \mathrm{Ker}(\mathrm{tr}_{P,G} : H^*(P, M) \to H^*(G, M)).$$

Thus we need to show that if $H^*(G, M)_{(p)} \oplus T(M)$ is Noetherian as a module over $H^*(G, R)_{(p)} \oplus T(R)$ then $H^*(G, M)_{(p)}$ is Noetherian as a module over $H^*(G, R)_{(p)}$.

Now by Lemma 3.7.9 (ii) of Volume I, we have

$$\mathrm{tr}_{P,G}(\beta).\alpha = \mathrm{tr}_{P,G}(\beta.\mathrm{res}_{G,P}(\alpha))$$

so that

$$T(R).H^+(G, M)_{(p)} \subseteq T(M).$$

We also have

$$H^+(G, R)_{(p)}.H^+(G, M)_{(p)} \subseteq H^+(G, M)_{(p)}.$$

Thus if A is an $H^+(G, R)_{(p)}$-submodule of $H^+(G, M)_{(p)}$ then

$$H^+(P, R).A = (H^+(G, R)_{(p)} \oplus T(R)).A$$
$$= H^+(G, R)_{(p)}.A \oplus T(R).A \subseteq H^+(G, M)_{(p)} \oplus T(M).$$

It follows that if A is a proper submodule of A' then $H^+(P, R).A$ is a proper submodule of $H^+(P, R).A'$, and so infinite ascending chains in $H^+(G, M)_{(p)}$ give rise to infinite ascending chains in $H^+(P, M)$.

THE CASE $G = P$ IS A p-GROUP.   We proceed by induction on the order of P. If $P = 1$ the theorem is trivial. If $P > 1$, then let Z be a cyclic subgroup of order p in $Z(P)$, and examine the spectral sequence of the central extension $1 \to Z \to P \to P/Z \to 1$. Set

$$E_2(R) = H^*(P/Z, H^*(Z, R)), \quad E_2(M) = H^*(P/Z, H^*(Z, M)).$$

Then $E_2(R)$ has a ring structure over which $E_2(M)$ is a module (see Section 3.9).

Now recall that the vertical edge homomorphism is just the restriction map, so that the following diagram commutes:

$$
\begin{array}{ccccc}
H^*(P, R) & \xrightarrow{\operatorname{res}_{P,Z}} & H^0(P, H^*(Z, R)) & = & H^*(Z, R) \\
\downarrow & & \| & & \\
E_\infty^{0*}(R) & \lhook\joinrel\longrightarrow & E_2^{0*}(R) & &
\end{array}
$$

Let $\sigma \in H^2(Z, R)$ be the element described in Corollary 3.5.4 of Volume I, with the property that cup product with σ induces an isomorphism

$$H^r(Z, -) \to H^{r+2}(Z, -)$$

for all r and for all coefficients. Then setting $n = |P : Z|$, we have

$$\sigma^n = \operatorname{res}_{P,Z} \operatorname{norm}_{Z,P}(\sigma)$$

so that the element $\xi = \sigma^n$ is an element of degree $2n$ in $\operatorname{Im}(\operatorname{res}_{P,Z})$, i.e., in $E_\infty^{0,2n}(R)$.

Set

$$E_2'(M) = \bigoplus_{0 \leq q \leq 2n-1} E_2^{*q}(M)$$

so that

$$E_2(M) = \bigoplus_{N \geq 0} \xi^n.E_2'(M).$$

Since each $H^q(Z, M)$ is a subquotient of M (Corollary 3.5.2 of Volume I), it is Noetherian over R, and so by the inductive hypothesis each $E_2^{*q}(M) = H^*(P/Z, H^q(Z, M))$ is Noetherian over $E_2^{*0}(R) = H^*(P/Z, R)$. Thus $E_2'(M)$ is Noetherian over $E_2^{*0}(R)$. It now follows from the Hilbert basis theorem that $E_2(M)$ is Noetherian over $E_2^{*0}(R)[\xi]$.

Now recall that

$$E_1(M) = Z_1(M) \supseteq Z_2(M) \supseteq \cdots \supseteq B_2(M) \supseteq B_1(M) = 0$$

with $E_2(M) = Z_2(M)/B_2(M)$. The product rule (Theorem 3.9.3) shows that each $Z_r(M)$ and each $B_r(M)$ is a submodule of $E_2(M)$ as an $E_2^{*0}(R)[\xi]$-module (note that if each $Z_r(M)$ and $B_r(M)$ were an $E_2(R)$-module the proof would be easier to word, but it seems that there is no reason why this should be the case). Since $E_\infty(M) = Z_\infty(M)/B_\infty(M)$ is a subquotient of $E_2(M)$, it is a Noetherian $E_2^{*0}(R)[\xi]$-module. Now there is a surjective ring homomorphism

$$E_2^{*0}(R)[\xi] \to E_\infty^{*0}(R)[\xi]$$

whose kernel acts trivially on $E_\infty(M)$. Thus $E_\infty(M)$ is Noetherian as an $E_\infty^{*0}(R)[\xi]$-module, and hence Noetherian as an $E_\infty(R)$-module.

Now $H^*(P, M)$ has a filtration whose filtered quotients are

$$E_\infty^{pq}(M) \cong F^p H^{p+q}(P, M)/F^{p+1} H^{p+q}(P, M),$$

so we need to "unfilter" this statement. Suppose $A_1 \subset A_2 \subset \cdots \subset H^*(P, M)$ is an infinite ascending chain of submodules for the action of $H^*(P, R)$. Let $F^p A_i = A_i \cap F^p H^*(P, M)$ and

$$B_i = \bigoplus_{p \geq 0} F^p A_i/F^{p+1} A_i \subseteq E_\infty(M).$$

If x is an element of A_{i+1} not in A_i, then for some p, $x \in F^p A_{i+1}$ but $x \notin F^{p+1} A_{i+1} + F^p A_i$ and so $x + F^{p+1} A_{i+1}$ is not in the image of the inclusion

$$F^p A_i/F^{p+1} A_i \hookrightarrow F^p A_{i+1}/F^{p+1} A_{i+1}.$$

Thus B_{i+1} properly contains B_i and so we have an infinite ascending chain of $E_\infty(R)$-submodules of $E_\infty(M)$. We know that this is impossible, and so this completes the proof of Theorem 4.2.1. $\square$

COROLLARY 4.2.4. *If G is finite and M and N are RG-modules such that $\mathrm{Hom}_R(M, N)$ is a Noetherian R-module, then $\mathrm{Ext}^*_{RG}(M, N)$ is Noetherian as a module over the ring $H^*(G, R)$ via cup product, and also as a module over the ring $\mathrm{Ext}^*_{RG}(M, M)$ by Yoneda composition. If $\mathrm{Hom}_R(M, M)$ is a Noetherian R-module then the ring $\mathrm{Ext}^*_{RG}(M, M)$ is Noetherian as a module over its centre $Z\mathrm{Ext}^*_{RG}(M, M)$.*

PROOF. We have

$$H^*(G, \mathrm{Hom}_R(M, N)) \cong \mathrm{Ext}^*_{RG}(M, N)$$

(see Proposition 3.1.8 of Volume I), so that by Theorem 4.2.1, $\mathrm{Ext}^*_{RG}(M, N)$ is Noetherian as a module over $H^*(G, R)$. According to Section 3.2 of Volume I, the cup product action of $H^*(G, R)$ on $\mathrm{Ext}^*_{RG}(M, N)$ factors as the map

$$H^*(G, R) = \mathrm{Ext}^*_{RG}(R, R) \xrightarrow{\otimes M} \mathrm{Ext}^*_{RG}(M, M)$$

followed by Yoneda composition. Therefore $\mathrm{Ext}^*_{RG}(M, N)$ is Noetherian as a module over $\mathrm{Ext}^*_{RG}(M, M)$. Finally, the image of even degree elements under the above map lie in the centre of $\text{Ext}^*_{RG}(M, M)$ by Section 3.2 of Volume I, and so $\text{Ext}^*_{RG}(M, M)$ is Noetherian as a module over its centre. $\square$

COROLLARY 4.2.5. *If H is a subgroup of a finite group G and R is a Noetherian ring, then $H^*(H, R)$ is finitely generated as a module over $H^*(G, R)$ (via the restriction map).*

PROOF. The Eckmann–Shapiro isomorphism (see Corollary 3.3.2 of Volume I)

$$H^*(H, R) \cong H^*(G, R_H \uparrow^G)$$

is an isomorphism of modules over $H^*(G, R)$. Since $R_H \uparrow^G$ is a Noetherian R-module, the result follows directly from Theorem 4.2.1. $\square$

Note that Evens' Theorem applies just as well to hypercohomology (see Section 2.7 of Volume I) of chain complexes of RG-modules. In this context, a chain complex of RG-modules is thought of as Noetherian as an R-module if it is non-zero in only finitely many degrees, and each non-zero module is Noetherian.

THEOREM 4.2.6. *Suppose that $\mathbf{C}$ and $\mathbf{D}$ are chain complexes of RG-modules. If the complex $\text{Hom}_R(\mathbf{C}, \mathbf{D})$ is a Noetherian R-module (and in particular non-zero in only finitely many degrees) then $\text{Ext}^*_{RG}(\mathbf{C}, \mathbf{D})$ is Noetherian as a module over $H^*(G, R)$ via cup product, and also as a module over $\text{Ext}^*_{RG}(\mathbf{C}, \mathbf{C})$ by Yoneda decomposition.*

PROOF. The proof is exactly the same as the proof in the module case. Namely, the Lyndon–Hochschild–Serre spectral sequence works just as well for hypercohomology as for cohomology, and we have

$$H^*(G, \text{Hom}_R(\mathbf{C}, \mathbf{D})) \cong \text{Ext}^*_{RG}(\mathbf{C}, \mathbf{D}).$$ $\square$

CÓROLLARY 4.2.7. *Suppose that R is a Noetherian ring with the property that for each prime p dividing $|G|$, either p is invertible in R or R/pR is Artinian. If $\mathbf{C}$ is a finite chain complex of finitely generated RG-modules with $\text{Ext}^n_{RG}(\mathbf{C}, \mathbf{C}) = 0$ for all n sufficiently large, then $\mathbf{C}$ has a finite projective resolution.*

PROOF. If $\mathbf{D}$ is another finite chain complex of finitely generated RG-modules, then $\text{Hom}_R(\mathbf{C}, \mathbf{D})$ is also a finite chain complex of finitely generated RG-modules, and so by the above theorem $\text{Ext}^*_{RG}(\mathbf{C}, \mathbf{D})$ is Noetherian as a module over the ring $\text{Ext}^*_{RG}(\mathbf{C}, \mathbf{C})$. Since $\text{Ext}^n_{RG}(\mathbf{C}, \mathbf{C}) = 0$ for all n sufficiently large, we deduce the same of $\text{Ext}^n_{RG}(\mathbf{C}, \mathbf{D})$.

Now for each p dividing $|G|$, $(R/pR)G$ is Artinian, and hence has only finitely many simple modules. So we may choose n_0 such that for all $n > n_0$, all p dividing $|G|$, and all simple $(R/pR)G$-modules S, $\text{Ext}^n_{RG}(\mathbf{C}, S) = 0$. We claim that for all $n > n_0$ and all finitely generated RG-modules M, $\text{Ext}^n_{RG}(\mathbf{C}, M) = 0$. For suppose $\text{Ext}^n_{RG}(\mathbf{C}, M) \neq 0$. By Proposition 3.6.17

of Volume I, multiplication by $|G|$ annihilates $\text{Ext}^n_{RG}(\mathbf{C}, M)$. The long exact sequences in Ext coming from the short exact sequences of modules

$$0 \to |G|.M \to M \to M/|G|.M \to 0$$

and

$$0 \to \text{Ker}(|G| \text{ on } M) \to M \to |G|.M \to 0$$

give

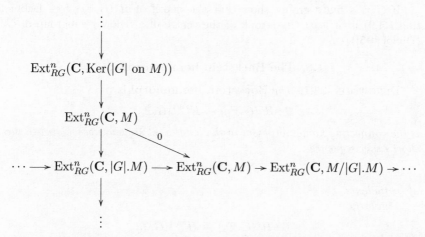

and so either $\text{Ext}^n_{RG}(\mathbf{C}, \text{Ker}(|G| \text{ on } M))$ or $\text{Ext}^n_{RG}(\mathbf{C}, M/|G|.M)$ is non-zero. Thus we may assume that multiplication by $|G|$ on M is zero. Using the long exact sequences coming from multiplication by the primes dividing $|G|$ in the same way, we may assume that multiplication by p on M is zero for some p dividing $|G|$. But now since $(R/pR)G$ is Artinian, this implies that M has finite composition length and so again using the long exact sequences coming from a composition series, we see that we may assume that M is simple. But we have chosen n so that $\text{Ext}^n_{RG}(\mathbf{C}, M)$ is zero in this case.

Now let $\mathbf{P}$ be a projective resolution of $\mathbf{C}$, bounded below, in which each P_n is a finitely generated RG-module. Choose n_0 as above, and so that also $C_n = 0$ for $n > n_0$. Then for $n > n_0$,

$$\cdots \to P_{n+1} \to P_n \to P_{n-1} \to \cdots$$

is exact. Let $M = \text{Im}(P_n \to P_{n-1})$. Then

$$\text{Hom}_{RG}(P_{n-1}, M) \to \text{Hom}_{RG}(P_n, M) \to \text{Hom}_{RG}(P_{n+1}, M) \to \cdots$$

is exact, since $\text{Ext}^n_{RG}(\mathbf{C}, M) = 0$. The natural homomorphism $P_n \to M$ goes to zero in $\text{Hom}_{RG}(P_{n+1}, M)$, and so it is a composite $P_n \to P_{n-1} \to M$. It follows that M is a summand of P_{n-1}, so that

$$0 \to P_{n-1}/M \to P_{n-2} \to \cdots \to 0$$

is a finite projective resolution of $\mathbf{C}$. $\square$

EXERCISE. Let $1 \to Z \to P \to P/Z \to 1$ be a central extension with P a p-group and Z cyclic of order p, and set $n = |P : Z|$. Let σ be a non-zero element of $H^2(Z, \mathbb{F}_p) \cong E_2^{0,2}$ in the spectral sequence

$$H^*(P/Z, H^*(Z, \mathbb{F}_p)) \Rightarrow H^*(P, \mathbb{F}_p).$$

Show that σ^n is a non zero-divisor on every page E_r^{**} of this spectral sequence, on which the differential vanishes. Deduce that $\mathrm{norm}_{Z,P}(\sigma)$ is a non zero-divisor in $H^*(P, \mathbb{F}_p)$.

If G is a finite group, show that the depth of $H^*(G, \mathbb{F}_n)$ (see Definition 5.4.9) is at least the p-rank of the centre of a Sylow p-subgroup of G (Duflot [95]).

4.3. The Bockstein homomorphism

DEFINITION 4.3.1. *The* **Bockstein homomorphism**

$$\beta : H^r(G, \mathbb{F}_p) \to H^{r+1}(G, \mathbb{F}_p)$$

is the connecting homomorphism in the long exact sequence associated to the short exact sequence

$$0 \to \mathbb{F}_p \to \mathbb{Z}/p^2 \to \mathbb{F}_p \to 0$$

of coefficients.
Similarly,

$$\hat{\beta} : H^r(G, \mathbb{F}_p) \to H^{r+1}(G, \mathbb{Z})$$

is the connecting homomorphism associated to the short exact sequence

$$0 \to \mathbb{Z} \xrightarrow{p} \mathbb{Z} \xrightarrow{\pi} \mathbb{F}_p \to 0$$

of coefficients.

LEMMA 4.3.2. $\beta^2 = 0 : H^r(G, \mathbb{F}_p) \to H^{r+2}(G, \mathbb{F}_p)$.

PROOF. We have $\beta = \pi_* \circ \hat{\beta}$, and $\hat{\beta} \circ \pi_* = 0$ since these are two adjacent maps in the long exact sequence. □

EXAMPLE. Let $G = \mathbb{Z}/p$, the cyclic group of order p. Then $H^r(G, \mathbb{Z})$ is isomorphic to $\mathbb{Z}/p$ for r even and zero for r odd (Section 3.5 of Volume I). Thus by the long exact sequence,

$$\hat{\beta} : H^r(G, \mathbb{F}_p) \to H^{r+1}(G, \mathbb{Z})$$

is an isomorphism for r odd and zero for r even, and hence also

$$\beta : H^r(G, \mathbb{F}_p) \to H^{r+1}(G, \mathbb{F}_p)$$

is an isomorphism for r odd and zero for r even.

It is worthwhile making explicit what the Bockstein of a degree one element of cohomology looks like, since we shall be dealing with these a lot in the next chapter. Since $H^1(G, \mathbb{F}_p) \cong \mathrm{Hom}(G, \mathbb{F}_p^+)$, an element x in degree one cohomology is always the inflation of an element $\bar{x} \in H^1(G/H, \mathbb{F}_p)$, where H

is the kernel of the corresponding homomorphism $\phi : G \to \mathbb{F}_p^+$. We thus have an isomorphism $\bar{\phi} : G/H \to \mathbb{F}_p^+$ corresponding to $\bar{x}$. Now G/H is cyclic of order p, and so $\beta(\bar{x}) \in H^2(G/H, \mathbb{F}_p) \cong \mathrm{Ext}^2_{\mathbb{F}_p(G/H)}(\mathbb{F}_p, \mathbb{F}_p)$ is represented by the exact sequence

$$0 \to \mathbb{F}_p \to \mathbb{F}_p(G/H) \to \mathbb{F}_p(G/H) \to \mathbb{F}_p \to 0$$

(cf. Corollary 3.5.4 of Volume I). In this sequence, the first map is given by taking $1 \in \mathbb{F}_p$ to the sum of the group elements in G/H. The second map takes $1 \in \mathbb{F}_p(G/H)$ to $1 - \bar{\phi}^{-1}(1) \in \mathbb{F}_p(G/H)$, and the third map takes $1 \in \mathbb{F}_p(G/H)$ to $1 \in \mathbb{F}_p$. Inflating to G, we see that $\beta(x)$ is represented by the sequence

$$0 \to \mathbb{F}_p \to (\mathbb{F}_p)_H \uparrow^G \to (\mathbb{F}_p)_H \uparrow^G \to \mathbb{F}_p \to 0. \qquad (*)$$

If k is a field of characteristic p, then

$$H^*(G, k) \cong k \otimes_{\mathbb{F}_p} H^*(G, \mathbb{F}_p)$$

and we regard $H^*(G, \mathbb{F}_p)$ as embedded in $H^*(G, k)$ via this isomorphism. If M is a kG-module, then we have a natural map

$$H^*(G, k) \cong \mathrm{Ext}^*_{kG}(k, k) \to \mathrm{Ext}^*_{kG}(M, M)$$

given by tensoring exact sequences with M. If $\zeta \in H^*(G, k)$, we write ζ_M for the image of ζ under this map. Tensoring the above sequence $(*)$ with M, we obtain a sequence

$$\beta(x)_M : \qquad 0 \to M \to M \downarrow_H \uparrow^G \to M \downarrow_H \uparrow^G \to M \to 0.$$

Finally, we remark that the above situation is symmetric. Namely, letting M' be the cokernel of η, or equivalently the kernel of η' (namely $\Omega_{G/H}(k) \otimes M = \Omega^{-1}_{G/H}(k) \otimes M$), we have an exact sequence

$$0 \to M' \to M \downarrow_H \uparrow^G \to M \downarrow_H \uparrow^G \to M' \to 0.$$

Since this sequence is obtained by tensoring M with the sequence

$$0 \to \Omega_{G/H}(k) \to k(G/H) \to k(G/H) \to \Omega_{G/H}(k) \to 0,$$

which represents $\beta(x)_{\Omega_{G/H}(k)}$, it is easy to see that the above sequence represents the element

$$\beta(x)_{M'} \in \mathrm{Ext}^2_{kG}(M', M').$$

Over a field k containing $\mathbb{F}_p$, we could extend the Bockstein map to a linear map $\beta : H^n(G, k) \to H^{n+1}(G, k)$, but this turns out not to be a good thing to do. Instead, we extend semilinearly via the Frobenius map. Namely, any element of $H^n(G, k)$ can be written as $\sum_i \lambda_i x_i$ with $\lambda_i \in k$ and $x_i \in H^n(G, \mathbb{F}_p)$. We then set

$$\beta(\sum_i \lambda_i x_i) = \sum_i \lambda_i^p \beta(x_i).$$

We shall only use this for degree one elements, where it may be justified by viewing the Bockstein $\beta(x)$ as a p-fold Massey product $\langle x, x, \ldots, x \rangle$ (or just as the cup square for $p = 2$). In Section 5.8 we shall see another related justification: this version of the Bockstein map commutes with restriction to shifted subgroups of an elementary abelian p-group.[1]

THE BOCKSTEIN SPECTRAL SEQUENCE.   The Bockstein spectral sequence is the spectral sequence associated to the exact couple

$$H^*(G, \mathbb{Z}) \xrightarrow{\quad p \quad} H^*(G, \mathbb{Z})$$
$$\overset{\hat{\beta}}{\diagdown} \quad \overset{\pi_*}{\diagup}$$
$$H^*(G, \mathbb{F}_p)$$

The differential $d_1 = \pi_* \circ \hat{\beta} = \beta$ is the Bockstein homomorphism. The higher differentials in this spectral sequence are the "higher Bocksteins" $d_n = \beta_n$. The map β_n is defined on the image of $H^r(G, \mathbb{Z}/p^n) \to H^r(G, \mathbb{Z}/p)$, and is given as follows. For such an element, the image of the connecting map

$$\hat{\beta}_n : H^r(G, \mathbb{F}_p) \to H^{r+1}(G, \mathbb{Z}/p^n)$$

associated to the short exact sequence of coefficients

$$0 \to \mathbb{Z}/p^n \to \mathbb{Z}/p^{n+1} \to \mathbb{F}_p \to 0$$

lies in the image of the map

$$H^{r+1}(G, \mathbb{F}_p) \to H^{r+1}(G, \mathbb{Z}/p^n)$$

given by the inclusion of $\mathbb{F}_p$ as the elements of additive order p in $\mathbb{Z}/p^n$, and so we pull back to an element of $H^{r+1}(G, \mathbb{F}_p)$. It is only well defined modulo the image of β_{n-1}. Thus the stage at which an element vanishes in this spectral sequence tells us the order of the p-torsion associated to this element. In particular the spectral sequence converges to

$$[H^*(G, \mathbb{Z})/(p\text{-torsion})] \otimes_{\mathbb{Z}} \mathbb{F}_p.$$

Of course, for a finite group this is zero except in degree zero.

The Bockstein spectral sequence should therefore be regarded as a tool for comparing integral and mod p cohomology.

MULTIPLICATIVE PROPERTIES.

LEMMA 4.3.3. *The Bockstein of a cup product is given by*

$$\beta(xy) = \beta(x)y + (-1)^{\deg(x)} x\beta(y).$$

PROOF. Let $\mathbf{P}$ be a projective resolution of $\mathbb{Z}$ as a $\mathbb{Z}G$-module, and let $\Delta : \mathbf{P} \to \mathbf{P} \otimes \mathbf{P}$ be a diagonal approximation (see Section 3.2 of Volume I). Suppose $x, y \in H^*(G, \mathbb{F}_p)$ are represented by cocycles $\hat{x}, \hat{y} : \mathbf{P} \to \mathbb{F}_p$. Choose cochains $\hat{\xi}, \hat{\eta} : \mathbf{P} \to \mathbb{Z}/p^2$ whose reductions modulo p are $\hat{x}, \hat{y}$, so that

---

[1]It is very easy to make mistakes in the discussion of the Bockstein map over fields larger than $\mathbb{F}_p$. In the first edition of this book, the Frobenius twist was omitted, resulting in problems for Section 5.8.

$\delta\hat{\xi} = p.\widehat{\beta(x)}$ and $\delta\hat{\eta} = p.\widehat{\beta(y)}$. We chase $\widehat{xy} = (\hat{x} \otimes \hat{y}) \circ \Delta$ explicitly through the connecting homomorphism

$$\text{Hom}_{\mathbb{Z}G}(P_r, \mathbb{Z}/p^2) \longrightarrow \text{Hom}_{\mathbb{Z}G}(P_r, \mathbb{F}_p)$$
$$\downarrow \delta$$
$$\text{Hom}_{\mathbb{Z}G}(P_{r+1}, \mathbb{F}_p) \longrightarrow \text{Hom}_{\mathbb{Z}G}(P_{r+1}, \mathbb{Z}/p^2)$$

We have

$$\delta(\widehat{\xi\eta}) = \delta(\hat{\xi} \otimes \hat{\eta}) \circ \Delta = (\delta\hat{\xi} \otimes \hat{\eta} + (-1)^{\deg(x)}\hat{\xi} \otimes \delta\hat{\eta}) \circ \Delta$$
$$= (p\widehat{\beta(x)} \otimes \hat{\eta} + (-1)^{\deg(x)}\hat{\xi} \otimes p\widehat{\beta(y)}) \circ \Delta$$
$$= (\widehat{\beta(x)} \otimes p\hat{\eta} + (-1)^{\deg(x)}p\hat{\xi} \otimes \widehat{\beta(y)}) \circ \Delta$$
$$= p((\widehat{\beta(x)y}) + (-1)^{\deg(x)}(\widehat{x\beta(y)}))$$

and hence

$$\beta(xy) = \beta(x)y + (-1)^{\deg(x)}x\beta(y). \qquad \square$$

What we would like to say next is that the Bockstein of a norm is a transfer. However, there is a slight technical difficulty here for elements in odd degree. Namely, the norm map goes from $H^r(H, \mathbb{F}_p)$ to $H^{nr}(G, \mathbb{F}_p^{(r)})$, where $\mathbb{F}_p^{(r)}$ is $\mathbb{F}_p$ for r even and the sign representation for the action of G on the cosets of H for r odd. So what we really need to consider is the *twisted Bockstein homomorphism*

$$\tilde{\beta} : H^q(G, \mathbb{F}_p^{(r)}) \to H^{q+1}(G, \mathbb{F}_p^{(r)})$$

associated to the exact sequence of coefficients

$$0 \to \mathbb{F}_p^{(r)} \to (\mathbb{Z}/p^2)^{(r)} \to \mathbb{F}_p^{(r)} \to 0.$$

Note that in case $p = 2$ this is a map on cohomology with trivial coefficients, which does not necessarily agree with the usual Bockstein.

LEMMA 4.3.4. *Suppose $x \in H^r(H, \mathbb{F}_p)$. Then the twisted Bockstein*

$$\tilde{\beta}(1 \wr x) \in H^{nr+1}(\Sigma_n \wr H, \mathbb{F}_p^{(r)})$$

is given by the formula

$$\tilde{\beta}(1 \wr x) = \text{Tr}_{H \times (\Sigma_{n-1} \wr H), \Sigma_n \wr H}(\beta(x) \otimes (1 \wr x)).$$

PROOF. Let **P** be a projective resolution of $\mathbb{Z}$ as a $\mathbb{Z}G$-module. Then x is represented by a cocycle $\hat{x} : P_r \to \mathbb{F}_p$. Choose a cochain $\hat{\xi} : P_r \to \mathbb{Z}/p^2$ whose reduction modulo p is $\hat{x}$. Then $1 \wr x$ is represented by the cocycle

$$1 \wr \hat{x} : (P^{\otimes n})_{nr} \to (P_r)^{\otimes n} \to \mathbb{F}_p$$

which lifts to

$$1 \wr \hat{\xi} : (P^{\otimes n})_{nr} \to \mathbb{Z}/p^2.$$

We have

$$\delta(1 \wr \hat{\xi}) = \sum_{j=0}^{n-1}(-1)^{jr}\hat{\xi}^{\otimes j} \otimes \delta\hat{\xi} \otimes \hat{\xi}^{\otimes(n-j-1)}$$

$$= p.\sum_{j=0}^{n-1}(-1)^{jr}\hat{\xi}^{\otimes j} \otimes \widehat{\beta(x)} \otimes \hat{\xi}^{\otimes(n-j-1)}.$$

This expression differs from

$$p.\mathrm{Tr}_{H\times\Sigma_{n-1}\wr H, \Sigma_n\wr H}(\widehat{\beta(x)} \otimes \hat{\xi}^{\otimes(n-1)})$$

by a coboundary. This completes the proof of the formula.   □

PROPOSITION 4.3.5. *If $|G:H| = n$ and $x \in H^r(H, \mathbb{F}_p)$ then*

$$\tilde{\beta}\,\mathrm{norm}_{H,G}(x) \in H^{nr+1}(G, \mathbb{F}_p^{(r)})$$

is a transfer from H.

PROOF. This follows from the lemma, since by the Mackey formula, the restriction to G of a transfer from $H \times \Sigma_{n-1} \wr H$ to $\Sigma_n \wr H$ is a transfer from $G \cap (H \times \Sigma_{n-1} \wr H) = H$ to G.   □

4.4. Steenrod operations

In this section we investigate the operations Sq^i ($p = 2$) and P^i (p odd) in group cohomology. This is a special case of a more general construction on cohomology of topological spaces. In a sense, the Kan–Thurston theorem [146] says that it is no loss of generality only to consider group cohomology, but the price we have to pay for our lack of generality is that some of the proofs are harder, in particular the proofs that the zeroth Steenrod operation is the identity and that the negative Steenrod operations are zero. The standard reference for Steenrod operations in topology is Steenrod and Epstein [253].

The idea of the construction is as follows. As indicated in Section 3.2 of Volume I, the diagonal approximation on a resolution may be chosen to be strictly co-associative, but only cocommutative up to homotopy. This lack of strict cocommutativity means that when we norm a restriction we do not end up with a power of what we started with (while the transfer of a restriction *is* a multiple of what we started with). Thus the norm from G to $\mathbb{Z}/p \times G$ should be an interesting operation, even though the transfer is not.

Now by the Künneth Theorem (3.5.6 of Volume I),

$$H^r(\mathbb{Z}/p \times G, \mathbb{F}_p) = \bigoplus_j H^j(\mathbb{Z}/p, \mathbb{F}_p) \otimes H^{r-j}(G, \mathbb{F}_p),$$

and from Section 3.5 of Volume I, we know that each $H^j(\mathbb{Z}/p, \mathbb{F}_p)$ is one dimensional over $\mathbb{F}_p$. We choose a basis vector $a_j \in H^j(\mathbb{Z}/p, \mathbb{F}_p)$ in the following manner. The element a_1 is chosen arbitrarily, we set $a_2 = -\beta(a_1)$, $a_{2j} = (a_2)^j$ and $a_{2j+1} = a_1(a_2)^j$, so that $\beta(a_{2j-1}) = -a_{2j}$ and $\beta(a_{2j}) = 0$.

DEFINITION 4.4.1. *We define operations*

$$D_j : H^r(G, \mathbb{F}_p) \to H^{pr-j}(G, \mathbb{F}_p)$$

as follows. If $x \in H^r(G, \mathbb{F}_p)$, *set*

$$\mathrm{norm}_{G,\mathbb{Z}/p \times G}(x) = \sum_j a_j \otimes D_j(x)$$

$$\in H^{pr}(\mathbb{Z}/p \times G) = \bigoplus_j H^j(\mathbb{Z}/p, \mathbb{F}_p) \otimes H^{pr-j}(G, \mathbb{F}_p).$$

REMARK. Since the transfer from G to $\mathbb{Z}/p \times G$ is zero (every element is a restriction, so its transfer is given by multiplying by p, by Proposition 3.6.17 of Volume I) it follows from Proposition 4.1.2 (iii) that the norm from G to $\mathbb{Z}/p \times G$ is additive. So the D_j are also additive maps.

Note that for r odd, the exact sign of $D_j(x)$ depends on the order we choose for the cosets of G in $\mathbb{Z}/p \times G$. We have a particular choice in mind, which we discuss at the end of Section 4.5.

DEFINITION 4.4.2. *If* $p = 2$ *and* $x \in H^r(G, \mathbb{F}_2)$, *we define*

$$Sq^i(x) = D_{r-i}(x) \in H^{r+i}(G, \mathbb{F}_2).$$

For p odd, it turns out that most of the D_j are either the zero map, or obtainable from others by composing with the Bockstein homomorphism, as will be seen in the next section. We therefore make the following definition:

DEFINITION 4.4.3. *If* p *is odd and* $x \in H^r(G, \mathbb{F}_p)$, *we define*

$$P^i(x) = (-1)^{i+mr(r+1)/2}(m!)^{-r} D_{(p-1)(r-2i)}(x) \in H^{r+2(p-1)i}(G, \mathbb{F}_p)$$

where $m = (p-1)/2$.

REMARK. The choice of the constant multiplier of $(-1)^{i+mr(r+1)/2}(m!)^{-r}$ ensures that $P^0 = 1$, as we shall see in the next section.

Note that $m!$ is a fourth root of unity modulo p. Namely, by Wilson's theorem,

$$(m!).(-1)^m(m!) \equiv (p-1)! \equiv -1 \pmod{p}$$

so that

$$(m!)^2 \equiv (-1)^{m+1} \equiv -\left(\frac{-1}{p}\right) \pmod{p}$$

(the expression on the right hand side is the Legendre symbol).

The principal properties of the Steenrod operations are listed in the following two theorems, whose proofs will occupy the next two sections.

THEOREM 4.4.4 ($p = 2$). *The Steenrod operations*

$$Sq^i : H^r(G, \mathbb{F}_2) \to H^{r+i}(G, \mathbb{F}_2)$$

satisfy the following axioms:

(i) Sq^i is an additive map and a natural transformation of functors on the category of groups.

(ii) $Sq^0 = 1$.

(iii) If $\deg(x) = r$ then $Sq^r(x) = x^2$.

(iv) If $\deg(x) < r$ then $Sq^r(x) = 0$.

(v) (Cartan formula)

$$Sq^k(xy) = \sum_{i+j=k} Sq^i(x)Sq^j(y).$$

(vi) $Sq^1 = \beta : H^r(G, \mathbb{F}_2) \to H^{r+1}(G, \mathbb{F}_2)$.

(vii) (Adem relations) If $0 < a < 2b$ then

$$Sq^a Sq^b = \sum_{j=0}^{[a/2]} \binom{b-1-j}{a-2j} Sq^{a+b-j} Sq^j.$$

THEOREM 4.4.5 (p odd). The Steenrod operations

$$P^i : H^r(G, \mathbb{F}_p) \to H^{r+2(p-1)i}(G, \mathbb{F}_p)$$

satisfy the following axioms:

(i) P^i is an additive map and a natural transformation of functors on the category of groups.

(ii) $P^0 = 1$.

(iii) If $\deg(x) = 2r$ then $P^r(x) = x^p$.

(iv) If $\deg(x) < 2r$ then $P^r(x) = 0$.

(v) (Cartan formula)

$$P^k(xy) = \sum_{i+j=k} P^i(x)P^j(y).$$

(vi) (Adem relations) If $a < pb$ then

$$P^a P^b = \sum_{j=0}^{[a/p]} (-1)^{a+j} \binom{(p-1)(b-j)-1}{a-pj} P^{a+b-j} P^j.$$

If $a \le b$ then

$$P^a \beta P^b = \sum_{j=0}^{[a/p]} (-1)^{a+j} \binom{(p-1)(b-j)}{a-pj} \beta P^{a+b-j} P^j$$

$$+ \sum_{j=0}^{[(a-1)/p]} (-1)^{a+j-1} \binom{(p-1)(b-j)-1}{a-pj-1} P^{a+b-j} \beta P^j.$$

4.5. Proof of the properties

In this section we shall prove most of Theorems 4.4.4 and 4.4.5 in parallel. The main differences between the two cases come from the difference in the multiplicative structure of cohomology of cyclic groups (Proposition 3.5.5 of

Volume I) and symmetric groups of degree p (example at end of Section 3.7 of Volume I).

Our approach is essentially the same as in Steenrod and Epstein [**253**], with minor modifications. For the benefit of the reader who wishes to compare, we shall indicate the parallel results in [**253**].

Property (i) of the Steenrod operations follows from the remark following the definition of the maps D_j, together with the following lemma.

LEMMA 4.5.1 (cf. [**253**] VII.3.3). *The maps D_j, and hence also Sq^i and P^i, are natural transformations of functors on the category of groups.*

PROOF. If $\phi : G \to G'$ is a group homomorphism, then the following diagram commutes.

$$
\begin{array}{ccccc}
H^*(G',\mathbb{F}_p) & \xrightarrow{\;\wr-\;} & H^*(\Sigma_p \wr G',\mathbb{F}_p) & \xrightarrow{\;\text{res}\;} & H^*(\mathbb{Z}/p \times G',\mathbb{F}_p) \\
\downarrow{\scriptstyle \phi^*} & & \downarrow{\scriptstyle 1\wr\phi} & & \downarrow{\scriptstyle (\mathrm{id}\times\phi)^*} \\
H^*(G,\mathbb{F}_p) & \xrightarrow{\;\wr-\;} & H^*(\Sigma_p \wr G,\mathbb{F}_p) & \xrightarrow{\;\text{res}\;} & H^*(\mathbb{Z}/p \times G,\mathbb{F}_p)
\end{array}
$$

The composite of each of the horizontal rows is the norm map.   □

Property (iv) of the Steenrod operations follows from the fact that $D_j = 0$ for $j < 0$, while property (iii) follows from the next lemma.

LEMMA 4.5.2 (cf. [**253**] VII.3.4). $D_0(x) = x^p$.

PROOF. We have

$$
D_0(x) = \mathrm{res}_{\mathbb{Z}/p\times G,G}\mathrm{norm}_{G,\mathbb{Z}/p\times G}(x) = x^p
$$

by the Mackey Formula 4.1.2 (v).   □

For p odd, we appear not to have used all the D_j in our definition of the Steenrod operations. However, the following two lemmas show that the only non-zero maps D_j are multiples of the P^i or βP^i for some i.

LEMMA 4.5.3 (cf. [**253**] VII.3.5). *For p odd and $x \in H^r(G,\mathbb{F}_p)$,*
(i) *if r is even then $D_j(x) = 0$ unless $j = 2k(p-1)$ or $j = 2k(p-1) - 1$ for some k; and*
(ii) *if r is odd then $D_j(x) = 0$ unless $j = (2k+1)(p-1)$ or $j = (2k+1)(p-1) - 1$ for some k.*

PROOF. The map

$$
\mathrm{norm}_{G,\mathbb{Z}/p\times G} : H^r(G,\mathbb{F}_p) \to H^{pr}(\mathbb{Z}/p \times G,\mathbb{F}_p)
$$

may be factored as follows

$$H^r(G, \mathbb{F}_p) \qquad\qquad x$$

$$\downarrow \qquad\qquad\qquad \downarrow$$

$$H^{pr}(\Sigma_p \wr G, \mathbb{F}_p^{(r)}) \qquad\qquad 1 \wr x$$

$$\downarrow \text{res}$$

$$H^{pr}(\Sigma_p \times G, \mathbb{F}_p^{(r)}) \xrightarrow{\qquad\text{res}\qquad} H^{pr}(\mathbb{Z}/p \times G, \mathbb{F}_p)$$

$$\| \qquad\qquad\qquad\qquad\qquad\qquad \|$$

$$\bigoplus_j H^j(\Sigma_p, \mathbb{F}_p^{(r)}) \otimes H^{pr-j}(G, \mathbb{F}_p) \longrightarrow \bigoplus_j H^j(\mathbb{Z}/p, \mathbb{F}_p) \otimes H^{pr-j}(G, \mathbb{F}_p).$$

Now by the example at the end of Section 3.6 of Volume I, the map

$$\text{res}_{\Sigma_p, \mathbb{Z}/p} : H^j(\Sigma_p, \mathbb{F}_p^{(r)}) \to H^j(\mathbb{Z}/p, \mathbb{F}_p)$$

is injective, and its image has a basis consisting of the a_j where
(i) $j = 2k(p-1)$ or $2k(p-1) - 1$ for some k if r is even, and
(ii) $j = (2k+1)(p-1)$ or $(2k+1)(p-1) - 1$ for some k if r is odd.
Thus $D_j(x) = 0$ unless j has one of these values. □

The following lemma shows that for $p = 2$, $Sq^{2i+1} = \beta Sq^{2i}$, so that in particular property (vi) follows from property (ii).

LEMMA 4.5.4 (cf. [253] VII.4.6 and errata). *We have* $\beta \circ D_0 = 0$, $\beta \circ D_{2j} = D_{2j-1}$, *and* $\beta \circ D_{2j-1} = 0$.

PROOF. We apply Lemma 4.3.4, and notice that
(i) when $p = 2$, $\tilde{\beta}$ and β agree on $H^*(\mathbb{Z}/2, \mathbb{F}_2)$ and hence on $H^*(\mathbb{Z}/2 \times G, \mathbb{F}_2)$, and
(ii) when p is odd, $\mathbb{Z}/p$ consists of even permutations, so that $\tilde{\beta} = \beta$ on $H^*(\mathbb{Z}/p \times G, \mathbb{F}_p)$.

Since the transfer map from G to $\mathbb{Z}/p \times G$ is zero, it now follows that

$$\beta \, \text{norm}_{G, \mathbb{Z}/p \times G}(x) = 0.$$

Since by definition

$$\text{norm}_{G, \mathbb{Z}/p \times G}(x) = \sum_j a_j \otimes D_j(x),$$

by Lemma 4.3.3 we have

$$\sum_j ((-1)^j a_j \otimes \beta D_j(x) + \beta(a_j) \otimes D_j(x)) = 0.$$

The result now follows by comparing coefficients of a_j, since $\beta(a_{2j-1}) = -a_{2j}$ and $\beta(a_{2j}) = 0$. □

The Cartan formula (property (v) of the Steenrod operations) follows from the next lemma, together with Lemma 4.5.3 in case p is odd. It is an easy exercise to show that the signs work out right.

LEMMA 4.5.5 (cf. [**253**] VII.4.7).
(i) *If $p = 2$ then*

$$D_k(xy) = \sum_{i+j=k} D_i(x)D_j(y).$$

(ii) *If p is odd then*

$$D_{2k}(xy) = (-1)^{\frac{p(p-1)}{2}\deg(x)\deg(y)} \sum_{i+j=k} D_{2i}(x)D_{2j}(y).$$

PROOF. By the multiplicativity of norm (Proposition 4.1.2 (i)) we have

$$\sum_k a_k \otimes D_k(xy) = \mathrm{norm}_{G,\mathbb{Z}/p \times G}(xy)$$

$$= (-1)^{\frac{p(p-1)}{2}\deg(x)\deg(y)} \mathrm{norm}_{G,\mathbb{Z}/p \times G}(x)\mathrm{norm}_{G,\mathbb{Z}/p \times G}(y)$$

$$= (-1)^{\frac{p(p-1)}{2}\deg(x)\deg(y)} (\sum_i a_i \otimes D_i(x))(\sum_j a_j \otimes D_j(y)).$$

(i) If $p = 2$ then $a_i a_j = a_{i+j}$ and the formula follows by comparing coefficients of a_k.
(ii) If p is odd then $a_i a_j = a_{i+j}$ when either i or j is even, while $a_i a_j = 0$ if i and j are odd. So the formula follows by comparing coefficients of a_{2k}. □

It is somewhat more subtle to show that Sq^0 (resp. P^0) is the identity operation, and that $Sq^n = 0$ (resp. $P^n = 0$) for n negative. This is really because Steenrod operations are defined more generally than for group algebras, for example they are defined for cocommutative Hopf algebras, Lie algebras, and so on, and in general these properties do not hold. So we need to make particular use of the fact that we are working with groups. There are (at least) two possible approaches here. One approach is to use the fact that group cohomology may be defined as cohomology of Eilenberg–Mac Lane spaces, and Steenrod operations are defined for all topological spaces. The required properties are true in this context, as is explained at the end of Section 2.2. For further details see Steenrod and Epstein [**253**]. The approach we shall use is to construct specific chain maps for the standard resolution, and use these to check the required properties.

We recall that if $\mathbf{T}$ is a projective resolution of $\mathbb{F}_p$ as an $\mathbb{F}_p(\mathbb{Z}/p)$-modules, and $\mathbf{C}(G)$ denotes the standard resolution of $\mathbb{F}_p$ as an $\mathbb{F}_pG$-module (see Section 3.4 of Volume I) then the norm map $\mathrm{norm}_{G,\mathbb{Z}/p \times G}$ may be computed using a map of complexes of $\mathbb{Z}/p \times G$-modules

$$\psi : \mathbf{T} \otimes \mathbf{C}(G) \to \mathbf{C}(G)^{\otimes p}$$

lifting the identity map on $\mathbb{F}_p$. Namely if $\hat{\zeta} : C_r(G) \to \mathbb{F}_p$ is an r-cocycle representing $\zeta \in H^r(G, \mathbb{F}_p)$ then the composite of ψ with $1 \wr \hat{\zeta} : (\mathbf{C}(G)^{\otimes p})_{pr} \to \mathbb{F}_p^{(r)}$ is a pr-cocycle on $\mathbb{Z}/p \times G$ representing $\mathrm{norm}_{G,\mathbb{Z}/p \times G}(\zeta)$. We shall construct an explicit map ψ and use it to calculate Sq^0 and Sq^{-n} (resp. P^0 and P^{-n}).

We first deal with the case $p = 2$, because this is easier than the odd case. In this case we take $\mathbf{T}$ to be the resolution given in Section 3.5 of Volume I consisting of one copy of $\mathbb{F}_2(\mathbb{Z}/2)$ in each degree j, with generator α_j. We then take the map ψ given by

$$\psi(\alpha_n \otimes (x_0, \dots, x_r)) = \sum_{0 \le i_1 < \cdots < i_{n+1} \le r} (x_0, x_1, \dots, x_{i_1}, x_{i_2}, x_{i_2+1}, \dots,$$

$$x_{i_3}, x_{i_4}, \dots, x_r) \otimes (x_{i_1}, x_{i_1+1}, \dots, x_{i_2}, x_{i_3}, \dots, x_{i_4}, x_{i_5}, \dots, x_{i_{n+1}})$$

if n is even, and

$$\psi(\alpha_n \otimes (x_0, \dots, x_r)) = \sum_{0 \le i_1 < \cdots < i_{n+1} \le r} (x_0, x_1, \dots, x_{i_1}, x_{i_2}, x_{i_2+1}, \dots,$$

$$x_{i_3}, x_{i_4}, \dots, x_{i_{n+1}}) \otimes (x_{i_1}, x_{i_1+1}, \dots, x_{i_2}, x_{i_3}, \dots, x_{i_4}, x_{i_5}, \dots, x_r)$$

if n is odd. Note that for $n = 0$ this is the Alexander–Whitney map, which confirms that for $\deg(\zeta) = r$ we have $Sq^r(\zeta) = \zeta^2$. For $n = r$ we have

$$\psi(\alpha_r \otimes (x_0, \dots, x_r)) = (x_0, \dots, x_r) \otimes (x_0, \dots, x_r)$$

so that $Sq^0(\zeta) = \zeta$. For $n > r$ we have

$$\psi(\alpha_n \otimes (x_0, \dots, x_r)) = 0$$

so that the Sq^i with i negative give the zero operation.

For general p we must make a more complicated construction which generalises the above construction for $p = 2$. We construct an exact sequence $\mathbf{X}$ of free modules for $\mathbb{F}_p(\mathbb{Z}/p)$ as follows. In degree n, the basis elements are sequences $(\lambda_1, \dots, \lambda_{p+n})$ such that each $\lambda_i \in \{1, \dots, p\}$, $\lambda_i \ne \lambda_{i+1}$, and each $j \in \{1, \dots, p\}$ appears as some λ_i. The generator of $\mathbb{Z}/p$ acts via

$$(\lambda_1, \dots, \lambda_{p+n}) \mapsto (\lambda_1 + 1, \dots, \lambda_{p+n} + 1),$$

where the addition is taken modulo p. The differential in this complex is given by

$$d(\lambda_1, \dots, \lambda_{p+n}) = \sum_{i=1}^{p+n} (-1)^i (\lambda_1, \dots, \lambda_{i-1}, \lambda_{i+1}, \dots, \lambda_{p+n}),$$

where such a sequence is to be thought of as the zero element if it does not satisfy the required properties. The map

$$h(\lambda_1, \dots, \lambda_{p+n}) = \begin{cases} (1, \lambda_1, \dots, \lambda_{p+n}) & \text{if } \lambda_1 \ne 1 \\ 0 & \text{if } \lambda_1 = 1 \end{cases}$$

satisfies $d \circ h + h \circ d = 1$, so that it is a homotopy from the identity map to the zero map (see Section 2.3 of Volume I). Thus $\mathbf{X}$ is an exact sequence of free $\mathbb{F}_p(\mathbb{Z}/p)$-modules. Its homology in degree one is $(p-1)!$ copies of the trivial module.

Let $\mathbf{T}$ be the resolution of $\mathbb{F}_p$ as an $\mathbb{F}_p(\mathbb{Z}/p)$-module described in Section 3.5 of Volume I, consisting of one copy of $\mathbb{F}_p(\mathbb{Z}/p)$ in each degree j. Then the map $T_0 \to X_0$ sending the generator to the sequence $(1, \dots, p)$

extends to a chain map $\phi : \mathbf{T} \to \mathbf{X}$, unique up to chain homotopy. So for example ϕ can be chosen to take the generator in degree one to $(1, \ldots, p, 1)$, but in higher degrees it is rather hard to write down an explicit chain map, except in the case $p = 2$. In this case $\mathbf{X}$ has two basis elements in each degree, and ϕ is an isomorphism.

We now construct a canonical map

$$\xi : \mathbf{X} \otimes \mathbf{C}(G) \to \mathbf{C}(G)^{\otimes p}$$

$$(\lambda_1, \ldots, \lambda_{n+p}) \otimes (x_0, \ldots, x_r) \mapsto \sum_{0=i_0 \leq i_1 \leq \cdots \leq i_{n+p}=r} (-1)^\sigma u_1 \otimes \cdots \otimes u_p$$

where u_j is defined as follows. If $\lambda_{k_1}, \lambda_{k_2}, \ldots$ are the λ_k which are equal to j, then

$$u_j = (x_{i_{k_1}-1}, x_{i_{k_1}-1+1}, \ldots, x_{i_{k_1}}, x_{i_{k_2}-1}, \ldots, x_{i_{k_2}}, \ldots).$$

The sign σ is defined as follows. Each sequence $(x_{i_{k-1}}, \ldots, x_{i_k})$ represents an element in degree $i_k - i_{k-1}$. Take the permutation which takes these sequences making up $u_1 \otimes \cdots \otimes u_p$ and puts them in their natural order, express it as a product of transpositions, and for each transposition multiply by minus one to the power of the product of the degrees.

The above sum should only be taken over those sequences satisfying $i_{k_j-1} < i_{k_j-1}$, which is the same as saying that each u_j has no repeated index.

Thus for example if $p = 3$, $n = 2$ and $(\lambda_1, \ldots, \lambda_{n+p}) = (3, 1, 3, 1, 2)$ then

$$d(3, 1, 3, 1, 2) = (1, 3, 1, 2) - (3, 1, 3, 2).$$

If $r = 2$, the only permitted sequence of i's is $i_0 = 0$, $i_1 = 0$, $i_2 = 1$, $i_3 = 2$, $i_4 = 2$, $i_5 = 2$. The corresponding term is $(x_0, x_1, x_2) \otimes (x_2) \otimes (x_0, x_1, x_2)$. The corresponding permutation sends (x_0, x_1), (x_2), (x_2), (x_0), (x_1, x_2) to (x_0), (x_0, x_1), (x_1, x_2), (x_2), (x_2), so no pair of elements of odd degree has been swapped and the sign is plus.

A straightforward but tedious combinatorial argument shows that ξ is a chain map of modules for $\mathbb{Z}/p \times G$ (even over the integers), and hence the map

$$\mathbf{T} \otimes \mathbf{C}(G) \to \mathbf{C}(G)^{\otimes p}$$

$$t \otimes x \mapsto \xi(\phi(t) \otimes x)$$

is a chain map of modules for $\mathbb{Z}/p \times G$ which may be used to compute the norm. Again for $n = 0$ this is the iterated Alexander–Whitney map (because of the choice of ϕ_0), which confirms that for $\deg(\zeta) = r$ we have $P^r(\zeta) = \zeta^p$.

If $r > pr - n$ then there are no available sequences of i's and so

$$\xi((\lambda_1, \ldots, \lambda_{n+p}) \otimes (x_0, \ldots, x_r)) = 0.$$

It follows that $D_n(\zeta) = 0$ for an element ζ of degree r with $n > (p-1)r$, and so $P^i = 0$ for i negative.

Similarly if $r = pr - n$, every possibility for the i's gives $(x_0, \ldots, x_r)^{\otimes p}$, and so for every sequence of λ_i's, $\xi((\lambda_1, \ldots, \lambda_{n+p}) \otimes (x_0, \ldots, x_r))$ is either zero or $\pm(x_0, \ldots, x_r)^{\otimes p}$. It follows that for some constant θ_r depending only on p and r, we have

$$D_{(p-1)r}(\zeta) = \theta_r.\zeta$$

for ζ an element of degree r, so that P^0 is some multiple of the identity map.

To evaluate θ_r, it suffices to work with any particular group G. If we choose for G a cyclic group of order p, then the calculation at the end of Section 4.1 shows that $\theta_{2r} = (-1)^r$, while $\theta_{2r+1} = (-1)^r \theta_1$. Since for an element of odd degree the exact sign of the norm depends on the order chosen for the coset representatives of G in $\mathbb{Z}/p \times G$, we are free to choose the sign of θ_1 as we wish, once and for all. We calculate θ_1 up to sign as follows. Choose for G an elementary abelian group of order p^2, so that there are two elements of cohomology in degree one whose product is non-zero. Applying the formula for the norm of a product to these elements, we see that $(-1)^{p(p-1)/2}\theta_1^2 = \theta_2 = -1$. Since $(\frac{p-1}{2})!^2 = -(-1)^{(p-1)/2}$ we may take $\theta_1 = (\frac{p-1}{2})!$.

With these values of θ_r it is now easy to check that P^0 is the identity map. It only remains to prove the Adem relations, and this we do in the next section.

4.6. Adem relations

The Adem relations essentially correspond to the fact that the following square commutes.

$$
\begin{array}{ccc}
H^*(G, \mathbb{F}_p) & \xrightarrow{\quad \text{norm} \quad} & H^*(1 \times \mathbb{Z}/p \times G, \mathbb{F}_p) \\
\downarrow{\scriptstyle \text{norm}} & & \downarrow{\scriptstyle \text{norm}} \\
H^*(\mathbb{Z}/p \times 1 \times G, \mathbb{F}_p) & \xrightarrow{\quad \text{norm} \quad} & H^*(\mathbb{Z}/p \times \mathbb{Z}/p \times G, \mathbb{F}_p)
\end{array}
$$

So we first prove a symmetric form of the Adem relations, due to Bullett and Macdonald [53], which goes as follows. For $p = 2$ we denote by Sq_t the formal power series $\sum_{i \geq 0} t^i Sq^i$, where t is an indeterminate. For p odd we denote by $P(t)$ the formal power series $\sum_{i \geq 0} t^i P^i$.

THEOREM 4.6.1. (i) *For $p = 2$ we have*

$$Sq_{t^2+st}Sq_{s^2} = Sq_{s^2+st}Sq_{t^2}.$$

(ii) *For p odd the formal power series*

$$P(t^p + st^{p-1} + \cdots + s^{p-1}t).P(s^p)$$

and

$$s.[\beta, P(t^p + st^{p-1} + \cdots + s^{p-1}t)].P(s^p)$$

are symmetric in s and t, where $[\beta, P] = \beta P - P\beta$.

We shall treat the cases $p = 2$ and p odd separately. Although most of the proof is exactly parallel, it helps to understand the case $p = 2$ first because there are fewer technical complications.

CASE (I) : $p = 2$. Let $H^*(\mathbb{Z}/2 \times \mathbb{Z}/2, \mathbb{F}_2) = \mathbb{F}_2[x_1, x_2]$ with $\deg(x_1) = \deg(x_2) = 1$. If $\zeta \in H^r(G, \mathbb{F}_2)$ then

$$\mathrm{norm}_{1 \times \mathbb{Z}/2 \times G, \mathbb{Z}/2 \times \mathbb{Z}/2 \times G} \mathrm{norm}_{G, 1 \times \mathbb{Z}/2 \times G}(\zeta) = \mathrm{norm}(\sum_k x_2^{r-k} Sq^k(\zeta))$$

$$= \sum_k \mathrm{norm}(x_2)^{r-k} \mathrm{norm}\, Sq^k(\zeta) = \sum_{j,k}(x_1 x_2 + x_2^2)^{r-k} x_1^{r+k-j} Sq^j Sq^k(\zeta)$$

$$= x_1^r x_2^r (x_1 + x_2)^r \sum_{j,k}(x_2 + x_1^{-1} x_2^2)^{-k} x_1^{-j} Sq^j Sq^k(\zeta)$$

$$= x_1^r x_2^r (x_1 + x_2)^r Sq_{x_1^{-1}} Sq_{(x_2 + x_1^{-1} x_2^2)^{-1}}(\zeta).$$

We make the substitution $x_1^{-1} = t(s + t)$, $x_2^{-1} = s(s + t)$ so that $(x_2 + x_1^{-1} x_2^2)^{-1} = s^2$ to deduce that $Sq_{t^2 + st} Sq_{s^2}$ is symmetric in s and t. This proves part (i) of Theorem 4.6.1.

We now put $s = 1$ (no information is lost by this substitution since the relation is homogeneous) to obtain

$$Sq_{t^2 + t} Sq_1 = Sq_{t+1} Sq_{t^2}.$$

Separating out terms which increase cohomological degree by n, we have

$$\sum_{a+b=n}(t^2 + t)^a Sq^a Sq^b = \sum_j (t+1)^{n-j} t^{2j} Sq^{n-j} Sq^j.$$

Thus $Sq^a Sq^b$ is equal to the residue at $t^2 + t = 0$ (or in particular at $t = 0$) of

$$\left(\sum_j (t+1)^{a+b-j} t^{2j} Sq^{a+b-j} Sq^j \, d(t^2 + t) \right) / (t^2 + t)^{a+1}$$

$$= \sum_j (t+1)^{b-j-1} t^{2j-a-1} Sq^{a+b-j} Sq^j \, dt,$$

namely the coefficient of t^{-1} in

$$\sum_j (t+1)^{b-j-1} t^{2j-a-1} Sq^{a+b-j} Sq^j,$$

which is

$$\sum_j \binom{b-1-j}{a-2j} Sq^{a+b-j} Sq^j.$$

This completes the proof of the Adem relations for $p = 2$.

CASE (II) : p ODD. Let $H^*(\mathbb{Z}/p \times \mathbb{Z}/p, \mathbb{F}_p)$ have generators y_1 and y_2 in degree one and $x_1 = -\beta(y_1)$, $x_2 = -\beta(y_2)$ in degree two. If $\zeta \in H^r(G, \mathbb{F}_p)$ then $P^i(\zeta) \in H^{r+2i(p-1)}(G, \mathbb{F}_p)$, and

$$\text{norm}_{G,1\times\mathbb{Z}/p\times G}(\zeta)$$
$$= (-1)^{mr(r+1)/2}(m!)^r \sum_i (-1)^i [x_2^{m(r-2i)} P^i(\zeta) + x_2^{m(r-2i)-1} y_2 \beta P^i(\zeta)]$$

where $m = (p-1)/2$ as usual. We have

$$\text{norm}_{1\times\mathbb{Z}/p,\mathbb{Z}/p\times\mathbb{Z}/p}(x_2) = x_2^p - x_1^{p-1}x_2$$
$$\text{norm}_{1\times\mathbb{Z}/p,\mathbb{Z}/p\times\mathbb{Z}/p}(y_2) = (m!)^{-1}x_1^{(p-3)/2}(x_1y_2 - y_1x_2)$$

(see the example at the end of Section 4.1 and the discussion of the constants θ_r at the end of the last section), and so

$$\text{norm}_{1\times\mathbb{Z}/p\times G,\mathbb{Z}/p\times\mathbb{Z}/p\times G}\text{norm}_{G,1\times\mathbb{Z}/p\times G}(\zeta) =$$
$$(-1)^{mr(r+1)/2}(m!)^r \sum_{i,j}(-1)^{i+j}[\text{norm}(x_2)^{m(r-2i)}(-1)^{mr(r+1)/2}(m!)^{r+2i(p-1)}$$
$$\{x_1^{m(r+2i(p-1)-2j)} P^j P^i(\zeta) + x_1^{m(r+2i(p-1)-2j)-1} y_1 \beta P^j P^i(\zeta)\} +$$
$$(-1)^{m(r+1)}\text{norm}(x_2)^{m(r-2i)-1}\text{norm}(y_2)(-1)^{m(r+1)(r+2)/2}(m!)^{r+2i(p-1)+1}$$
$$\{x_1^{m(r+2i(p-1)+1-2j)} P^j \beta P^i(\zeta) + x_1^{m(r+2i(p-1)+1-2j)-1} y_1 \beta P^j \beta P^i(\zeta)\}]$$

$$= (-1)^{(m+1)r} x_1^{mr} x_2^{mr} (x_2^{p-1} - x_1^{p-1})^{mr} \sum_{i,j}(-1)^{i+j}\{(x_2^p x_1^{1-p} - x_2)^{-2mi}$$
$$x_1^{-2mj} P^j P^i(\zeta) + (x_2^p x_1^{1-p} - x_2)^{-2mi} x_1^{-2mj-1} y_1 \beta P^j P^i(\zeta)$$
$$+ (x_2^p x_1^{1-p} - x_2)^{-2mi-1} x_1^{-2mj-1}(x_1y_2 - y_1x_2) P^j \beta P^i(\zeta)$$
$$+ (x_2^p x_1^{1-p} - x_2)^{-2mi-1} x_1^{-2mj-1} y_2 y_1 \beta P^j \beta P^i(\zeta)\}.$$

The terms in the above expression not involving y_1 or y_2 show that the expression

(a) $\sum_{i,j}(-1)^{i+j}(x_2^p x_1^{1-p} - x_2)^{-2mi} x_1^{-2mj} P^j P^i$

is symmetric in x_1 and x_2.

Similarly if we multiply the coefficient of y_1 by x_1^p, multiply the coefficient of y_2 by x_2^p and add, the resulting expression

(b) $-\sum_{i,j}(x_2^p x_1^{1-p} - x_2)^{-2mi} x_1^{-2m(j-1)}(\beta P^j P^i - P^j \beta P^i)$

is symmetric in x_1 and x_2.

We now make the substitution

$$x_1^{-1} = (-t)^{1/(p-1)}(s-t), \qquad x_2^{-1} = (-s)^{1/(p-1)}(t-s)$$

so that

$$x_1^{-2m} = -t(s-t)^{2m} = -(t^p + st^{p-1} + \cdots + s^{p-1}t)$$

and

$$(x_2^p x_1^{1-p} - x_2)^{-2m} = -s^p.$$

Making this substitution in (a) we see that

$$\sum_{i,j} s^{pi}(t^p + st^{p-1} + \cdots + s^{p-1}t)^j P^j P^i$$

is symmetric in s and t. Making the same substitution in (b) and then multiplying by $st(s^{p-1} + s^{p-2}t + \cdots + t^{p-1})$, we see that

$$\sum_{i,j} s^{pi+1}(t^p + st^{p-1} + \cdots + s^{p-1}t)^j [\beta, P^j] P^i$$

is symmetric in s and t. This completes the proof of part (ii) of Theorem 4.6.1.

We now put $s = 1$ and write u for $(t^{p-1} + t^{p-2} + \cdots + 1) = (1-t)^{p-1}$ to obtain

$$P(tu)P(1) = P(u)P(t^p)$$
$$[\beta, P(tu)]P(1) = t[\beta, P(u)]P(t^p),$$

so that

$$P(tu)\beta P(1) = ((1-t)\beta P(u) + tP(u)\beta)P(t^p).$$

Separating out terms increasing cohomological degree by $2n(p-1)$ in the first case and $2n(p-1)+1$ in the second, we have

$$\sum_{a+b=n} (tu)^a P^a P^b = \sum_j u^{n-j} t^{pj} P^{n-j} P^j$$
$$\sum_{a+b=n} (tu)^a P^a \beta P^b = \sum_j u^{n-j} t^{pj}((1-t)\beta P^{n-j}P^j + tP^{n-j}\beta P^j).$$

Thus $P^a P^b$ (resp. $P^a \beta P^b$) is equal to the residue at $t = 0$ of the right-hand side of the first (resp. second) equation multiplied by

$$\frac{d(tu)}{(tu)^{a+1}} = \frac{(1-t)^{p-2} dt}{(tu)^{a+1}}.$$

Hence $P^a P^b$ is equal to the coefficient of t^{-1} in

$$\sum_j (1-t)^{(p-1)(b-j)-1} t^{pj-a-1} P^{a+b-j} P^j$$

while $P^a \beta P^b$ is equal to the coefficient of t^{-1} in

$$\sum_j (1-t)^{(p-1)(b-j)-1} t^{pj-a-1}((1-t)\beta P^{a+b-j}P^j + tP^{a+b-j}\beta P^j).$$

This completes the proof of the Adem relations for p odd.

4.7. Serre's theorem on products of Bocksteins

In this section we shall prove a theorem of Serre, which states that if G is a p-group which is not elementary abelian, then there exist non-zero elements $x_1, \ldots, x_n \in H^1(G, \mathbb{F}_p)$, for some n, such that the product of the Bocksteins is zero

$$\beta(x_1).\beta(x_2).\ldots.\beta(x_n) = 0.$$

Of course, if G is elementary abelian, the Bocksteins of degree one elements form a polynomial subring of $H^*(G, \mathbb{F}_p)$ (Section 3.5 of Volume I) and so no such relation exists.

Serre's original proof [232] used Steenrod operations and Hilbert's Nullstellensatz, and in particular gave no bound on n. Later, Kroll [159] gave a proof using Chern classes of complex representations, which showed that it suffices to take each non-zero element of $H^1(G, \mathbb{F}_p)$ once only in the above relation. Serre, not to be outdone, then showed in [234] how to modify his original proof to show that it suffices to take one non-zero element from each one dimensional subspace of $H^1(G, \mathbb{F}_p)$. Okuyama and Sasaki [204] gave a proof which uses Evens' norm map instead of the Steenrod operations, and the proof we shall give is a modification by Evens of their proof, which gives a slight improvement on Serre's bound. This proof proceeds by induction, starting with the case of a group of order p^3.

LEMMA 4.7.1. *If $|G| = p^3$ and $|\Phi(G)| = p$ then there are non-zero elements $x_1, \ldots, x_{p+1}$ of $H^1(G, \mathbb{F}_p)$, such that*

$$\beta(x_1).\beta(x_2).\ldots.\beta(x_{p+1}) = 0.$$

PROOF. If G has a cyclic maximal subgroup H of order p^2, then there is a non-zero element $y \in H^1(H, \mathbb{F}_p)$ with $\beta(y) = 0$. The element y is the inflation of an element $\bar{y} \in H^1(H/\Phi(H), \mathbb{F}_p)$, and so by Proposition 4.1.2 (vi) and Proposition 4.1.4 we have

$$0 = \mathrm{norm}_{H,G}(\beta(y)) = \inf_{G/\Phi(G),G} \mathrm{norm}_{H/\Phi(H),G/\Phi(G)}(\beta(\bar{y}))$$

$$= \inf_{G/\Phi(G),G}(\beta(\bar{y}_2)^p - \beta(\bar{y}_1)^{p-1}\beta(\bar{y}_2)) = \beta(y_2)^p - \beta(y_1)^{p-1}\beta(y_2)$$

$$= \prod_{\lambda=0}^{p-1}(\beta(y_2 + \lambda y_1)).$$

Here, y_1 and y_2 are suitably chosen generators for $H^1(G, \mathbb{F}_p)$, which are inflations of generators $\bar{y}_1$ and $\bar{y}_2$ of $H^1(G/\Phi(G), \mathbb{F}_p)$.

If on the other hand G has no elements of order p^2, then G is generated by elements g and h of order p. Their commutator $z = [g, h]$ is central and of order p. Let H be the subgroup generated by h and z, so that H is elementary abelian of order p^2. Then $H^*(H, \mathbb{F}_p)$ has generators y_1 and y_2 in degree one, and $x_1 = \beta(y_1)$, $x_2 = \beta(y_2)$ in degree two, with $y_1^2 = y_2^2 = 0$ if p is odd, and $y_1^2 = x_1$, $y_2^2 = x_2$ if $p = 2$. We choose the notation so that conjugation by g

fixes x_2 and sends x_1 to $x_1 + x_2$. So by Proposition 4.1.2 we have

$$\mathrm{norm}_{H,G}(x_1) = \mathrm{norm}_{H,G}(x_1 + x_2)$$
$$= \mathrm{norm}_{H,G}(x_1) + \mathrm{norm}_{H,G}(x_2) - \mathrm{Tr}_{H,G}(\alpha)$$

for some $\alpha \in H^{2p}(H, \mathbb{F}_p)$, and hence

$$\mathrm{norm}_{H,G}(x_2) = \mathrm{Tr}_{H,G}(\alpha).$$

Choose generators $\bar{u}_1$, $\bar{u}_2$ in degree one and $\bar{v}_1 = \beta(\bar{u}_1)$, $\bar{v}_2 = \beta(\bar{u}_2)$ in degree two for $H^*(G/\Phi(G), \mathbb{F}_p)$, and write u_1, u_2, v_1, v_2 for their inflations in $H^*(G, \mathbb{F}_p)$. Choose the notation so that $\mathrm{res}_{G,H}(u_1) = 0$ and $\mathrm{res}_{G,H}(u_2) = y_2$. Then we have $\mathrm{res}_{G,H}(v_1) = 0$ and $\mathrm{res}_{G,H}(v_2) = x_2$, and so by Proposition 4.1.2 and Proposition 4.1.4 we have

$$0 = \mathrm{Tr}_{H,G}(\mathrm{res}_{G,H}(v_1).\alpha) = v_1.\mathrm{Tr}_{H,G}(\alpha) = v_1.\mathrm{norm}_{H,G}(x_2)$$

$$= \mathrm{inf}_{G/\Phi(G),G}(\bar{v}_1.\mathrm{norm}_{H/\Phi(G),G/\Phi(G)}\mathrm{res}_{G/\Phi(G),H/\Phi(G)}(\bar{v}_2))$$

$$= \mathrm{inf}_{G/\Phi(G),G}(\bar{v}_1.(\bar{v}_2^p - \bar{v}_1^{p-1}\bar{v}_2)) = v_1(v_2^p - v_1^{p-1}v_2) = \beta(u_1)\prod_{\lambda=0}^{p-1}\beta(u_2 + \lambda u_1).$$

$\square$

LEMMA 4.7.2. *If E is an elementary abelian p-group and $E' \leq E$, then for any $x \in H^1(E', \mathbb{F}_p)$, $\mathrm{norm}_{E',E}\beta(x)$ is a product of $|E : E'|$ Blocksteins of elements of $H^1(E, \mathbb{F}_p)$.*

PROOF. By transitivity and multiplicativity of the norm (Proposition 4.1.2), we may assume E' is maximal in E. Then x is the inflation to E' of some element $\bar{x} \in H^1(E'/N, \mathbb{F}_p)$, for some maximal subgroup N of E'. So by Proposition 4.1.2 (vi), it suffices to prove the lemma with $N = 1$, namely in the case $|E'| = p$, $|E| = p^2$. This case is dealt with in Proposition 4.1.4. $\square$

REMARK. In fact it is easy to see from the above argument (or by symmetry) that $\mathrm{norm}_{E',E}\beta(x)$ is exactly the product of the Blocksteins of those elements whose restriction is x.

THEOREM 4.7.3 (Serre, ...). *If G is a p-group which is not elementary abelian, then there exist non-zero elements*

$$x_1, \ldots, x_n \in H^1(G, \mathbb{F}_p),$$

with $n \leq ((p+1)/p^2)|G : \Phi(G)|$, such that

$$\beta(x_1).\beta(x_2).\ldots.\beta(x_n) = 0.$$

PROOF. If G is not elementary abelian, then $\Phi(G) \neq 1$, and so $\Phi(G)$ has a maximal subgroup N of index p, normal in G. If there is such a relation in G/N, then it inflates to such a relation in G, so without loss of generality $N = 1$, and $\Phi(G) \cong \mathbb{Z}/p$.

If $G/\Phi(G)$ is cyclic, then G is cyclic and $\beta(x) = 0$ for any $x \in H^1(G, \mathbb{F}_p)$. The result follows since $1 \leq ((p + 1)/p^2).p$. Otherwise, G contains a subgroup H of order p^3, containing $\Phi(G)$, and such that $\Phi(H) = \Phi(G)$. By Lemma 4.7.1, we can find elements $x_1, \ldots, x_{p+1} \in H^1(H, \mathbb{F}_p)$ such that

$$\beta(x_1).\beta(x_2). \ldots .\beta(x_{p+1}) = 0$$

in $H^*(H, \mathbb{F}_p)$. Each x_i is the inflation from $H/\Phi(G)$ of an element $\bar{x}_i$, and so by Proposition 4.1.2 we have

$$
\begin{aligned}
0 &= \mathrm{norm}_{H,G}(\beta(x_1). \ldots .\beta(x_{p+1})) \\
&= \mathrm{norm}_{H,G}\mathrm{inf}_{H/\Phi(G),H}(\beta(\bar{x}_1. \ldots .\beta(\bar{x}_{p+1})) \\
&= \mathrm{inf}_{G/\Phi(G),G}\mathrm{norm}_{H/\Phi(G),G/\Phi(G)}(\beta(\bar{x}_1. \ldots .\beta(\bar{x}_{p+1})) \\
&= \mathrm{inf}_{G/\Phi(G),G}(\mathrm{norm}\beta(\bar{x}_1). \ldots .\mathrm{norm}\beta(\bar{x}_{p+1})).
\end{aligned}
$$

Now by Lemma 4.7.2, each $\mathrm{norm}_{H/\Phi(G),G/\Phi(G)}\beta(\bar{x}_i)$ is a product of degree one elements of $H^*(G/\Phi(G), \mathbb{F}_p)$, and so the inflation is a product of Bocksteins of degree one elements of $H^*(G, \mathbb{F}_p)$. The number of Bocksteins involved in this expression is

$$(p + 1).|G/\Phi(G) : H/\Phi(G)| = (p + 1/p^2).|G : \Phi(G)|. \qquad \square$$

REMARK. The above proof easily shows that the elements x_i produced are in distinct one dimensional subspaces of $H^1(G, \mathbb{F}_p)$, although we have not taken the trouble to keep track of this in the argument.

4.8. Steenrod operations and spectral sequences

When calculating cohomology of groups, one often ends up applying spectral sequence techniques, and needing methods for determining the differentials. One such method involves the fact that the Steenrod operations commute with the transgressions. See Section 5.5 for an example of an application of this.

THEOREM 4.8.1. *In the Lyndon–Hochschild–Serre spectral sequence*

$$H^*(G/N, H^*(N, \mathbb{F}_p)) \Rightarrow H^*(G, \mathbb{F}_p),$$

suppose that $x \in E_n^{0,n-1}$ with $d_n(x) = y$ in $E_n^{n,0}$, and ϕ is a Steenrod operation of degree r. Then $\phi(x)$ survives to $E_{n+r}^{0,n+r-1}$, and $d_{n+r}(\phi(x)) = \phi(y)$ in $E_{n+r}^{n+r,0}$.

PROOF. This theorem is proved in the context of Steenrod operations in the cohomology of topological spaces in Serre [231], Section II.9 c. Namely, if $F \rightarrow E \rightarrow B$ is a fibration, then in the spectral sequence

$$H^*(B; H^*(F; \mathbb{F}_p)) \Rightarrow H^*(E; \mathbb{F}_p)$$

the transgression $d_n : E_n^{0,n-1} \rightarrow E_n^{n,0}$ has the following description. If $x \in H^{n-1}(F; \mathbb{F}_p)$ survives to give an element of $E_n^{0,n-1}$, then it is the image of an element $\hat{x}$ under the boundary homomorphism

$$\partial : H^n(E, F; \mathbb{F}_p) \to H^{n-1}(F; \mathbb{F}_p).$$

All such choices of $\hat{x}$ have the same image $\pi_*(\hat{x})$ under the projection map

$$\pi_* : H^n(E, F; \mathbb{F}_p) \to H^n(B; \mathbb{F}_p).$$

The element $\pi_*(\hat{x})$ survives to $E_n^{n,0}$, and is equal to $d_n(x)$. Since the Steenrod operations commute with both ∂ and π_*, the theorem follows.

To prove the theorem in the cohomology of groups, one can simply notice that this is a special case of the spectral sequence of a fibration (Remark (iv) at the end of Section 3.3). Alternatively, to provide an algebraic proof, one must introduce the relative version of group cohomology corresponding to $H^*(E, F; \mathbb{F}_p)$. This is not the same as the relative cohomology introduced in Section 3.9 of Volume I, and has less interesting algebraic properties, but it is not hard to describe algebraically. One then introduces Steenrod operations into this relative theory in the same way as in Section 4.4, and the proof carries over verbatim.   □

If p is odd, there is also an internal differential which may be calculated in a similar way. Namely, if $x \in E_{2n+1}^{0,2n}$ with $d_{2n+1}(x) = y$ in $E_{2n+1}^{2n+1,0}$, then since d_{2n+1} is a derivation we have $d_{2n+1}(x^j) = jx^{j-1}y$. In particular, $d_{2n+1}(x^p) = px^{p-1}y = 0$, and the above theorem states that $x^p = P^n(x)$ survives to $E_{2np+1}^{0,2np}$ and $d_{2np+1}(x^p) = P^n(y)$ in $E_{2np+1}^{2np+1,0}$. Kudo's transgression theorem [161] states that the element $x^{p-1}y$, which failed to get hit by d_{2n+1}, also survives to $E_{2n(p-1)+1}^{2n+1,2n(p-1)}$ (draw a picture), and

$$d_{2n(p-1)+1}(x^{p-1}y) = -\beta P^n(y) \in E_{2n(p-1)+1}^{2np+2}.$$

Further information about how Steenrod operations fit into spectral sequences can be found in Araki [15, 16], Sawka [227], Singer [236, 237], among others.

Turning now to the Eilenberg–Moore spectral sequence, we recall from Section 3.7 that given a central extension $1 \to Z \to G \to G/Z \to 1$, there is a spectral sequence

$$\mathrm{Tor}^{**}_{H^*(K(Z,2))}(H^*(\text{point}), H^*(G/Z)) \Rightarrow H^*(G).$$

In case $Z = \mathbb{Z}/2$, we have

$$H^*(K(\mathbb{Z}/2, 2), \mathbb{F}_2) = \mathbb{F}_2[u_0, u_1, \dots],$$

a polynomial ring in variables u_i of degree $2^i + 1$. We have $Sq^{2^i} u_i = u_{i+1}$, and the ring homomorphism

$$\mathbb{F}_2[u_0, u_1, \dots] \to H^*(G/Z, \mathbb{F}_2)$$

sends u_0 to the element ζ of $H^2(G, \mathbb{F}_2)$ classifying the central extension (Section 3.7 of Volume I) and

$$u_i \mapsto Sq^{2^{i-1}} Sq^{2^{i-2}} \cdots Sq^1 \zeta.$$

Thus in the E_2 page of the Eilenberg–Moore spectral sequence, all the differentials determined by the transgressions in the Lyndon–Hochschild–Serre spectral sequence have already happened. So one expects the Eilenberg–Moore spectral sequence to converge much more rapidly than the Lyndon–Hochschild–Serre spectral sequence. For an example, see the exercise at the end of Section 5.5.

CHAPTER 5

Varieties for modules and multiple complexes

5.1. Overview and historical background

All groups considered in this chapter will be finite. Recall from the last chapter (Section 4.2) that if R is a Noetherian ring then $H^*(G, R)$ is finitely generated as a ring over R. It is also graded commutative (Section 3.2 of Volume I) in the sense that

$$xy = (-1)^{\deg(x)\deg(y)} yx.$$

In particular, if $R = k$ is a field of characteristic p, then

(i) for $p = 2$, $H^*(G, k)$ is commutative;

(ii) for p odd, the subring $H^{\mathrm{ev}}(G, k)$ generated by elements of even degree is commutative, while elements of odd degree square to zero.

Thus we define

$$H^{\cdot}(G, k) = \begin{cases} H^*(G, k) & \text{if } p = 2 \\ H^{\mathrm{ev}}(G, k) & \text{if } p \text{ is odd} \end{cases}$$

so that $H^{\cdot}(G, k)$ is a finitely generated commutative graded ring over k. For any finitely generated commutative ring over k, the maximal ideals form the points of an affine algebraic variety. We denote by V_G the variety corresponding to $H^{\cdot}(G, k)$. Quillen [209, 210] (1971) investigated the variety V_G and showed that it is stratified by pieces coming from the elementary abelian subgroups of G. In particular, he showed that the dimension of V_G (i.e., the Krull dimension of $H^{\cdot}(G, k)$) is equal to the maximal rank $r_p(G)$ of an elementary abelian p-subgroup of G, as was conjectured by Atiyah (unpublished) and independently by Swan [257].

Five years later, Chouinard [81] (1976) proved that a kG-module M is projective if and only if the restriction $M \downarrow_E$ is projective for every elementary abelian subgroup E of G.

Another five years passed before Alperin and Evens [11] (1981) found a common generalisation of these theorems. They defined the *complexity* $c_G(M)$ of a finitely generated kG-module M to be the rate of growth of a minimal resolution of M. More precisely, if

$$\cdots \to P_2 \to P_1 \to P_0 \to M \to 0$$

is the minimal resolution, then $c_G(M)$ is defined to be the least integer s such that there is a constant $\kappa > 0$ with

$$\dim_k P_n \leq \kappa.n^{s-1} \quad \text{for} \quad n > 0.$$

153

Thus for example a module M has complexity zero if and only if its minimal resolution stops at some finite stage. Since projective modules are also injective (Proposition 3.1.2 of Volume I), this means that M is projective. Similarly, M has complexity one if and only if the modules in the minimal resolution have bounded dimension. We shall see in Section 5.10 that this happens if and only if the minimal resolution repeats with finite period (we then say that M is *periodic*).

Alperin and Evens proved that the complexity of a module M is equal to the maximal complexity of $M\downarrow_E$ as E ranges over the elementary abelian p-subgroups of G. The case of complexity zero is Chouinard's theorem, while the case $M = k$, the trivial module, is Quillen's theorem on the dimension of V_G, after a little re-interpretation.

Meanwhile, Carlson [**65, 69**] had been investigating certain varieties associated to a module M. The *cohomological variety* $V_G(M)$ is defined to be the subvariety of V_G defined by the ideal in $H^\cdot(G,k) = \mathrm{Ext}^\cdot_{kG}(k,k)$ of elements annihilating $\mathrm{Ext}^*_{kG}(M,M)$. The *rank variety* $V_E^\sharp(M)$ is only defined for an elementary abelian group E, and its definition does not involve cohomology. Carlson showed that $V_E(M)$ and $V_E^\sharp(M)$ both have dimension equal to the complexity of M. He conjectured that these varieties were equal, and showed how this would imply that $V_G(M)$ is well behaved for tensor products and restrictions of modules. Avrunin and Scott [**24**] proved Carlson's conjecture, and deduced that $V_G(M)$ is stratified by pieces coming from elementary abelian subgroups of G, in analogy with Quillen's theorem.

Our presentation will differ considerably from this historical order, and benefits from later simplifications by Carlson, Evens and others. The following portmanteau theorem gives some of the main properties of the varieties $V_G(M)$.

THEOREM 5.1.1. (i) *If M is a finitely generated kG-module, then the dimension of $V_G(M)$ is equal to the complexity $c_G(M)$. In particular $V_G(M) = \{0\}$ if and only if M is projective.*

(ii) $V_G(M_1 \oplus M_2) = V_G(M_1) \cup V_G(M_2)$.

(iii) $V_G(M_1 \otimes M_2) = V_G(M_1) \cap V_G(M_2)$.

(iv) $V_G(M) = V_G(M^*) = V_G(\Omega(M)) = V_G(\Omega^{-1}(M))$.

(v) *Denote by* $\mathrm{res}^*_{G,H} : V_H \to V_G$ *the map induced by* $\mathrm{res}_{G,H} : H^\cdot(G,k) \to H^\cdot(H,k)$. *Then* $V_H(M) = (\mathrm{res}^*_{G,H})^{-1}V_G(M)$.

(vi) $V_G(M) = \bigcup_{\substack{E \leq G \\ \text{elemab}}} \mathrm{res}^*_{G,E}V_E(M)$.

(vii) *If* $0 \neq \zeta \in \mathrm{Ext}^n_{kG}(k,k)$ *is represented by a map* $\hat\zeta : \Omega^n k \to k$ *(n even if p is odd) with kernel L_ζ then $V_G(L_\zeta) = V_G\langle\zeta\rangle$ is the hypersurface in V_G determined by ζ.*

(viii) *If $V_G(M) = V_1 \cup V_2$ with $V_1 \cap V_2 = \{0\}$ then $M \cong M_1 \oplus M_2$ with $V_G(M_1) = V_1$ and $V_G(M_2) = V_2$.*

The proofs may be found in Sections 5.7, 5.9 and 5.11.

5.2. Restriction to elementary abelian subgroups

In this section, we show that an element of $\mathrm{Ext}^*_{kG}(M, M)$ is nilpotent if and only if its restriction to every elementary abelian subgroup is nilpotent.

In the case where M is the trivial module, this theorem is due to Quillen [209, 210], see also Quillen and Venkov [218]. In the general case it is due to Carlson [66]. The proof uses Serre's Theorem 4.7.3 together with the following theorem. This theorem is due to Quillen and Venkov [218] in case M is the trivial module and Alperin and Evens [11] in the general case. The proof we give is due to Kroll [156].

THEOREM 5.2.1 (Quillen, Venkov, ...). *Suppose k is a field of characteristic p, M is a kG-module and H is a normal subgroup of index p in G. If ζ is an element of positive degree in $\mathrm{Ext}^*_{kG}(M, M)$ with $\mathrm{res}_{G,H}(\zeta) = 0$, then*

$$\zeta^2 = \beta(x).\zeta'$$

for some $\zeta' \in \mathrm{Ext}^*_{kG}(M, M)$, where x is the inflation to G of a non-zero element $\bar{x} \in H^1(G/H, \mathbb{F}_p)$.

PROOF. We have short exact sequences

$$0 \to M \xrightarrow{\eta} M{\downarrow}_H{\uparrow}^G \to M' \to 0$$

$$0 \to M' \to M{\downarrow}_H{\uparrow}^G \xrightarrow{\eta'} M \to 0$$

where η and η' are the natural maps described in Section 2.8 of Volume I, and $M' \cong \mathrm{coker}(\eta) \cong \mathrm{Ker}(\eta')$ is the tensor product of M with $\Omega_{G/H}(k)$ regarded as a kG-module of dimension $p-1$ with H in the kernel. Denote by $\rho \in \mathrm{Ext}^1_{kG}(M', M)$ and $\rho' \in \mathrm{Ext}^1_{kG}(M, M')$ the elements corresponding to these sequences.

Recall from Section 2.8 of Volume I that the composite

$$\mathrm{Ext}^n_{kG}(M, M) \xrightarrow{\eta_*} \mathrm{Ext}^n_{kG}(M, M{\downarrow}_H{\uparrow}^G) \xrightarrow{\cong} \mathrm{Ext}^n_{kH}(M{\downarrow}_H, M{\downarrow}_H)$$

of η_* with the Eckmann–Shapiro isomorphism is equal to $\mathrm{res}_{G,H}$, as also is the composite

$$\mathrm{Ext}^n_{kG}(M, M) \xrightarrow{(\eta')^*} \mathrm{Ext}^n_{kG}(M{\downarrow}_H{\uparrow}^G, M) \xrightarrow{\cong} \mathrm{Ext}^n_{kH}(M{\downarrow}_H, M{\downarrow}_H).$$

So the condition $\mathrm{res}_{G,H}(\zeta) = 0$ implies that $\eta_*(\zeta) = 0$ and $(\eta')^*(\zeta) = 0$.

The above short exact sequences give rise to long exact sequences

$$\cdots \to \mathrm{Ext}^{n-1}_{kG}(M, M') \to \mathrm{Ext}^n_{kG}(M, M) \xrightarrow{\eta_*} \mathrm{Ext}^n_{kG}(M, M{\downarrow}_H{\uparrow}^G) \to \cdots$$

$$\cdots \to \mathrm{Ext}^{n-1}_{kG}(M', M) \to \mathrm{Ext}^n_{kG}(M, M) \xrightarrow{(\eta')^*} \mathrm{Ext}^n_{kG}(M{\downarrow}_H{\uparrow}^G, M) \to \cdots$$

whose connecting homomorphisms are given by Yoneda composition with ρ, resp. ρ'. So the conditions $\eta_*(\zeta) = 0$ and $(\eta')^*(\zeta) = 0$ imply that there are elements $\zeta_0 \in \mathrm{Ext}^{n-1}_{kG}(M, M')$ and $\zeta'_0 \in \mathrm{Ext}^{n-1}_{kG}(M', M)$ such that $\zeta = \rho \circ \zeta_0 = \zeta'_0 \circ \rho'$. Now by the discussion following Lemma 4.3.2, the element

$\rho' \circ \rho \in \text{Ext}^2_{kG}(M', M')$ is equal to the element $\beta(x)_{M'}$ obtained by tensoring M' with the exact sequence representing $\beta(x)$. Since Yoneda product with $\beta(x)_{M'}$ equals cup product with $\beta(x)$, and hence commutes with other Yoneda compositions (see Section 3.2 of Volume I) we have

$$\zeta^2 = \zeta \circ \zeta = \zeta_0' \circ \rho' \circ \rho \circ \zeta_0 = \zeta_0' \circ \beta(x)_{M'} \circ \zeta_0 = \beta(x).(\zeta_0' \circ \zeta_0). \qquad \square$$

THEOREM 5.2.2 (Quillen, Carlson). *Suppose k is a field of characteristic p, and M is a kG-module. Then an element $\zeta \in \text{Ext}^*_{kG}(M, M)$ of positive degree is nilpotent (under Yoneda product) if and only if $\text{res}_{G,E}(\zeta)$ is nilpotent for all elementary abelian subgroups E of G.*

PROOF. Since restriction from G to a Sylow p-subgroup is injective on $\text{Ext}^*_{kG}(M, M)$ (Corollary 3.6.18 of Volume I), without loss of generality G is a p-group. We may suppose that G is not elementary abelian, so that by Serre's Theorem 4.7.3 we may choose elements $x_1, \ldots, x_n \in H^1(G, \mathbb{F}_p)$ with

$$\beta(x_1). \ldots .\beta(x_n) = 0.$$

Working by induction, we may assume that for each maximal subgroup H of G, $\text{res}_{G,H}(\zeta)$ is nilpotent. Denoting by H_i the maximal subgroup corresponding to x_i, we suppose that $\text{res}_{G,H_i}(\zeta^{r_i}) = 0$. Thus by the Quillen–Venkov Theorem 5.2.1, we have

$$\zeta^{2r_i} = \beta(x_i).\zeta_i'.$$

So letting $r = 2\sum_i r_i$, we have

$$\zeta^r = \zeta^{2r_1} \circ \cdots \circ \zeta^{2r_n} = (\beta(x_1).\zeta_1') \circ \cdots \circ (\beta(x_n).\zeta_n')$$
$$= \beta(x_1). \ldots .\beta(x_n).(\zeta_1' \circ \cdots \circ \zeta_n') = 0.$$

We have used here the fact that cup product with $\beta(x_i)$ commutes with Yoneda products (Corollary 3.2.2 of Volume I). $\qquad \square$

Using the finite generation of cohomology, we may now deduce Chouinard's Theorem.

LEMMA 5.2.3. *Suppose k is a field of characteristic p an M is a finitely generated kG-module. Then the following are equivalent:*
(i) *M is projective.*
(ii) *$\text{Ext}^n_{kG}(M, M) = 0$ for all $n > 0$.*
(iii) *$\text{Ext}^n_{kG}(M, M) = 0$ for all n large enough.*
(iv) *Every element of $\text{Ext}^*_{kG}(M, M)$ of positive degree is nilpotent.*

PROOF. Since $\text{Ext}^*_{kG}(M, M)$ is finitely generated as a module over its centre (Corollary 4.2.4), (iii) is equivalent to (iv). It is clear that (i) $\Rightarrow$ (ii) $\Rightarrow$ (iii), so we shall prove that (iii) $\Rightarrow$ (i).

Now by Corollary 4.2.4, if N is another finitely generated kG-module, then $\text{Ext}^*_{kG}(M, N)$ is finitely generated as a module over $\text{Ext}^*_{kG}(M, M)$. So if we have $\text{Ext}^n_{kG}(M, M) = 0$ for all n large enough, then the same is true of $\text{Ext}^n_{kG}(M, N)$ for any finitely generated module N. In particular, taking N to be a simple module S, this shows that the projective cover of S appears only finitely often in the minimal projective resolution of M. Since there are only finitely many simple modules, this means that the minimal projective resolution of M stops. By Corollary 3.6.5 of Volume I, this forces M to be projective.   □

THEOREM 5.2.4 (Chouinard). *A finitely generated kG-module M is projective if and only if $M \downarrow_E$ is projective for every elementary abelian subgroup E of G.*

PROOF. If M is projective, it is clear that $M \downarrow_E$ is projective. Conversely, if $M \downarrow_E$ is projective for every elementary abelian subgroup E of G, then $\mathrm{Ext}^n_{kE}(M, M) = 0$ for every E and $n > 0$. Thus by Theorem 5.2.2, every element of $\mathrm{Ext}^*_{kG}(M, M)$ of positive degree is nilpotent, and so by the lemma M is projective.   □

5.3. Poincaré series and complexity

Let V be a graded vector space of **finite type** over k. In other words,

$$V = \bigoplus_{r \geq 0} V_r$$

with each V_r a finite dimensional vector space over k. We define the **Poincaré series** of V to be

$$p(V, t) = \sum_{r \geq 0} t^r \dim_k V_r$$

as a formal power series in the indeterminate t. Thus for example if $V = k[x_1, \ldots, x_n]$ is a polynomial ring in generators x_i of degree one, then V_r is the rth symmetric power of the n dimensional vector space V_1, and so

$$p(V, t) = 1 + nt + n(n+1)t^2/2 + \cdots = 1/(1-t)^n.$$

As another example, if $V = k\{x_1, x_2\}$ is the free ring on two non-commuting generators x_1 and x_2 of degree one, then

$$p(V, t) = 1 + 2t + 4t^2 + 8t^3 + \cdots = 1/(1 - 2t).$$

The following proposition shows that if V is a finitely generated graded module over a finitely generated commutative graded ring of finite type, then $p(V, t)$ is a rational function of t whose poles are at roots of unity.

PROPOSITION 5.3.1 (Hilbert, Serre). *(see Atiyah–Macdonald [20], Theorem 11.1) Suppose that A is a commutative graded ring of finite type over k, finitely generated over A_0 by homogeneous elements $x_1, \ldots, x_s$ in degrees $k_1, \ldots, k_s$. Suppose V is a finitely generated graded A-module (i.e., we have $A_i.A_j \subseteq A_{i+j}$ and $A_i.V_j \subseteq V_{i+j}$). Then the Poincaré series $p(V, t)$ is of the form*

$$f(t) / \prod_{j=1}^{s} (1 - t^{k_j})$$

where $f(t)$ is a polynomial in t with integer coefficients.

PROOF. We work by induction on s. If $s = 0$ then $p(V, t)$ is a polynomial, so suppose $s > 0$. Denoting by K and L the kernel and cokernel of multiplication by x_s, we have an exact sequence

$$0 \to K_r \to V_r \xrightarrow{x_s} V_{r+k_s} \to L_{r+k_s} \to 0.$$

Now K and L are finitely generated graded modules for $A_0[x_1, \dots, x_{s-1}]$, and so by induction their Poincaré series have the given form. We have from the above exact sequence

$$t^{k_s} p(K, t) - t^{k_s} p(V, t) + p(V, t) - p(L, t) + g(t) = 0$$

where $g(t)$ is a polynomial of degree less than k_s with integer coefficients. Thus

$$p(V, t) = (p(L, t) - t^{k_s} p(K, t) - g(t))/(1 - t^{k_s})$$

has the given form. □

REMARK. It doesn't matter for the proof of the above proposition whether the ring A is strictly commutative or only commutative in the graded sense $xy = (-1)^{\deg(x)\deg(y)} yx$. Of course, in the latter case A is finitely generated as a module over the subring A^{ev} of elements of even degree, and so we may always work with A^{ev} if it makes us feel more comfortable. This will become a real issue when we come to study varieties, since most algebraic geometry texts are written for strictly commutative rings. The proofs usually carry over verbatim to the graded commutative case, but it is often not worth the trouble of saying so.

The following proposition shows that the rate of growth of the dimensions in a graded module is determined by the order of the pole of the Poincaré series at $t = 1$.

PROPOSITION 5.3.2. *Suppose*

$$p(t) = f(t) / \prod_{j=1}^{s} (1 - t^{k_j}) = \sum_{r \geq 0} a_r t^r$$

where $f(t)$ is a polynomial with integer coefficients and the a_r are non-negative integers. Let γ be the order of the pole of $p(t)$ at $t = 1$. Then

(i) *there exists a constant $\kappa > 0$ such that $a_n \leq \kappa . n^{\gamma-1}$ for $n > 0$, but*

(ii) *if $\gamma \geq 1$, there does not exist a constant $\kappa > 0$ such that $a_n \leq \kappa . n^{\gamma-2}$ for $n > 0$.*

PROOF. The hypothesis and conclusion remain unaltered if we replace $p(t)$ by $p(t).(1 + t + \dots + t^{k_j-1})$, and so without loss of generality each $k_j = 1$. So we may suppose $p(t) = f(t)/(1 - t)^\gamma$ with $f(1) \neq 0$. Suppose $f(t) = \alpha_m t^m + \dots + \alpha_0$. We have

$$a_n = \alpha_0 \binom{n + \gamma - 1}{\gamma - 1} + \alpha_1 \binom{n + \gamma - 2}{\gamma - 1} + \dots + \alpha_m \binom{n + \gamma - m - 1}{\gamma - 1}.$$

The condition $f(1) \neq 0$ implies that $\alpha_0 + \cdots + \alpha_m \neq 0$, so this expression is a polynomial of degree exactly $\gamma - 1$ in n. $\qquad\square$

DEFINITION 5.3.3. *Suppose V is a graded vector space of finite type whose Poincaré series $p(V, t)$ has the form $f(t)/\prod_{j=1}^{s}(1 - t^{k_j})$ where $f(t)$ is a polynomial with integer coefficients. Then we write $\gamma(V)$ for the order of the pole of $p(V, t)$ at $t = 1$. By the proposition, this measures the polynomial rate of growth of the V_r, so we call it the* **growth** *of V.*

If V is a finitely generated graded module over a commutative graded ring A, finitely generated and of finite type over k, then Proposition 5.3.1 shows that $\gamma(V)$ is well defined. Clearly if $0 \to V'' \to V' \to V \to 0$ is a short exact sequence of finitely generated graded A-modules then $\gamma(V')$ is equal to the maximum of $\gamma(V)$ and $\gamma(V'')$. In particular, any finitely generated graded module V is a quotient of a direct sum of finitely many copies of A by a graded ideal, and so we have $\gamma(V) \leq \gamma(A)$.

Now suppose M is a finitely generated kG-module with minimal resolution

$$\cdots \to P_2 \to P_1 \to P_0 \to M \to 0.$$

Then the multiplicity of the projective cover P_S of a simple kG-module S as a summand of P_r is equal to

$$\dim_k \mathrm{Hom}_{kG}(P_r, S) / \dim_k \mathrm{End}_{kG}(S).$$

Since the resolution is minimal, every homomorphism $P_r \to S$ is a cocycle, and every coboundary is zero, so that

$$\mathrm{Hom}_{kG}(P_r, S) \cong \mathrm{Ext}_{kG}^r(M, S).$$

Thus we have

$$\dim_k P_r = \sum_{S \text{ simple}} \dim_k P_S . \dim_k \mathrm{Ext}_{kG}^r(M, S) / \dim_k \mathrm{End}_{kG}(S).$$

Since $\mathrm{Ext}_{kG}^*(M, S)$ is a finitely generated graded module over the finitely generated graded commutative ring $H^*(G, k) = \mathrm{Ext}_{kG}^*(k, k)$, it follows from Proposition 5.3.1 that the Poincaré series $p(\mathrm{Ext}_{kG}^*(M, S), t)$ is of the form $f(t)/\prod_{j=1}^{s}(1 - t^{k_j})$, where $f(t)$ is a polynomial with integer coefficients and the k_j are the degrees of generators of $H^*(G, k)$. Thus by the above formula for the dimension of P_r, the graded vector space P_* also has a Poincaré series of this form.

DEFINITION 5.3.4. *The* **complexity** *$c_G(M)$ of a finitely generated kG-module M is the growth $\gamma(P_*)$ of the minimal resolution P_* of M.*

Now if N is another finitely generated kG-module, then $\mathrm{Ext}_{kG}^r(M, N)$ is a subquotient of $\mathrm{Hom}_{kG}(P_r, N)$, and this is a subspace of $\mathrm{Hom}_k(P_r, N)$. Thus

$$\dim_k \mathrm{Ext}_{kG}^r(M, N) \leq \dim_k P_r . \dim_k N$$

and so
$$\gamma(\mathrm{Ext}^*_{kG}(M,N)) \leq \gamma(P_*) = c_G(M).$$

PROPOSITION 5.3.5. *We have*
$$c_G(M) = \gamma(\mathrm{Ext}^*_{kG}(M,M)) = \gamma(Z\mathrm{Ext}^*_{kG}(M,M)) = \max_{S \text{ simple}} \gamma(\mathrm{Ext}^*_{kG}(M,S)).$$

PROOF. Since $\mathrm{Ext}^*_{kG}(M,M)$ is finitely generated as a graded module over its centre $Z\mathrm{Ext}^*_{kG}(M,M)$ (Corollary 4.2.4), we have
$$\gamma(\mathrm{Ext}^*_{kG}(M,M)) = \gamma(Z\mathrm{Ext}^*_{kG}(M,M)).$$
If S is a simple kG-module, then $\mathrm{Ext}^*_{kG}(M,S)$ is finitely generated as a module over $\mathrm{Ext}^*_{kG}(M,M)$, and so we have
$$\max_{S \text{ simple}} \gamma(\mathrm{Ext}^*_{kG}(M,S)) \leq \gamma(\mathrm{Ext}^*_{kG}(M,M)) \leq \gamma(P_*) = c_G(M)$$
$$\leq \max_{S \text{ simple}} \gamma(\mathrm{Ext}^*_{kG}(M,S))$$
by our formula for $\dim_k P_r$. □

LEMMA 5.3.6. *Suppose A is a finitely generated commutative graded ring of finite type over k. Then the nilpotent elements in A form an ideal J with $J^n = 0$ for some n. We have $\gamma(A) = \gamma(A/J)$.*

PROOF. Since A is Noetherian, J is finitely generated by nilpotent elements x_i. If $x_i^{n_i} = 0$ then $J^n = 0$ with $n = \sum_i n_i$.

Now each J^k/J^{k+1} is a finitely generated module for A/J, and so
$$\gamma(A/J) \geq \gamma(J^k/J^{k+1}).$$
Hence $\gamma(A/J) = \gamma(A/J^2) = \cdots = \gamma(A)$. □

THEOREM 5.3.7 (Alperin, Evens [11]). *The complexity $c_G(M)$ is equal to the maximum complexity $c_E(M)$ of the restriction of M to an elementary abelian subgroup $E \leq G$.*

PROOF. By Proposition 5.3.5 and Lemma 5.3.6 we have
$$c_G(M) = \gamma(\mathrm{Ext}^*_{kG}(M,M)) = \gamma(Z\mathrm{Ext}^*_{kG}(M,M)) = \gamma(Z\mathrm{Ext}^*_{kG}(M,M)/J).$$
Now by Theorem 5.2.2 the map
$$Z\mathrm{Ext}^*_{kG}(M,M)/J \to \bigoplus_{\substack{E \leq G \\ \text{elemab}}} Z\mathrm{Ext}^*_{kE}(M,M)/J$$
is injective, and so
$$c_G(M) \leq \gamma\Big(\bigoplus_{\substack{E \leq G \\ \text{elemab}}} Z\mathrm{Ext}^*_{kE}(M,M)/J\Big)$$
$$= \max_{\substack{E \leq G \\ \text{elemab}}} \gamma(Z\mathrm{Ext}^*_{kE}(M,M)/J) = \max_{\substack{E \leq G \\ \text{elemab}}} c_E(M).$$
Conversely, it is clear that $c_E(M) \leq c_G(M)$ for any such E. □

We now define the **Krull dimension** of a finitely generated commutative graded ring A of finite type over k to be its rate of growth $\gamma(A)$ as a graded vector space. We shall give other equivalent (but more general) definitions in Section 5.4; the definition we have given is designed to make the following theorem a direct corollary of the Alperin–Evens Theorem (of course, it came historically earlier).

THEOREM 5.3.8 (Quillen [**209, 210**]). *The Krull dimension of $H^\cdot(G,k)$ is equal to the p-rank $r_p(G)$ (i.e., the maximal rank of an elementary abelian p-subgroup of G).*

PROOF. The Krull dimension of $H^\cdot(G,k) = \operatorname{Ext}^*_{kG}(k,k)$ is equal to $c_G(k)$, by Proposition 5.3.5. By the Alperin–Evens Theorem, this is equal to the maximal value of $c_E(k)$ as E ranges over the elementary abelian p-subgroups of G. If E is an elementary abelian group of order p^n, then the Poincaré series of $H^*(E,k)$ is $1/(1-t)^n$ (see Section 3.5 of Volume I) and so $c_E(k) = \gamma(H^*(E,k)) = n$.   □

EXERCISE. Let E be an elementary abelian p-group of order p^r, and let k be a field of characteristic p. Denote by Λ the algebra $kE/J^2(kE)$ of dimension $r+1$. Show that

$$H^*(\Lambda, k) = \operatorname{Ext}^*_\Lambda(k,k) \cong k\{x_1, \dots, x_r\},$$

the free ring on non-commuting generators $x_1, \dots, x_r$ of degree one, so that

$$p(H^*(\Lambda,k),t) = 1/(1-rt).$$

5.4. Varieties and commutative algebra

In preparation for the theory of varieties for modules, we include here a brief summary of the commutative algebra and algebraic geometry we shall need.

Suppose A is a commutative ring, finitely generated over an algebraically closed field k. Then A has the form $k[x_1, \dots, x_n]/I_A$ for some ideal I_A in a polynomial ring $k[x_1, \dots, x_n]$. We wish to investigate the prime ideals of A. In order to do this, we should try to understand the prime ideals in $k[x_1, \dots, x_n]$, and then see which ones contain I_A.

We start by describing the maximal ideals. We regard $k[x_1, \dots, x_n]$ as the polynomial functions on an n-dimensional space $\mathbb{A}^n(k)$. This is really an affine space rather than a vector space, since if $(\lambda_1, \dots, \lambda_n)$ is a point in this space, we can just as well regard $k[x_1, \dots, x_n]$ as $k[x_1 - \lambda_1, \dots, x_n - \lambda_n]$; from this new point of view, this point becomes the origin.

Each point $(\lambda_1, \dots, \lambda_n) \in \mathbb{A}^n(k)$ determines a ring homomorphism

$$k[x_1, \dots, x_n] \to k$$
$$x_i \mapsto \lambda_i,$$

namely evaluation of the polynomial function at the point. The kernel is a maximal ideal in $k[x_1, \dots, x_n]$. Since polynomial functions separate points in $\mathbb{A}^n(k)$, distinct points give rise to distinct maximal ideals.

LEMMA 5.4.1 (Weak Nullstellensatz). *Every maximal ideal in the polynomial ring $k[x_1, \ldots, x_n]$ is determined by a point $(\lambda_1, \ldots, \lambda_n)$ in $\mathbb{A}^n(k)$ in the above way.*

PROOF. We shall prove by induction on n, that even if k is not algebraically closed, if $\mathcal{M}$ is a maximal ideal in $k[x_1, \ldots, x_n]$ then in the field $K = k[x_1, \ldots, x_n]/\mathcal{M}$ the x_i are algebraic over k. In the algebraically closed case, this forces the x_i to be in k, proving the lemma.

Since $k[x_1]$ is not a field (x_1 has no inverse), the statement is clear for $n = 1$. If $n > 1$ and (say) x_1 is transcendental over k, then $k(x_1)$ is a subfield of K, and by induction $x_2, \ldots, x_n$ are algebraic over $k(x_1)$. So there exists a polynomial $p(x_1)$ such that $p(x_1)x_2, \ldots, p(x_1)x_n$ are integral over $k[x_1]$. So given any element $f(x_1, \ldots, x_n) \in K$, for some $r > 0$ the element $p(x_1)^r f(x_1, \ldots, x_n)$ is integral over $k[x_1]$. Since $k[x_1]$ is integrally closed in $k(x_1)$, this means that if $f(x_1, \ldots, x_n) \in k(x_1)$ then $p(x_1)^r f(x_1, \ldots, x_n) \in k[x_1]$, so that $f(x_1, \ldots, x_n)$ has the form $q(x_1)/p(x_1)^r$. But $p(x_1)$ is independent of the element of $k(x_1)$ chosen, which is absurd because there are infinitely many irreducible polynomials in $k[x_1]$. $\qquad\square$

The above lemma is a special case of the following theorem.

THEOREM 5.4.2 (Hilbert's Nullstellensatz). *If I is an ideal in the polynomial ring $k[x_1, \ldots, x_n]$, we set*

$$V(I) = \{v \in \mathbb{A}^n(k) \mid f(v) = 0 \ \forall \ f \in I\}.$$

If $f \in k[x_1, \ldots, x_n]$ with $f(v) = 0$ for all $v \in V(I)$, then for some $r > 0$ we have $f^r \in I$.

PROOF. Suppose $I = (f_1, \ldots, f_s)$, and suppose $f(v) = 0$ for all $v \in I$. Let t be a new indeterminate. Then the polynomials $f_1, \ldots, f_s, 1 - f.t$ have no common zero in $\mathbb{A}^{n+1}(k)$. So by the weak Nullstellensatz the ideal they generate is not contained in any maximal ideal, so it is the whole ring $k[x_1, \ldots, x_n, t]$. So there are polynomials $g_1, \ldots, g_s, g \in k[x_1, \ldots, x_n, t]$ with

$$f_1.g_1 + \cdots + f_s.g_s + (1 - f.t)g = 1.$$

Substituting $t = 1/f$ and clearing denominators, we obtain polynomials $h_1, \ldots, h_s$ in $k[x_1, \ldots, x_n]$ with

$$f_1.h_1 + \cdots + f_s.h_s = f^r,$$

as required. $\qquad\square$

The effect this has is the following. If I is an ideal in $k[x_1, \ldots, x_n]$, we define the **radical** of I to be

$$\sqrt{I} = \{f \in k[x_1, \ldots, x_n] \mid \text{ for some } r > 0, \ f^r \in I\}.$$

A **radical ideal** is one with $I = \sqrt{I}$. Certainly $V(I) = V(\sqrt{I})$, and Hilbert's Nullstellensatz says that $f \in \sqrt{I}$ if and only if $f(v) = 0$ for all $v \in V(\sqrt{I})$.

COROLLARY 5.4.3. *There is a one–one inclusion reversing correspondence between sets in $\mathbb{A}^n(k)$ of the form $V(I)$ for some I (simultaneous sets of zeroes of some polynomials), and radical ideals $I = \sqrt{I}$.*   □

The sets $V(I)$ form the closed sets of a topology on $\mathbb{A}^n(k)$ called the **Zariski topology**. This is because

$$V(I_1.I_2) = V(I_1) \cup V(I_2), \quad V(\sum_\alpha I_\alpha) = \bigcap_\alpha V(I_\alpha)$$

so that a finite union $\cdots$ arbitrary intersection of closed sets is closed. Note that this topology is very badly separated. For example, every non-empty open set is dense, so that in particular the space is not Hausdorff.

A closed set V in the Zariski topology is said to be **irreducible** if whenever we write $V = V_1 \cup V_2$ as a union of two closed sets then either $V = V_1$ or $V = V_2$. This is equivalent to the corresponding ideal being *prime*. Since $k[x_1, \ldots, x_n]$ is Noetherian, every closed set is a finite union of irreducible closed sets.

Now we go back to the situation of a general commutative ring A, finitely generated over k. If we write $A = k[x_1, \ldots, x_n]/I_A$, then the maximal ideals in A are in one–one correspondence with the maximal ideals in $k[x_1, \ldots, x_n]$ containing I_A, and hence with the points of $V(I_A)$. We thus write

$$\max(A) = V(I_A) \subseteq \mathbb{A}^n(k)$$

for the set of maximal ideals of A. It inherits the Zariski topology from $\mathbb{A}^n(k)$. Namely, radical ideals in A correspond to radical ideals in $k[x_1, \ldots, x_n]$ containing I_A, and hence to closed subsets of $V(I_A)$. We thus make the following definition.

DEFINITION 5.4.4. *Suppose that A is a commutative ring, finitely generated over an algebraically closed field k. Denote by $\max(A)$ the set of maximal ideals of A. If I is an ideal in A, we set $V(I) \subseteq \max(A)$ to be the set of maximal ideals containing I. The **Zariski topology** on $\max(A)$ is the topology whose closed sets are of the form $V(I)$, for some ideal I. The set $\max(A)$ with this topology is called the **maximal ideal spectrum** of A. An object of this form is called an **affine variety**.*

Giving a presentation $A = k[x_1, \ldots, x_n]/I_A$ corresponds to giving an embedding of the "abstract" affine variety $\max(A)$ as a closed set in $\mathbb{A}^n(k)$.

We regard elements $a \in A$ as "polynomial functions" on $\max(A)$ as follows. If $\alpha \in \max(A)$ corresponds to a ring homomorphism $\phi_\alpha : A \to k$, we set $a(\alpha) = \phi_\alpha(a)$. Hilbert's Nullstellensatz tells us that the elements giving the zero function are precisely the nilpotent elements in A.

If $\phi : A \to B$ is a homomorphism of commutative rings (both finitely generated over k), then we have a map of affine varieties

$$\phi^* : \max(B) \to \max(A)$$

given as follows. If $\alpha \in \max(B)$ corresponds to $\phi_\alpha : B \to k$ then $\phi^*(\alpha) \in \max(A)$ corresponds to $\phi_\alpha \circ \phi : A \to k$.

REMARK. For more general commutative rings (not necessarily finitely generated over an algebraically closed field), it is not true that the inverse image of a maximal ideal under a ring homomorphism is again a maximal ideal. However, the inverse image of a prime ideal is always prime, and so we define spec(A) to be the set of prime ideals of A. If I is an ideal in A, then $V(I)$ is the set of prime ideals containing I. These form the closed sets of the Zariski topology on spec(A). The set spec(A) with this topology is called the **prime ideal spectrum** of A. Notice that max(A) can be recovered as the set of closed points in spec(A). The remaining points should be thought of as "generic points" of irreducible subvarieties of max(A).

Since all the rings we wish to consider are finitely generated over algebraically closed fields, we shall always use max(A) rather than spec(A).

GRADED RINGS AND PROJECTIVE VARIETIES. Now suppose A is a finitely generated commutative graded ring of finite type over an algebraically closed field k,

$$A = \bigoplus_{r \geq 0} A_r$$

with each A_r finite dimensional over k. Then A_0 is a finite dimensional commutative k-algebra, and hence is a finite sum of local rings whose residue field is k. The corresponding idempotents split A as a direct sum of graded rings, so we may as well suppose A_0 is a local ring. Since nilpotent elements do not affect the maximal (or prime) ideal spectrum, we may thus assume $A_0 = k$. Thus $A = k[x_1, \ldots, x_n]/I_A$, where say $\deg(x_i) = n_i > 0$ and I_A is a **homogeneous ideal**; namely if $a = \sum a_j \in I_A$ with $\deg(a_j) = j$ then each $a_j \in I_A$. Note that $k[x_1, \ldots, x_n]/I$ inherits a grading from $k[x_1, \ldots, x_n]$ if and only if I is homogeneous.

Thus A has a distinguished maximal ideal A^+, consisting of the elements of positive degree in A. We regard this as the origin in max(A). If $\lambda \in k$ is a scalar, we define "dilation by λ" as an operation on max(A) as follows. We have a ring homomorphism $m_\lambda : A \to A$ which multiplies an element of degree r by λ^r. The map $m_\lambda^* : \max(A) \to \max(A)$ is dilation by λ. If α is not the origin, then the set of dilations $\lambda\alpha$ of α, $\lambda \in k$, forms a subvariety which should be thought of as the line through the origin, containing α. The corresponding ideal in A is the largest homogeneous ideal contained in the maximal ideal α.

EXERCISE. Show that an ideal I of A is homogeneous if and only if the dilation by λ of an element of I is in I for all $\lambda \in k$.

If we regard max(A) as a closed subset of $\mathbb{A}^n(k)$ via the presentation

$$A = k[x_1, \ldots, x_n]/I_A$$

with I_A homogeneous as above, then $\max(A)$ is a union of lines through the origin in $\mathbb{A}^n(k)$, i.e., a "homogeneous" closed set. So if A is a graded ring as above, we say $\max(A)$ is a **homogeneous affine variety**.

In this situation, it is sometimes better to think in terms of projective varieties. The points in $\text{proj}(A)$ are the lines through the origin in $\max(A)$; namely the maximal homogeneous ideals of A (properly contained in A^+). If I is a homogeneous ideal in A, we write $\bar{V}(I)$ for the subset of $\text{proj}(A)$ consisting of the maximal homogeneous ideals containing I. These form the closed sets of the Zariski topology on $\text{proj}(A)$. A set of the form $\text{proj}(A)$ with this topology is called a **projective variety**. The projective variety $\text{proj}\, k[x_1, \dots, x_n]$ is written $\mathbb{P}^{n-1}(k)$ and is called $(n-1)$-dimensional **projective space**. There are obvious maps $\max(A) \setminus \{0\} \twoheadrightarrow \text{proj}(A)$ and $\mathbb{A}^n(k) \setminus \{0\} \twoheadrightarrow \mathbb{P}^{n-1}(k)$ making the following diagram commute.

$$
\begin{array}{ccccc}
\max(A) \setminus \{0\} & = & V(I_A) \setminus \{0\} & \subseteq & \mathbb{A}^n(k) \setminus \{0\} \\
\downarrow & & \downarrow & & \downarrow \\
\text{proj}(A) & = & \bar{V}(I_A) & \subseteq & \mathbb{P}^{n-1}(k).
\end{array}
$$

The most general sort of variety we shall consider is a **quasiprojective variety**, namely an open subset of a projective variety, with the inherited topology. For example, $\mathbb{A}^{n-1}(k)$ is the open subset of $\mathbb{P}^{n-1}(k)$ complementing the closed set $x_n = 0$. Thus every affine variety is also a quasiprojective variety, as also is every projective variety.

One can think of quasiprojective varieties, and also of open subsets in affine varieties, in terms of localisation. If I is an ideal in A, then the set S of elements which are not zero divisors modulo I is closed under multiplication. So we can form the localisation $A[S^{-1}]$ whose elements are formal quotients a/s ($a \in A$, $s \in S$) with $a/s = a'/s'$ if and only if for some $s'' \in S$ we have $s''(s'a - sa') = 0$. The maximal ideals of $A[S^{-1}]$ correspond to the maximal ideals of A not containing I, and so

$$
\max(A[S^{-1}]) = \max(A) \setminus V(I).
$$

Similarly if A is graded and I is homogeneous, then $A[S^{-1}]$ is graded (in positive and negative degrees), and letting $\text{proj}(A[S^{-1}])$ be the spectrum of homogeneous ideals we have

$$
\text{proj}(A[S^{-1}]) = \text{proj}(A) \setminus \bar{V}(I).
$$

Note that $\text{proj}(A[S^{-1}])$ is a quasiprojective, not a projective variety.

DIMENSION. There are many different concepts of dimensions of rings; for an extensive discussion of this topic see McConnell and Robson [**184**], Part II. In the case of a finitely generated commutative ring over a field, all these definitions agree. The following lemma is the starting point of our discussion of dimension.

LEMMA 5.4.5 (Noether Normalisation Lemma). *Suppose that k is a field and A is a finitely generated commutative algebra over k. Then there exist elements $y_1, \ldots, y_n \in A$ generating a polynomial subalgebra $k[y_1, \ldots, y_n] \subseteq A$ over which A is finitely generated as a module.*

If A is graded then $y_1, \ldots, y_n$ may be chosen to be homogeneous elements.

PROOF. See Matsumura [**179**] (14G). □

The integer n in the above lemma is called the **dimension** of A. If A is an integral domain, this is the same as the transcendence degree of the field of fractions of A as an extension of k.

If A_* is a finitely generated graded ring of finite type over k then $y_1, \ldots, y_n$ may be chosen to be homogeneous of positive degree, so that the rate of growth is

$$\gamma(A_*) = \gamma(k[y_1, \ldots, y_n]) = n.$$

Thus the above definition of dimension agrees with the definition of Krull dimension given in the last section.

The usual definition of the Krull dimension of a ring is as follows. If there is a chain of prime ideals of length n

$$\mathfrak{p}_0 \supset \mathfrak{p}_1 \supset \cdots \supset \mathfrak{p}_n$$

in A but none of length $n+1$, then the Krull dimension of A is n. If there are arbitrarily long chains of prime ideals, A has infinite Krull dimension. In case A is a finitely generated commutative ring over a field k, the Krull dimension is equal to the dimension as defined above. This is proved in Matsumura [**179**] Chapter 5 §14. In summary, we have the following theorem:

THEOREM 5.4.6. *If A is a finitely generated commutative ring over a field k, then the Krull dimension $\dim(A)$ (defined in terms of chains of prime ideals) is equal to the integer n given in the Noether normalisation lemma. We also have the following:*

(i) *If A is an integral domain then $\dim(A)$ is equal to the transcendence degree of the field of fractions of A.*

(ii) *If A_* is a finitely generated commutative graded ring of finite type over k then $\dim(A_*) = \gamma(A_*)$.* □

FINITE MAPS. Suppose $\phi : A \to B$ is a map of finitely generated commutative algebras over k. We say

$$\phi^* : \max(B) \to \max(A)$$

is a **finite map** if B is finitely generated as a module over the image of ϕ (or equivalently B is integral over the image of ϕ). If ϕ^* is finite, then every point in $\max(A)$ has only finitely many inverse images in $\max(B)$. To see this, let $B = k[x_1, \ldots, x_n]/I_B$ so that $\max(B) \subseteq \mathbb{A}^n(k)$. Then each x_j satisfies an equation of the form $x_j^m + \phi(a_{m-1})x_j^{m-1} + \cdots + \phi(a_0) = 0$ with $a_i \in A$. So on the pre-image of a point in $\max(A)$, the coordinate x_j can take on only finitely many values.

We say ϕ^* is a **dominant map** if the kernel of ϕ is a nilpotent ideal in A. In general a dominant map of varieties is not necessarily surjective, but its image is a dense open subset. However, a finite dominant map is always surjective because of the following theorem:

THEOREM 5.4.7. *Let $A \subset B$ be commutative rings with B integral over A, and let $\mathfrak{p}$ be a prime ideal of A. Then there exists a prime ideal $\mathfrak{p}'$ of B such that $\mathfrak{p}' \cap A = \mathfrak{p}$.*

PROOF. See Atiyah and Macdonald [**20**] Theorem 5.10. □

Thus, for example, we can view the Noether Normalisation Lemma as saying that any affine variety of dimension n can be expressed as a finite branched cover of $\mathbb{A}^n(k)$, and any projective variety of dimension $n - 1$ can be expressed as a finite branched cover of $\mathbb{P}^{n-1}(k)$.

Suppose $A \subseteq B$ are finitely generated commutative algebras over a field k of characteristic p, and suppose there exists some power p^a of p such that the p^ath power of any element of B lies in A. Writing $B^{[p^a]}$ for the subring consisting of p^ath powers of elements of B, the inclusions $B^{[p^a]} \subseteq A \subseteq B$ give rise to finite dominant maps

$$\max(B) \to \max(A) \to \max(B^{[p^a]}).$$

The composite of these surjective maps is equal to the Frobenius morphism on $\max(B)$, which at the level of the Zariski topology is a homeomorphism. It follows that the map

$$\max(B) \twoheadrightarrow \max(A)$$

is also a homeomorphism in the Zariski topology. Such a map is called an **inseparable isogeny** or an **F-isomorphism**.

Finally, we discuss finite group actions on varieties.

PROPOSITION 5.4.8 (Hilbert–Noether). *Suppose a finite group G acts as automorphisms on a finitely generated commutative k-algebra A. Then the fixed point subalgebra A^G is a finitely generated k-algebra over which A is integral. The variety $\max(A^G)$ is the quotient of $\max(A)$ by the action of G.*

PROOF. If a is an element of A, then a satisfies the monic equation $\prod_{g \in G}(x - g(a)) = 0$ with coefficients in A^G. This proves that A is integral over A^G. Let B be the subalgebra of A^G generated by the coefficients of the monic polynomials satisfied by a finite set of k-algebra generators of A. Then B is a finitely generated commutative k-algebra, and hence Noetherian. So A is a finitely generated B-module, and hence so is A^G. Thus A^G is a finitely generated k-algebra.

Now the map of varieties

$$\max(A) \twoheadrightarrow \max(A^G)$$

is a finite dominant map, and hence surjective. If $\mathcal{M}$ and $\mathcal{M}'$ are maximal ideals in A which are not G-conjugate, then there exists an element $a \in A$

with $a \in \mathcal{M}$ but $a \notin g(\mathcal{M}')$ for all $g \in G$. So $\prod_{g \in G} g(a)$ is an element of A^G lying in $\mathcal{M}$ but not in $\mathcal{M}'$. It follows that the preimage in $\max(A^G)$ is a single G-orbit of points in $\max(A)$. $\qquad\square$

COHEN–MACAULAY RINGS. We end this section with a brief review of Cohen–Macaulay rings and modules. For further details, see Zariski and Samuel [286], Appendix 6, Matsumura [179] Chapter 6, and Serre [233], Chapter 4.

DEFINITION 5.4.9. *Suppose that* $A = \bigoplus_{n \geq 0} A_n$ *is a finitely generated graded (anti-) commutative k-algebra, and* $M = \bigoplus_{n \geq 0} M_n$ *is a finitely generated graded A-module. A sequence* $\zeta_1, \ldots, \zeta_r$ *of homogeneous elements of degree* $n_1, \ldots, n_r$ *in A is said to be a* **regular sequence** *for M if for each $i = 1, \ldots, r$ the map*

$$M_n/M_n \cap (\zeta_1, \ldots, \zeta_{i-1})M \to M_{n+n_i}/M_{n+n_i} \cap (\zeta_1, \ldots, \zeta_{i-1})M$$

induced by multiplication by ζ_i is injective. Note that we are not asking that $M_n/M_n \cap (\zeta_1, \ldots, \zeta_{i-1})M$ *be non-zero.*

The **depth** *of M is the length of the longest regular sequence. M is said to be* **Cohen–Macaulay** *if its depth is equal to its Krull dimension. The ring A is said to be Cohen–Macaulay if it is Cohen–Macaulay as a module over itself. This is equivalent to the condition that there is a polynomial subring $k[\zeta_1, \ldots, \zeta_r] \subseteq A$ generated by homogeneous elements ζ_i, such that A is a finitely generated free module over $k[\zeta_1, \ldots, \zeta_r]$.*

The following theorem is proved in Serre [233], p. IV–20, Theorem 2; see also Stanley [251], Proposition 3.1. Actually the proof given in Serre [233] is for strictly commutative rings, but the proof carries over verbatim to the graded commutative case.

THEOREM 5.4.10. *Suppose that A is a finitely generated graded commutative k-algebra. Then the following are equivalent.*

(i) *There exists a polynomial subring* $k[\zeta_1, \ldots, \zeta_r] \subseteq A$ *generated by homogeneous elements* ζ_i, *such that A is a finitely generated free module over* $k[\zeta_1, \ldots, \zeta_r]$.

(ii) *For every polynomial subring* $k[\zeta_1, \ldots, \zeta_r] \subseteq A$ *generated by homogeneous elements* ζ_i, *such that A is finitely generated as a module over* $k[\zeta_1, \ldots, \zeta_r]$ *(i.e., $\zeta_1, \ldots, \zeta_r$ is a homogeneous set of parameters for A), A is a free module.*

One can recognise regular sequences in polynomial rings using the following theorem.

THEOREM 5.4.11 (Macaulay). *Suppose $I = (\zeta_1, \ldots, \zeta_r)$ is an ideal in a polynomial ring $A = k[x_1, \ldots, x_n]$. Then $\zeta_1, \ldots, \zeta_r$ is a regular sequence for A if and only if the Krull dimension of A/I is equal to $n - r$.*

PROOF. See Zariski and Samuel [286], Appendix 6, Theorem 2. $\qquad\square$

EXERCISE. Show that if A and B are commutative rings, finitely generated over an algebraically closed field k, then

$$\max(A \otimes_k B) \cong \max(A) \times \max(B).$$

5.5. Example: extraspecial 2-groups

Quillen [211] calculated the mod 2 cohomology of extraspecial 2-groups. We shall give a simplified version of his computation [36], as an example of how to apply the commutative algebra discussed in the last section to the calculation of cohomology rings. This is also a good example to keep in mind when reading about the Quillen stratification in the next section.

We are interested in examining the Lyndon–Hochschild–Serre spectral sequence of a central extension

$$1 \to N \to G \to E \to 1$$

with N cyclic of order two and E an elementary abelian 2-group of order 2^n. Strictly speaking, for G to be extraspecial one requires $N = Z(G)$ in the above situation. However, we shall abuse terminology by using the term extraspecial even when this condition does not hold.

We start with some preliminaries on quadratic forms. Recall that a quadratic form on a vector space k of characteristic two is a map $q : V \to k$ with the property that

$$q(x + y) = q(x) + q(y) + b(x, y)$$

with $b : V \times V \to k$ a symmetric bilinear form. A linear subspace W of V is said to be isotropic if $q(w) = 0$ for all $w \in W$. If k' is an extension field of k, we extend q to a quadratic form on $k' \otimes_k V$, also denoted q by abuse of notation, by setting

$$q\Big(\sum_i \lambda_i x_i\Big) = \sum_i \lambda_i^2 q(x_i) + \sum_{i<j} \lambda_i \lambda_j b(x_i, x_j).$$

LEMMA 5.5.1. *Suppose that k is a perfect field of characteristic 2. Let q be a quadratic form defined on a vector space E over $\mathbb{F}_2$. Let $V = k \otimes_{\mathbb{F}_2} E$, and denote by $F : V \to V$ the Frobenius morphism, corresponding to squaring elements of k.*

Let h be the codimension in E of a maximal isotropic subspace E'. If v is a vector in V with the property that the linear span of $v, F(v), \ldots, F^{r-1}(v)$ is isotropic, with $r \geq h$, then the linear span of $v, F(v), \ldots, F^r(v)$ is also isotropic.

PROOF. Let $W = k \otimes_{\mathbb{F}_2} E'$, so that W is a maximal F-stable isotropic subspace of V. Since $r \geq h$, there is a linear relation of the form

$$w + \mu_s F^s(v) + \mu_{s+1} F^{s+1}(v) + \cdots + \mu_t F^t(v) = 0$$

with $w \in W$, $0 \leq s \leq t \leq r$ and μ_s, μ_t non-zero. Since k is perfect, F is invertible and so without loss of generality $s = 0$. Then by applying F^{r-t}

and adding if necessary, we obtain such a relation with $t = r$. Applying $q(-)$ and $b(w, -)$ to the above relation, we obtain

$$b(w, \mu_0 v + \cdots + \mu_r F^r(v)) + q(\mu_0 v + \cdots + \mu_r F^r(v)) = 0$$
$$b(w, \mu_0 v + \cdots + \mu_r F^r(v)) = 0.$$

Subtracting and using the fact that $q(F^i(v)) = 0$ and $b(F^i(v), F^j(v)) = 0$ for $|i - j| < r$, we obtain

$$\mu_0 \mu_r b(v, F^r(v)) = 0.$$

Since μ_0 and μ_r are non-zero, we deduce that $b(v, F^r(v)) = 0$, which completes the proof of the lemma.   □

Now, keeping the notation of the lemma, assume that k is algebraically closed, and denote by $k[V]$ the coordinate ring of V; namely the ring of polynomial functions on V. Thus F can be thought of as the k-linear ring homomorphism on $k[V]$ sending each element of E in degree one to its square in degree two. So for example we can regard $q(v)$ as an element of $k[V]$ given by a degree two polynomial, and $b(v, F^i(v))$ as an element of $k[V]$ given by a polynomial of degree $2^i + 1$.

PROPOSITION 5.5.2. *The sequence*

$$q(v), b(v, F(v)), b(v, F^2(v)), \ldots, b(v, F^{h-1}(v))$$

is a regular sequence in $k[V]$; and $b(v, F^r(v))$ is in the radical of the ideal they generate, for $r \geq h$. The variety defined by this ideal is the union of the F-stable isotropic subspaces of V; namely the union over the isotropic subspaces $E' \leq E$ of $k \otimes_{\mathbb{F}_2} E'$.

PROOF. For $r = 0$ set $J_r = \{0\}$, and for $1 \leq r \leq \infty$ we denote by J_r the ideal in $k[V]$ generated by the sequence of elements $q(v), b(v, F(v)), \ldots, b(v, F^{r-1}(v))$. Since

$$q\left(\sum_{i=1}^{r-1} \lambda_i F^i(v)\right) = \sum_{i=1}^{r-1} \lambda_i^2 q(v)^{2^i} + \sum_{1 \leq i < j \leq r-1} \lambda_i \lambda_j b(v, F^{j-i}(v))^{2^i},$$

it follows that a vector $v \in V$ lies in the subvariety $V(J_r)$ if and only if the linear span of $v, F(v), \ldots, F^{r-1}(v)$ is isotropic. So by the lemma, we have

$$V(J_h) = V(J_{h+1}) = \cdots = V(J_\infty),$$

so that the new generators are in the radical of the ideal generated by the previous ones by Hilbert's Nullstellensatz (Theorem 5.4.2). Now $V(J_\infty)$ is just the union of the F-stable isotropic subspaces of V. Some elementary linear algebra shows that an isotropic subspace of V is F-stable if and only if it is of the form $k \otimes_{\mathbb{F}_2} E'$ for some isotropic subspace $E' \leq E$. So $V(J_\infty)$ is a subvariety of codimension h in V. Since J_h is generated by exactly h elements, it follows that these elements form a regular sequence (see Theorem 5.4.11).   □

Now returning to the above central extension

$$1 \to N \to G \to E \to 1,$$

if we write elements of N and E additively, we can identify N with $\mathbb{F}_2$ and E with a vector space over $\mathbb{F}_2$. We define a quadratic form $q : E \to N$ by taking $q(x)$ to be the square of a preimage $\tilde{x}$ of x in G. Since commutators are central and self-inverse, we have

$$(\tilde{x}\tilde{y})^2 = \tilde{x}^2 \tilde{y}(\tilde{y}^{-1}\tilde{x}^{-1}\tilde{y}\tilde{x})\tilde{y} = \tilde{x}^2 \tilde{y}^2[\tilde{x}, \tilde{y}],$$

so the associated symmetric bilinear form $b : E \times E \to N$ is given by the commutator map on preimages in G.

We have $H^*(E, \mathbb{F}_2) = \mathbb{F}_2[x_1, \dots, x_n]$ with $\deg(x_i) = 1$. Let k be an algebraically closed field of characteristic two, and let $V = k \otimes_{\mathbb{F}_2} E$, a vector space of dimension n over k. Then $H^*(E, k) = k[x_1, \dots, x_n] = k[V]$, the coordinate ring of V regarded as an affine variety over k. The element of $H^2(E, \mathbb{F}_2)$ classifying the above central extension is $q(v)$. This is regarded as a polynomial function on V of degree two as follows. Let $e_1, \dots, e_n$ be a basis of E over $\mathbb{F}_2$. Then if $v = \sum_{i=1}^n x_i e_i$, we have

$$q(v) = q(\sum_{i=1}^n x_i e_i) = \sum_{i=1}^n q(e_i) x_i^2 + \sum_{1 \leq i < j \leq n} b(e_i, e_j) x_i x_j.$$

Thus the $q(e_i)$ and $b(e_i, e_j)$ are the coefficients of $q(v)$ as a quadratic polynomial in $k[V]$ with coefficients in $\mathbb{F}_2$.

Recall that the Frobenius map F on $k[V]$ is defined to be the k-linear ring homomorphism sending x_i to x_i^2. Thus we have (see Section 4.4)

$$Sq^1 q(v) = \sum_{1 \leq i < j \leq n} b(e_i, e_j)(x_i x_j^2 + x_j x_i^2) = b(v, F(v)),$$

as a polynomial of degree three in $k[V]$. Similarly for $r \geq 1$ we have

$$Sq^{2^{r-1}} Sq^{2^{r-2}} \cdots Sq^1 q(v) = \sum_{1 \leq i < j \leq n} b(e_i, e_j)(x_i x_j^{2^r} + x_j x_i^{2^r}) = b(v, F^r(v)).$$

THEOREM 5.5.3 (Quillen [**208**]). *Let G be as above. Then*

$$H^*(G, \mathbb{F}_2) = \mathbb{F}_2[x_1, \dots, x_n]/(q(v), b(v, F(v)), \dots, b(v, F^{h-1}(v))) \otimes \mathbb{F}_2[\zeta].$$

Here, $x_1, \dots, x_n$ are degree one elements inflated from $H^*(E, \mathbb{F}_2)$. The integer h is the codimension in V of a maximal F-stable isotropic subspace; in other words, 2^h is the index in G of a maximal elementary abelian subgroup. The element ζ is any element of degree 2^h which restricts non-trivially to $H^*(N, \mathbb{F}_2)$.

The elements $b(v, F^j(v))$, for $j \geq h$, are in the ideal generated by the regular sequence $q(v), b(v, F(v)), \dots, b(v, F^{h-1}(v))$.

PROOF. We examine the Lyndon–Hochschild–Serre spectral sequence

$$H^*(E, H^*(N, \mathbb{F}_2)) \Rightarrow H^*(G, \mathbb{F}_2).$$

The E_2 page of this spectral sequence is

$$H^*(E, \mathbb{F}_2) \otimes H^*(N, \mathbb{F}_2) = \mathbb{F}_2[x_1, \ldots, x_n] \otimes \mathbb{F}_2[z]$$

with $x_1, \ldots, x_n \in E_2^{1,0}$ and $z \in E_2^{0,1}$. Since $q(v) \in H^2(E, \mathbb{F}_2)$ is the element corresponding to the central extension, we have $d_2(z) = q(v) \in E_2^{2,0}$. Since Steenrod operations commute with transgressions (Section 4.8), we have, for $r \geq 1$,

$$d_{2^r+1}(z^{2^r}) = d_{2^r+1}(Sq^{2^{r-1}} Sq^{2^{r-2}} \cdots Sq^1(z))$$
$$= Sq^{2^{r-1}} Sq^{2^{r-2}} \cdots Sq^1 q(v) = b(v, F^r(v)).$$

Now by Proposition 5.5.2, the elements $q(v), b(v, F(v)), \ldots, b(v, F^{h-1}(v))$ form a regular sequence in $H^*(E, \mathbb{F}_2)$, and so by induction on r we have, for $1 \leq r \leq h$ and $2^{r-1} + 1 < j \leq 2^r + 1$,

$$E_j^{**} = \mathbb{F}_2[x_1, \ldots, x_n]/(q(v), b(v, F(v)), \ldots, b(v, F^{r-1}(v))) \otimes \mathbb{F}_2[z^{2^r}].$$

To complete the calculation, it suffices to show that the element $z^{2^h} \in E_2^{0,2^h}$ is a universal cycle. Let H be a maximal elementary abelian subgroup of G, of index 2^h. Then z is a restriction of an element $\hat{z} \in H^1(H, \mathbb{F}_2)$. One then easily checks using the Mackey formula for Evens' norm map (Section 4.1), that $\text{norm}_{H,G}(\hat{z})$ is an element of $H^*(G, \mathbb{F}_2)$ restricting to $z^{2^h} \in H^*(N, \mathbb{F}_2)$. Thus z^{2^h} is a universal cycle. In particular, for $j \geq h$, $d_{2^j+1}(z^{2^j}) = 0$, and so $b(v, F^j(v))$ is in the ideal generated by $q(v), b(v, F(v)), \ldots, b(v, F^{h-1}(v))$. $\quad\square$

EXERCISES. 1. Let G be as above. Show that the E_2 page of the Eilenberg–Moore spectral sequence (see Sections 3.7 and 4.8) converging to $H^*(G, \mathbb{F}_2)$ is equal to the E_∞ page, and has

$$\mathbb{F}_2[x_1, \ldots, x_n]/(q(v), b(v, F(v)), \ldots, b(v, F^{h-1}(v)))$$

on the vertical axis.

2. (Avrunin and Carlson) Let G be the nilpotent 2-group of class two and order 2^{2n} generated by elements $g_1, \ldots, g_n$ subject to the relations $g_i^4 = 1$, $[g_i, g_{i+1}] = g_{i+1}^2$, and $[g_i, g_j] = 1$ if $j \geq i+2$. Let N be the central subgroup of order 2^n generated by $g_1^2, \ldots, g_n^2$, and let $E = G/N$. Show that the Lyndon–Hochschild–Serre spectral sequence for the central extension $1 \to N \to G \to E \to 1$ satisfies $E_3 = E_\infty$. Show that there is an element x of $H^1(G, \mathbb{F}_p)$ with $x^n \neq 0$ but $x^{n+1} = 0$.

5.6. The Quillen stratification

The basic object of study here is the maximal ideal spectrum

$$V_G = \max H^{\cdot}(G, k).$$

Here, $H^{\cdot}(G, k)$ denotes $H^*(G, k)$ if $\text{char}(k) = 2$, and the subring $H^{ev}(G, k)$ of elements of even degree if $\text{char}(k) \neq 2$ (in this case, elements of odd degree square to zero). Thus $H^{\cdot}(G, k)$ is a commutative graded ring, and so V_G is a homogeneous affine variety. We denote by $\bar{V}_G$ the projective variety of one smaller dimension

$$\bar{V}_G = \text{proj}\, H^{\cdot}(G, k).$$

If H is a subgroup of G then we have a restriction map

$$\text{res}_{G,H} : H^{\cdot}(G, k) \to H^{\cdot}(H, k),$$

and hence a corresponding map of varieties $\text{res}^*_{G,H} : V_H \to V_G$.

PROPOSITION 5.6.1. *We have*

$$V_G = \bigcup_{\substack{E \leq G \\ \text{elemab}}} \text{res}^*_{G,E} V_E.$$

PROOF. By Theorem 5.2.2, an element of $H^{\cdot}(G, k) = \text{Ext}^{\cdot}_{kG}(k, k)$ is nilpotent if and only if it is nilpotent on restriction to every elementary abelian subgroup $E \leq G$. Thus the map

$$\bigoplus \text{res}_{G,E} : H^{\cdot}(G, k) \to \bigoplus H^{\cdot}(E, k)$$

has nilpotent kernel. Also, by Corollary 4.2.5, the right-hand side is finitely generated as a module over the image. Thus

$$\bigcup \text{res}^*_{G,E} : \bigcup V_E \to V_G$$

is a finite dominant map, so that by Theorem 5.4.7 it is surjective. $\qquad\square$

The Quillen Stratification Theorem [**209, 210**] is a refinement of the above proposition, giving a decomposition of V_G into disjoint pieces corresponding to the conjugacy classes of elementary abelian subgroups of G.

Recall from Section 3.5 of Volume I that if E is an elementary abelian p-group of rank n, then

$$H^*(E, k) \cong k[x_1, \dots, x_n]$$

with $\deg(x_i) = 1$ if $p = 2$, while

$$H^*(E, k) \cong \Lambda(x_1, \dots, x_n) \otimes k[y_1, \dots, y_n]$$

with $\deg(x_i) = 1$, $\deg(y_i) = 2$ and $\beta(x_i) = y_i$ if p is odd. In either case, we see that V_E is a k-vector space of dimension equal to the p-rank of E. We define

$$V_E^+ = V_E \setminus \bigcup_{E' < E} \text{res}^*_{E,E'} V_{E'}.$$

Thus V_E^+ is V_E with the hyperplanes defined over $\mathbb{F}_p$ removed. Let

$$\sigma_E = \prod_{0 \neq \zeta \in H^1(E, \mathbb{F}_p)} \beta(\zeta) \in H^{\cdot}(E, k).$$

Then since the subvariety of V_E defined by $\beta(\zeta)$ is $\text{res}^*_{E,E'} V_{E'}$, where E' is the maximal subgroup corresponding to ζ, the subvariety defined by the above element σ_E is $\bigcup_{E' < E} \text{res}^*_{E,E'} V_{E'}$. So the points of V_E^+ correspond to maximal ideals of $H^{\cdot}(E, k)$ not containing σ_E, and hence to the maximal ideals of $H^{\cdot}(E, k)[\sigma_E^{-1}]$. So we have

$$V_E^+ = \max H^{\cdot}(E, k)[\sigma_E^{-1}].$$

We define $V_{G,E} = \operatorname{res}_{G,E}^* V_E \subseteq V_G$, and $V_{G,E}^+ = \operatorname{res}_{G,E}^* V_E^+$.

The following lemma produces enough elements of cohomology to describe the structures of these varieties.

LEMMA 5.6.2. *If E is an elementary abelian p-subgroup of G with $|N_G(E) : E| = p^a h$ and $(p, h) = 1$ then the following hold:*

(i) *If $y \in H^{\cdot}(E, k)$ is invariant under the action of $N_G(E)$ then there exists an element $y' \in H^{\cdot}(G, k)$ with*

$$\operatorname{res}_{G,E}(y') = (\sigma_E.y)^{p^a}.$$

(ii) *There exists an element $\rho_E \in H^{\cdot}(G, k)$ such that $\operatorname{res}_{G,E}(\rho_E) = (\sigma_E)^{p^a}$, and such that if E is not contained in any conjugate of an elementary abelian p-subgroup $E' \leq G$ then $\operatorname{res}_{G,E'}(\rho_E) = 0$.*

PROOF. (i) Without loss of generality y is homogeneous. Let

$$z = \operatorname{norm}_{E,G}(1 + \sigma_E.y).$$

Then the Mackey formula 4.1.2 (v) shows that

$$\operatorname{res}_{G,E}(z) = (1 + \sigma_E.y)^{p^a h} = (1 + (\sigma_E.y)^{p^a})^h$$
$$= 1 + h(\sigma_E.y)^{p^a} + \text{terms of higher degree.}$$

So we take for y' the homogeneous part of z of degree $p^a . \deg(\sigma_E.y)$, divided by h.

(ii) We take $y = 1$ in part (i), and write ρ_E for the element y' obtained in this case. Since $\operatorname{res}_{E,F}(\sigma_E) = 0$ for all proper subgroups F of E, the Mackey formula 4.1.2 (v) shows that $\operatorname{res}_{G,E'}(\rho_E) = 0$ unless E is conjugate to a subgroup of E'.   □

THEOREM 5.6.3 (Quillen Stratification [**209, 210**]). *The variety V_G is the disjoint union of locally closed subvarieties $V_{G,E}^+$, one for each conjugacy class of elementary abelian subgroups $E \leq G$. Let $W_G(E) = N_G(E)/C_G(E)$. Then $W_G(E)$ acts on V_E and V_E^+, and $\operatorname{res}_{G,E}^*$ induces an inseparable isogeny*

$$V_E^+/W_G(E) \to V_{G,E}^+$$

(see the end of Section 5.4).

PROOF. By the lemma we have

$$\{(H^{\cdot}(E, k)[\sigma_E^{-1}])^{W_G(E)}\}^{[p^a]} \subseteq \operatorname{res}_{G,E} H^{\cdot}(G, k)[\rho_E^{-1}] \subseteq (H^{\cdot}(E, k)[\sigma_E^{-1}])^{W_G(E)}$$

and so we have an inseparable isogeny

$$\max(H^{\cdot}(E, k)[\sigma_E^{-1}])^{W_G(E)} \to \max \operatorname{res}_{G,E} H^{\cdot}(G, k)[\rho_E^{-1}].$$

By Proposition 5.4.8 the left-hand side is the quotient of V_E^+ by the action of $W_G(E)$, while the right-hand side is $V_{G,E}^+$. Thus we have an inseparable isogeny

$$V_E^+/W_G(E) \to V_{G,E}^+.$$

If E is not conjugate to a subgroup of E' then by part (ii) of the lemma we have $\text{res}_{G,E'}(\rho_E) = 0$, and so ρ_E is a function which is everywhere non-zero on $V_{G,E}^+$ but everywhere zero on $V_{G,E'}^+$. It follows that the $V_{G,E}^+$ are disjoint. Since V_G is the union of the $\text{res}_{G,E}^* V_E$ by Proposition 5.6.1, it is the union of the $V_{G,E}^+$. □

A more succinct way to write the Quillen stratification theorem is as follows. We form a category $\mathcal{C}_G$ whose objects are the elementary abelian subgroups of G, and where a morphism from E to E' consists of an element $g \in G$ such that $gEg^{-1} \subseteq E'$. Then $E \mapsto V_E$ is a covariant functor on $\mathcal{C}_G$, and so it makes sense to talk of the limit $\varinjlim_E V_E$. This is the maximal ideal spectrum of $\varprojlim_E H^{\cdot}(E,k)$.

COROLLARY 5.6.4. *The natural map*

$$\varinjlim_E V_E \to V_G$$

is an inseparable isogeny. Equivalently, the natural map

$$q_G : H^{\cdot}(G,k) \to \varprojlim_E H^{\cdot}(E,k)$$

is a finite map, whose kernel is nilpotent, and with the property that there is some power p^a of p such that the p^ath power of any element in the right hand side lies in the image of q_G. □

REMARK. It is an interesting question to ask when the map q_G is an isomorphism of rings. Gunawardena, Lannes and Zarati [**121**] have shown that one can tell whether q_G is an isomorphism just by looking at $H^{\cdot}(G,k)$ as a module over the Steenrod algebra. Using this, they have shown that for $p = 2$ and G a symmetric group $\mathcal{S}_n$ the map q_G is an isomorphism. This is false for p odd, but it seems that the reason is in some sense that the definition of $H^{\cdot}(G,k)$ is wrong for p odd. A better definition would be to take the subring consisting of all elements which are annihilated by all elements of the Steenrod algebra involving the Bockstein operation. For example, with this definition, we have $H^{\cdot}(G_1 \times G_2, k) \cong H^{\cdot}(G_1, k) \otimes H^{\cdot}(G_2, k)$, so that if E is elementary abelian then $H^{\cdot}(E,k)$ is just the polynomial subring, and the exterior part disappears. It is still an open question as to whether, with this definition, the map q_G is an isomorphism for $G = \mathcal{S}_n$ and p odd.

EXERCISE. Show that in Theorem 5.6.3, the action of $W_G(E)$ on V_E^+ is free. (Hint: show that if $1 \neq w \in W_G(E)$ then the 1-eigenspace of w on V_E is contained in $\text{res}_{E,E'}^*(V_{E'})$ for some $E' < E$)

5.7. Varieties for modules

If M is a kG-module, we associate to M a subvariety $V_G(M)$ of V_G as follows. There is a natural map

$$\Phi_M : H^{\cdot}(G,k) = \operatorname{Ext}^{\cdot}_{kG}(k,k) \xrightarrow{\otimes M} \operatorname{Ext}^*_{kG}(M,M)$$

whose image is central, by Corollary 3.2.2 of Volume I. We denote the kernel of this map by $I_G(M)$. This is a homogeneous ideal in $H^{\cdot}(G,k)$, and therefore determines a closed homogeneous subvariety $V_G(M) = \max H^{\cdot}(G,k)/I_G(M)$ of V_G. The corresponding projective variety is denoted by $\bar{V}_G(M) \subseteq \bar{V}_G$.

If M' is another kG-module, we write $I_G(M',M)$ for the annihilator in the ring $\operatorname{Ext}^{\cdot}_{kG}(k,k)$ of the module $\operatorname{Ext}^*_{kG}(M',M)$, with the cup product action. We write $V_G(M',M)$ for the closed homogeneous subvariety of V_G defined by this ideal, so that in particular $V_G(M) = V_G(M,M)$. Since this cup product action factors as Φ_M followed by Yoneda composition, or as $\Phi_{M'}$ followed by Yoneda composition (Proposition 3.2.1 of Volume I), we have

$$I_G(M',M) \supseteq I_G(M') + I_G(M)$$
$$V_G(M',M) \subseteq V_G(M') \cap V_G(M).$$

If $0 \to M' \to M'' \to M''' \to 0$ is a short exact sequence of kG-modules, then the long exact Ext sequence (Section 2.5 of Volume I) shows that

$$I_G(M'',M) \supseteq I_G(M',M).I_G(M''',M)$$
$$V_G(M'',M) \subseteq V_G(M',M) \cup V_G(M''',M).$$

Combining these statements, we obtain the following:

PROPOSITION 5.7.1. *If M is a finitely generated kG-module then*

$$V_G(M) = \bigcup_S V_G(S,M),$$

where S runs over the simple kG-modules, or over the composition factors of M. In particular, if G is a p-group then $V_G(M) = V_G(k,M)$, the subvariety of V_G defined by the annihilator of $H^*(G,M)$.

PROOF. The containment of the right-hand side in the left-hand side follows from the first of the above inequalities, while the reverse containment follows from the second.   □

REMARK. If M is a finitely generated kG-module, then by Corollary 4.2.4, the ring $\operatorname{Ext}^*_{kG}(M,M)$ is finitely generated as a module over the image of Φ_M. Thus if we set

$$\tilde{V}_G(M) = \max Z\operatorname{Ext}^*_{kG}(M,M),$$

the maximal ideal spectrum of the centre of $\operatorname{Ext}^*_{kG}(M,M)$, then by Theorem 5.4.7 the map

$$\Phi^*_M : \tilde{V}_G(M) \to V_G(M)$$

is a finite surjective map.

In some sense, we should not have to stop at the centre. Since the ring $\text{Ext}^*_{kG}(M, M)$ is finitely generated as a module over its centre, it is an affine polynomial identity (P.I.) algebra, see McConnell and Robson [**184**], Chapter 13, Corollary 1.13. There is a well developed theory of maximal ideal spectra for such rings, see for example Procesi's book [**207**].

We begin with some properties of the varieties $V_G(M)$ which are easy consequences of our previous discussions.

PROPOSITION 5.7.2. *If M is a finitely generated kG-module then the dimension of $V_G(M)$ is equal to the complexity $c_G(M)$. In particular $V_G(M) = \{0\}$ if and only if M is projective.*

PROOF. Since $\text{Ext}^*_{kG}(M, M)$ is finitely generated as a module over the ring $H^{\cdot}(G, k)$ (Corollary 4.2.4), we have by Proposition 5.3.5 and Theorem 5.4.6

$$c_G(M) = \gamma(\text{Ext}^*_{kG}(M, M)) = \gamma(H^{\cdot}(G, k)/I_G(M)) = \dim V_G(M). \qquad \square$$

PROPOSITION 5.7.3. *We have*

$$V_G(M) = V_G(M^*) = V_G(\Omega(M)) = V_G(\Omega^{-1}(M)).$$

PROOF. This follows from the isomorphisms

$$\text{Ext}^*_{kG}(M, M) \cong \text{Ext}^*_{kG}(M^*, M^*) \cong \text{Ext}^*_{kG}(\Omega(M), \Omega(M))$$
$$\cong \text{Ext}^*_{kG}(\Omega^{-1}(M), \Omega^{-1}(M))$$

as modules for $\text{Ext}^{\cdot}_{kG}(k, k)$. $\qquad \square$

PROPOSITION 5.7.4. *We have*

$$V_G(M) = \bigcup_{\substack{E \le G \\ \text{elemab}}} \text{res}^*_{G,E} V_E(M).$$

PROOF. According to Theorem 5.2.2, an element of $\text{Ext}^*_{kG}(M, M)$ is nilpotent if and only if it is nilpotent on restriction to every elementary abelian subgroup $E \le G$. Applying this to elements in the image of Φ_M, we have

$$\sqrt{I_G(M)} = \bigcap_{\substack{E \le G \\ \text{elemab}}} \text{res}^{-1}_{G,E} \sqrt{I_E(M)}.$$

The subvariety of V_G determined by $\text{res}^{-1}_{G,E} \sqrt{I_E(M)}$ is the same as the subvariety defined by $\text{res}^{-1}_{G,E} I_E(M)$, namely $\text{res}^*_{G,E} V_E(M)$. $\qquad \square$

PROPOSITION 5.7.5. *We have*

$$V_G(M_1 \oplus M_2) = V_G(M_1) \cup V_G(M_2).$$

PROOF. Tensoring with $M_1 \oplus M_2$ factors as follows:

$$\mathrm{Ext}^{\cdot}_{kG}(k,k) \xrightarrow{(\otimes M_1, \otimes M_2)} \mathrm{Ext}^*_{kG}(M_1, M_1) \oplus \mathrm{Ext}^*_{kG}(M_2, M_2)$$
$$\hookrightarrow \mathrm{Ext}^*_{kG}(M_1 \oplus M_2, M_1 \oplus M_2)$$

and so

$$I_G(M_1 \oplus M_2) = I_G(M_1) \cap I_G(M_2)$$
$$V_G(M_1 \oplus M_2) = V_G(M_1) \cup V_G(M_2).$$ □

Next, we develop a version of the Quillen stratification theorem for the varieties $V_G(M)$, due to Avrunin and Scott [24]. We begin with a technical lemma.

LEMMA 5.7.6. *Suppose H is a normal subgroup of index p in G, and x is the inflation to G of a non-zero element $\bar{x} \in H^1(G/H, \mathbb{F}_p)$. If $\zeta \in H^*(G,k)$ with $\mathrm{res}_{G,H}(\zeta) \in I_H(M)$, then some power of ζ lies in the ideal generated by $I_G(M)$ and the Bockstein $\beta(x)$.*

PROOF. If $\mathrm{res}_{G,H}(\zeta) \in I_H(M)$ then $\mathrm{res}_{G,H}(\zeta_M) = 0$, and so by Theorem 5.2.1, $\zeta^2_M = \beta(x).\xi$ for some $\xi \in \mathrm{Ext}^*_{kG}(M,M)$. Now by Corollary 4.2.4 $\mathrm{Ext}^*_{kG}(M,M)$ is finitely generated as a module over $\mathrm{Ext}^{\cdot}_{kG}(k,k)$, and so ξ satisfies some monic polynomial of the form

$$\xi^n + (a_{n-1})_M \xi^{n-1} + \cdots + (a_0)_M = 0$$

with the $a_i \in \mathrm{Ext}^{\cdot}_{kG}(k,k)$. Multiplying this equation by $\beta(x)^n$, we obtain

$$\zeta^{2n}_M + (a_{n-1})_M \beta(x) \zeta^{2n-2}_M + \cdots + (a_0)_M \beta(x)^n = 0.$$

Thus the element

$$\zeta^{2n} + a_{n-1} \beta(x) \zeta^{2n-2} + \cdots + a_0 \beta(x)^n$$

is in $I_G(M)$. □

PROPOSITION 5.7.7. *If $E' \leq E$ are elementary abelian p-groups and M is a kE-module, then regarding $V_{E'}$ as embedded in V_E via $\mathrm{res}^*_{E,E'}$, we have*

$$V_{E'}(M) = V_{E'} \cap V_E(M).$$

PROOF. Arguing by induction, we may suppose that $|E : E'| = p$. If $\zeta \in H^{\cdot}(E,k)$ lies in $I_E(M)$ then $\mathrm{res}_{E,E'}(\zeta)$ lies in $I_{E'}(M)$. Conversely, if $\mathrm{res}_{E,E'}(\zeta)$ lies in $I_{E'}(M)$ then by the lemma, some power of ζ lies in the ideal generated by $\beta(x)$ and $I_E(M)$, where x is the inflation to E of a non-zero element of $H^1(E/E', k)$. So we have

$$\sqrt{\mathrm{res}^{-1}_{E,E'} I_{E'}(M)} = \sqrt{\langle \mathrm{Ker}(\mathrm{res}_{E,E'}), I_E(M) \rangle}.$$

Taking varieties, we obtain the required equality. □

We now define

$$V_E^+(M) = V_E^+ \cap V_E(M)$$
$$V_{G,E}(M) = \mathrm{res}_{G,E}^* V_E(M) \subseteq V_G(M)$$
$$V_{G,E}^+(M) = \mathrm{res}_{G,E}^* V_E^+(M).$$

THEOREM 5.7.8 (Avrunin and Scott [**24**]). *Suppose M is a finitely generated kG-module. Then the variety $V_G(M)$ is the disjoint union of locally closed subvarieties $V_{G,E}^+(M)$, one for each conjugacy class of elementary abelian subgroups $E \leq G$. The group $W_G(E)$ acts on $V_E(M)$ and $V_E^+(M)$, and $\mathrm{res}_{G,E}^*$ induces an inseparable isogeny*

$$V_E^+(M)/W_G(E) \to V_{G,E}^+(M).$$

PROOF. It follows from the above proposition that

$$V_E^+(M) = V_E(M) \setminus \bigcup_{E' < E} \mathrm{res}_{E,E'}^* V_{E'}(M)$$

so that $V_E(M)$ is the disjoint union over $E' \leq E$ of locally closed subvarieties $V_{E'}^+(M)$. It now follows from Proposition 5.7.4 that

$$V_G(M) = \bigcup_{\substack{E \leq G \\ \text{elemab}}} \mathrm{res}_{G,E}^* V_E^+(M).$$

It follows from Theorem 5.6.3 that the subvarieties $V_{G,E}^+(M) = \mathrm{res}_{G,E}^* V_E^+(M)$ are disjoint and that $\mathrm{res}_{G,E}^*$ induces an inseparable isogeny

$$V_E^+(M)/W_G(E) \to V_{G,E}^+(M). \qquad \square$$

COROLLARY 5.7.9. *If H is a subgroup of G and M is a kG-module then*

$$V_H(M) = (\mathrm{res}_{G,H}^*)^{-1} V_G(M).$$

PROOF. Clearly $\mathrm{res}_{G,H}^* V_H(M) \subseteq V_G(M)$. Conversely, by the theorem, the diagram

$$
\begin{array}{ccc}
V_E^+(M)/W_H(E) & \longrightarrow\!\!\!\!\!\rightarrow & V_E^+(M)/W_G(E) \\
\downarrow \scriptstyle\sim & & \downarrow \scriptstyle\sim \\
V_{H,E}^+(M) & \xrightarrow{\ \mathrm{res}_{G,H}^*\ } & V_{G,E}^+(M)
\end{array}
$$

shows that $V_{H,E}^+(M) = (\mathrm{res}_{G,H}^*)^{-1} V_{G,E}^+(M).$ $\qquad \square$

PROPOSITION 5.7.10. *There is a natural isomorphism $V_{G_1 \times G_2} \cong V_{G_1} \times V_{G_2}$. If M_1 is a kG_1-module and M_2 is a kG_2-module then the image of $V_{G_1 \times G_2}(M_1 \otimes M_2)$ under this isomorphism is $V_{G_1}(M_1) \times V_{G_2}(M_2)$.*

PROOF. This follows directly from the Künneth Theorem (3.5.6 of Volume I; the Tor term there is zero, as we are working over a field). $\qquad \square$

THEOREM 5.7.11. *If M_1 and M_2 are kG-modules then*

$$V_G(M_1 \otimes M_2) = V_G(M_1) \cap V_G(M_2).$$

PROOF. The interior tensor product $M_1 \otimes M_2$ as a kG-module is the restriction from $G \times G$ to the diagonal $G = \Delta(G)$ of the exterior product $M_1 \otimes M_2$ as a $k(G \times G)$-module. So by Proposition 5.7.10 and Corollary 5.7.9 we have

$$V_G(M_1 \otimes M_2) = (\mathrm{res}_{G \times G, \Delta(G)}^*)^{-1} V_{G \times G}(M_1 \otimes M_2)$$
$$= (\mathrm{res}_{G \times G, \Delta(G)}^*)^{-1} V_G(M_1) \times V_G(M_2) = V_G(M_1) \cap V_G(M_2)$$

(Note that in general if X and Y are two subsets of a set Z then the inverse image of $X \times Y$ under the diagonal map $Z = \Delta(Z) \to Z \times Z$ is equal to $X \cap Y$). $\qquad\square$

COROLLARY 5.7.12. *If $V_G(M_1) \cap V_G(M_2) = \{0\}$ then $\mathrm{Ext}_{kG}^n(M_1, M_2) = 0$ for all $n > 0$.*

PROOF. By Proposition 5.7.3 and Theorem 5.7.11 we have $V_G(M_1^* \otimes M_2) = \{0\}$, and hence by Proposition 5.7.2 $M_1^* \otimes M_2$ is projective. Hence

$$\mathrm{Ext}_{kG}^n(M_1, M_2) \cong \mathrm{Ext}_{kG}^n(k, M_1^* \otimes M_2) = 0. \qquad\square$$

5.8. Rank varieties

In this section, we give an alternative description of $V_E(M)$ in case E is elementary abelian. This is Jon Carlson's notion [69] of the rank variety $V_E^\sharp(M)$. Carlson conjectured that $V_E(M) = V_E^\sharp(M)$. He proved the inclusion of the rank variety in the cohomology variety and the equality of the dimensions. Carlson's conjecture was proved by Avrunin and Scott [24].

We start with a discussion of shifted subgroups. Let $E = \langle g_1, \dots, g_r \rangle$ be an elementary abelian p-group of rank r (order p^r), and let k be an algebraically closed field of characteristic p. Then the linear subspace $V_E^\sharp$ of $J = J(kE)$ spanned by $g_1 - 1, \dots, g_r - 1$ has dimension r, and maps isomorphically onto J/J^2.

If

$$0 \neq v = \lambda_1(g_1 - 1) + \cdots + \lambda_r(g_r - 1)$$

is any element of $V_E^\sharp$, then $v^p = 0$, and so $1+v$ is a unit in kE with $(1+v)^p = 1$. If $v_1, \dots, v_r$ form a basis for $V_E^\sharp$, say $v_i = \sum_{j=1}^r \lambda_{ij}(g_j - 1)$ with (λ_{ij}) a nonsingular matrix, then there is an induced automorphism of kE given by

$$g_i \mapsto 1 + \sum_{j=1}^r \lambda_{ij}(g_j - 1).$$

If (μ_{ij}) is the inverse matrix, then the inverse of this automorphism is given by

$$g_i \mapsto 1 + \sum_{j=1}^{r} \mu_{ij}(g_j - 1).$$

It follows that given linearly independent elements $v_1, \ldots, v_s$ of $V_E^\sharp$, the subgroup

$$E' = \langle 1 + v_1, \ldots, 1 + v_s \rangle \subseteq (kE)^\times$$

has the property that its group algebra kE' is a subalgebra of kE over which kE is free as a module. A subgroup $E' \subseteq (kE)^\times$ obtained in this way is called a **shifted subgroup** of E. Thus for example a cyclic shifted subgroup is one generated by an element $g = 1 + v \in kE$ of order p, where $0 \neq v \in V_E^\sharp$. Any shifted subgroup can be extended to a maximal shifted subgroup, which has the same rank r as E. It should be noted that the inclusion $kE' \hookrightarrow kE$ of the group algebra of a shifted subgroup is not in general a homomorphism of Hopf algebras. So we need to be careful when restricting tensor products, or cup products in cohomology.

DEFINITION 5.8.1. *If M is a finitely generated kE-module, we define the* **rank variety** *of M to be*

$$V_E^\sharp(M) = \{0\} \cup \{v \in V_E^\sharp \mid M \!\downarrow_{\langle 1+v \rangle} \text{ is not free}\}.$$

LEMMA 5.8.2. *There is a natural isomorphism $V_E \cong V_E^\sharp$.*

PROOF. According to Proposition 2.4.3 of Volume I, we have a natural isomorphism

$$\mathrm{Ext}_{kE}^1(k, k) \cong \mathrm{Hom}_k(V_E^\sharp, k).$$

Now the structure of $\mathrm{Ext}_{kE}^*(k, k)$ is given in Section 3.5 of Volume I. If $p = 2$ then $\mathrm{Ext}_{kE}^*(k, k)$ is a polynomial algebra generated by $\mathrm{Ext}_{kE}^1(k, k)$, and the non-zero elements of $\mathrm{Ext}_{kE}^1(k, k)$ are precisely the linear maps from $V_E = \max \mathrm{Ext}_{kE}^*(k, k)$ to k. So V_E is the dual vector space of $\mathrm{Ext}_{kE}^1(k, k)$ and is hence isomorphic to $V_E^\sharp$. If $p > 2$ then we have the Bockstein map

$$\beta : \mathrm{Ext}_{kE}^1(k, k) \to \mathrm{Ext}_{kE}^2(k, k).$$

Modulo nilpotent elements, $\mathrm{Ext}_{kE}^*(k, k)$ is a polynomial algebra on the image of β, and we now use the same argument as before. $\qquad\square$

REMARK. For $p = 2$ the isomorphism of Lemma 5.8.2 is an isomorphism of varieties. For $p > 2$, it is semilinear via the Frobenius morphism (see Section 4.3) so that while it is a bijection, it is not invertible as a map of varieties.

In terms of lines through the origin, the above isomorphism may be described as follows. A line through the origin in $V_E^\sharp$ is represented by a cyclic shifted subgroup $T = \langle 1 + v \rangle$. The kernel of $\mathrm{res}_{E,T}$ is a maximal *graded* ideal $\mathcal{M}_T \leq \mathrm{Ext}^{\cdot}_{kE}(k,k)$ and hence corresponds to a line through the origin in V_E.

To see this, we need to verify that the Bockstein map commutes with restriction to a cyclic shifted subgroup. There are a number of ways of doing this, but we have chosen to follow Carlson [69] because it is explicit and convincing. We assume that p is odd for this purpose, because for $p = 2$ the Bockstein map is just the cup square, and anyway the isomorphism of Lemma 5.8.2 does not involve the Bockstein in this case.

We use the explicit resolution for k as a kE-module which begins

$$0 \to \Omega^2 k \to P_1 \xrightarrow{\delta} P_0 \to k \to 0$$

where $P_0 = kE$ and P_1 is the free kE-module with free basis $\gamma_1, \ldots, \gamma_r$. The differential $\delta : P_1 \to P_0$ is given by $\delta(\gamma_i) = g_i - 1$ ($1 \leq i \leq r$). The kernel $\Omega^2 k$ of δ is generated by elements

$$b_i = (g_i - 1)^{p-1} \gamma_i \qquad\qquad 1 \leq i \leq r$$
$$c_{ij} = (g_i - 1)\gamma_j - (g_j - 1)\gamma_i \qquad\qquad 1 \leq i < j \leq r.$$

The b_i correspond to the Bocksteins of the degree one elements, while the c_{ij} correspond to products of degree one elements. Let

$$0 \neq v = \lambda_1(g_1 - 1) + \cdots + \lambda_r(g_r - 1)$$

be an element of $V_E^{\sharp}$, so that $T = \langle 1 + v \rangle$ is a cyclic shifted subgroup. Then we have a diagram

$$
\begin{array}{ccccccccc}
0 & \longrightarrow & k & \longrightarrow & kT & \xrightarrow{\ v\ } & kT & \longrightarrow & k & \longrightarrow & 0 \\
& & \downarrow & & \downarrow & & \downarrow & & \| \\
0 & \longrightarrow & \Omega^2 k & \longrightarrow & P_1 & \xrightarrow{\ \delta\ } & P_0 & \longrightarrow & k & \longrightarrow & 0
\end{array}
$$

where the middle map in the top row is given by multiplication by v, the vertical map $kT \to P_0$ is the inclusion, and the vertical map $kT \to P_1$ sends the identity element to $\sum_{i=1}^{r} \lambda_i \gamma_i$. It is easily checked that this commutes. The inclusion of k into kT on the top row sends the identity element to v^{p-1}, whose image in P_1 is equal to

$$\left[\sum_{i=1}^{r} \lambda_i(g_i - 1) \right]^{p-1} \left[\sum_{i=1}^{r} \lambda_i \gamma_i \right].$$

The fact that the diagram commutes means that this lies in $\Omega^2 k$, and we claim that modulo $\mathrm{Rad}(\Omega^2 k)$ we have

$$\left[\sum_{i=1}^{r} \lambda_i(g_i - 1) \right]^{p-1} \left[\sum_{i=1}^{r} \lambda_i \gamma_i \right] \equiv \sum_{i=1}^{r} \lambda_i^p (g_i - 1)^{p-1} \gamma_i = \sum_{i=1}^{r} \lambda_i^p b_i.$$

This formula confirms that the version of the Bockstein defined by extending semilinearly from $\mathbb{F}_p$ to k via the Frobenius map (see Section 4.3) commutes with restriction to cyclic shifted subgroups. To prove the formula, we argue by induction on the rank. It is clearly true if $r = 1$. For $r > 1$, we expand the left hand side out as

$$\left[\sum_{i=1}^{r-1} \lambda_i(g_i - 1) + \lambda_r(g_r - 1)\right]^{p-1} \left[\sum_{i=1}^{r-1} \lambda_i\gamma_i + \lambda_r\gamma_r\right].$$

Taking the first term from each bracket gives

$$\sum_{i=1}^{r-1} \lambda_i^p (g_i - 1)^{p-1}\gamma_i$$

by the inductive hypothesis. Using the fact that $\binom{p-1}{j} \equiv (-1)^j \pmod{p}$, the remaining terms give

$$\left[\sum_{j=0}^{p-2}(-1)^j \left(\sum_{i=1}^{r-1} \lambda_i(g_i - 1)\right)^j (\lambda_r(g_r - 1))^{p-1-j}\right]\left[\sum_{i=1}^{r-1} \lambda_i\gamma_i\right]$$

$$+ \left[\sum_{j=0}^{p-1}(-1)^j \left(\sum_{i=1}^{r-1} \lambda_i(g_i - 1)\right)^j (\lambda_r(g_r - 1))^{p-1-j}\right]\lambda_r\gamma_r.$$

Using the fact that

$$\lambda_r(g_r - 1)\left(\sum_{i=1}^{r-1} \lambda_i\gamma_i\right) - \left(\sum_{i=1}^{r-1} \lambda_i(g_i - 1)\right)\lambda_r\gamma_r$$

$$= \sum_{i=1}^{r-1} \lambda_i\lambda_r\left[(\lambda_r - 1)\gamma_i - (\lambda_i - 1)\gamma_r\right] \in \Omega^2 k,$$

we see that modulo $\mathrm{Rad}(\Omega^2 k)$, the term in the first sum indexed by j cancels with the term in the second sum indexed by $j+1$. We are left with the term with $j = 0$ in the second sum, namely $\lambda_r^p(g_r - 1)^{p-1}\gamma_r$, as required.

We shall see later in this section that under the identification of Lemma 5.8.2, the subvariety $V_E(M)$ is identified with $V_E^\sharp(M)$. For now, let us indicate why $V_E^\sharp(M)$ is a closed homogeneous subvariety of $V_E^\sharp$. We may test whether $M\downarrow_{\langle 1+v\rangle}$ is free as follows. By Jordan canonical form, the rank of the matrix representing the action of v on M is always at most $(p-1)\dim_k(M)/p$, with equality if and only if $M\downarrow_{\langle 1+v\rangle}$ is free. Thus $V_E^\sharp(M)$ is defined by the vanishing of certain minors in the matrix representation, and these are homogeneous equations in the variables $\lambda_1, \ldots, \lambda_r$. We illustrate this with an example.

CARLSON'S FAVOURITE EXAMPLE. Let $E = \langle g_1, g_2, g_3\rangle$ be an elementary abelian group of order 8 and $\mathrm{char}\, k = 2$. For each triple of parameters $(a, b, c) \in k^3$, we have a kE-module $M_{a,b,c}$ given in terms of matrices as follows.

$$g_1 \mapsto \begin{pmatrix} 1 & 0 & 0 & 0 \\ 0 & 1 & 0 & 0 \\ 1 & 0 & 1 & 0 \\ 0 & 1 & 0 & 1 \end{pmatrix} \qquad g_2 \mapsto \begin{pmatrix} 1 & 0 & 0 & 0 \\ 0 & 1 & 0 & 0 \\ a & 0 & 1 & 0 \\ 0 & b & 0 & 1 \end{pmatrix} \qquad g_3 \mapsto \begin{pmatrix} 1 & 0 & 0 & 0 \\ 0 & 1 & 0 & 0 \\ 0 & c & 1 & 0 \\ 1 & 0 & 0 & 1 \end{pmatrix}$$

Then $g_1 - 1$, $g_2 - 1$ and $g_3 - 1$ form a basis for $V_E^\sharp$, and the element

$$v = \lambda_1(g_1 - 1) + \lambda_2(g_2 - 1) + \lambda_3(g_3 - 1)$$

is represented by the matrix

$$\begin{pmatrix} 0 & 0 & 0 & 0 \\ 0 & 0 & 0 & 0 \\ \lambda_1 + \lambda_2 a & \lambda_3 c & 0 & 0 \\ \lambda_3 & \lambda_1 + \lambda_2 b & 0 & 0 \end{pmatrix}.$$

Thus the restriction of $M_{a,b,c}$ to $\langle 1 + v \rangle$ is free if and only if the rank of this matrix is two, namely if and only if

$$\begin{vmatrix} \lambda_1 + \lambda_2 a & \lambda_3 c \\ \lambda_3 & \lambda_1 + \lambda_2 b \end{vmatrix} \neq 0.$$

It follows that $V_E^\sharp(M_{a,b,c})$ is given by the homogeneous quadratic equation in $\mathbb{A}^3(k)$

$$(X_1 + aX_2)(X_1 + bX_2) = cX_3^2.$$

The main obstruction to proving the equality of the varieties $V_E(M)$ and $V_E^\sharp(M)$ is that if E' is a shifted subgroup of E then the following diagram does not necessarily commute.

$$\begin{array}{ccc} \mathrm{Ext}^*_{kE}(k, k) & \xrightarrow{\otimes M} & \mathrm{Ext}^*_{kE}(M, M) \\ \downarrow{\scriptstyle \mathrm{res}_{E,E'}} & & \downarrow{\scriptstyle \mathrm{res}_{E,E'}} \\ \mathrm{Ext}^*_{kE'}(k, k) & \xrightarrow{\otimes M} & \mathrm{Ext}^*_{kE'}(M, M) \end{array}$$

The problem is that tensoring with M involves the Hopf algebra structure of kE, and therefore the definition of the map is changed by moving the group basis.

EXAMPLE. (Carlson [69]) Let E be the Klein four group and let $M = V_{n,\alpha}$ be the indecomposable module of dimension $2n$ parametrized by $\alpha \in \mathbb{P}^1(k)$ as described in Section 4.3 of Volume I. Let ζ_α be the corresponding element in degree one cohomology, and let E_α be the corresponding cyclic shifted subgroup. Then for $n = 1$ and α not equal to 0, 1 or ∞, $\mathrm{res}_{E,E_\alpha}(\zeta_\alpha \otimes \mathrm{Id}_M) \neq 0$ even though $\mathrm{res}_{E,E_\alpha}(\zeta_\alpha) = 0$.

REMARK. Niwasaki [203] has shown that the above diagram almost commutes, in the sense that it commutes on the subring of $\mathrm{Ext}^*_{kE}(k, k)$ generated by the Bocksteins of degree one elements. He has based an alternative proof of Theorem 5.8.3 on this observation.

Although it is true that the action of $\mathrm{Ext}^{\cdot}_{kE}(k,k)$ on $\mathrm{Ext}^*_{kE}(k,M)$ by Yoneda composition commutes with restriction to shifted subgroups, the following example shows that we cannot get around the problem just by arguing directly with $\mathrm{Ext}^*_{kE}(k,M)$ instead of $\mathrm{Ext}^*_{kE}(M,M)$.[1]

EXAMPLE. Let k be a field of characteristic two, and let

$$E = \langle g_1, g_2, g_3 \rangle \cong (\mathbb{Z}/2)^3.$$

Let M be the four dimensional kE-module given in terms of matrices by

$$g_1 \mapsto \begin{pmatrix} 1 & 0 & 0 & 0 \\ 1 & 1 & 0 & 0 \\ 0 & 0 & 1 & 0 \\ 0 & 0 & 1 & 1 \end{pmatrix} \quad g_2 \mapsto \begin{pmatrix} 1 & 0 & 0 & 0 \\ 0 & 1 & 0 & 0 \\ 1 & 0 & 1 & 0 \\ 0 & 1 & 0 & 1 \end{pmatrix} \quad g_3 \mapsto \begin{pmatrix} 1 & 0 & 0 & 0 \\ 0 & 1 & 0 & 0 \\ 0 & 0 & 1 & 0 \\ 1 & 0 & 0 & 1 \end{pmatrix}$$

Then M is periodic with period one, so each $\mathrm{Ext}^n_{kE}(k,M)$ is isomorphic to $\mathrm{Hom}_{kE}(k,M)$, and is one dimensional, given by the map $\zeta : k \to M$ corresponding to the last basis element. This map does not factor through a projective module, even though its restriction to every cyclic shifted subgroup does.

THEOREM 5.8.3 (Avrunin, Scott [24]). *Under the isomorphism $V_E \cong V_E^{\sharp}$ given in Lemma 5.8.2, the subvariety $V_E(M)$ corresponds to $V_E^{\sharp}(M)$.*

PROOF. By Proposition 5.7.1, $V_E(M)$ is equal to the variety determined by the annihilator $I_E(k,M)$ in $H^{\cdot}(E,k)$ of $H^*(E,M)$. The action can be considered as coming from Yoneda composition, so it in no way depends on the Hopf algebra structure of kE. So we can replace E by a shifted subgroup E' of maximal rank so that $kE = kE'$ without changing $V_E(M)$ or $V_E^{\sharp}(M)$. Furthermore, the above discussion of the Bockstein map shows that the isomorphism between V_E and $V_E^{\sharp}$ described in Lemma 5.8.2 is not affected by this replacement. So for $0 \neq v \in V_E^{\sharp}$, we can extend $T = \langle 1+v \rangle$ to a cyclic shifted subgroup E' of maximal rank in kE. Then by Proposition 5.7.7,

$$V_T(M) = V_T \cap V_{E'}(M) = V_T \cap V_E(M).$$

So $M \downarrow_{\langle 1+v \rangle}$ is not free if and only if $V_T(M) \neq \{0\}$, namely if and only if $V_T \subseteq V_E(M)$. □

The following corollary was proved directly by Dade [87], and another direct proof may be found in Carlson [69].

COROLLARY 5.8.4. *A kE-module M is free if and only if $M \downarrow_T$ is free for each cyclic shifted subgroup T of E.*

---

[1] As was erroniously attempted in the first edition of this book!

PROOF. If M is free then certainly $M \downarrow_T$ is free. Conversely, if $M \downarrow_T$ is free for each cyclic shifted subgroup T of E, then $V_E^{\sharp}(M) = \{0\}$. So by the Avrunin–Scott theorem $V_E(M) = \{0\}$ and hence by Proposition 5.7.2 M is projective, which for a p-group is the same as free. $\qquad \square$

COROLLARY 5.8.5. *If p does not divide* $\dim(M)$ *then* $V_G(M) = V_G$.

PROOF. This follows from Proposition 5.7.4 and Theorem 5.8.3, since if p does not divide $\dim(M)$, the restriction of M to a cyclic shifted subgroup of an elementary abelian subgroup of G is never free.

This corollary also follows immediately from Theorem 3.1.9 of Volume I; namely if p does not divide $\dim(M)$ then $V_G(M) \supseteq V_G(M \otimes M^*) \supseteq V_G(k) = V_G$. $\qquad \square$

5.9. The modules L_ζ

If $0 \neq \zeta \in H^n(G, k) = \mathrm{Ext}^n_{kG}(k, k) \cong \mathrm{Hom}_{kG}(\Omega^n k, k)$ then ζ corresponds to a surjective map $\hat{\zeta} : \Omega^n k \to k$. We write L_ζ for the kernel of $\hat{\zeta}$. If $\zeta = 0 \in H^n(G, k)$, then we define $L_\zeta = \Omega^n k \oplus \Omega k$. The reason for this definition is as follows. If H is a subgroup of G, then by Schanuel's Lemma, 1.5.3 of Volume I, we have $\Omega^n_G(k) \downarrow_H \cong \Omega^n_H(k) \oplus P$ with P a projective kH-module. As long as $\mathrm{res}_{G,H}(\zeta)$ is still non-zero, we have

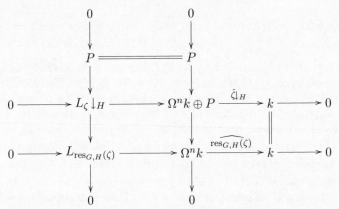

so that since P is also injective (Proposition 3.1.2 of Volume I) it follows that

$$L_\zeta \downarrow_H \cong L_{\mathrm{res}_{G,H}(\zeta)} \oplus (\text{projective}).$$

If, however, $\mathrm{res}_{G,H}(\zeta) = 0$, then $\hat{\zeta} \downarrow_H$ has $\Omega^n k$ in its kernel and so

$$L_\zeta \downarrow_H \cong \Omega^n k \oplus \Omega k \oplus (\text{projective}).$$

The importance of the modules L_ζ is that the variety $V_G(L_\zeta)$ may be explicitly computed.

PROPOSITION 5.9.1 (Carlson [**70**]). *The variety $V_G(L_\zeta)$ is the hypersurface $V_G\langle\zeta\rangle$ in V_G determined by ζ (in other words, the set of maximal ideals containing ζ).*

PROOF. By Theorem 5.7.8 and the above discussion, it suffices to prove the proposition for $G = E$ elementary abelian. By Theorem 5.8.3 it suffices to prove that $V_E^\sharp(L_\zeta) = V_E^\sharp\langle\zeta\rangle$. In other words, we must show that if $0 \neq u \in J(kE)/J^2(kE)$ and $x = 1 + u$, then $L_\zeta \downarrow_{\langle x\rangle}$ is projective if and only if $\mathrm{res}_{G,\langle x\rangle}(\zeta) \neq 0$.

If p is odd and $n = \deg(\zeta)$ is odd, then $\zeta^2 = 0$ so that $V_E\langle\zeta\rangle = V_E$, while $\dim L_\zeta = \dim \Omega^n k - 1$ is congruent to -2 modulo p, so that $L_\zeta \downarrow_{\langle x\rangle}$ is never projective. So we may assume that if p is odd then n is even. Thus $\Omega^n(k) \downarrow_{\langle x\rangle} \cong k \oplus P$ with P projective. If $\mathrm{res}_{G,\langle x\rangle}(\zeta) = 0$ then the map $\hat\zeta \downarrow_{\langle x\rangle}: k \oplus P \to k$ has the first summand in its kernel and so $L_\zeta \downarrow_{\langle x\rangle} \cong k \oplus \Omega k \oplus (\text{projective})$. If $\mathrm{res}_{G,\langle x\rangle}(\zeta) \neq 0$ then the map $\hat\zeta \downarrow_{\langle x\rangle}: k \oplus P \to k$ splits and so $L_\zeta \downarrow_{\langle x\rangle} \cong P$. $\qquad\square$

COROLLARY 5.9.2. *Every closed homogeneous subvariety of V_G is of the form $V_G(M)$ for some finitely generated kG-module M.*

PROOF. If V is a closed homogeneous subvariety of V_G then the corresponding ideal $I(V) \subseteq H^\cdot(G,k)$ is generated by homogeneous elements $\zeta_1, \ldots, \zeta_s$. Let $M = L_{\zeta_1} \otimes \cdots \otimes L_{\zeta_s}$. Then by Theorem 5.7.11 and Proposition 5.9.1 we have

$$V_G(M) = V_G(L_{\zeta_1} \otimes \cdots \otimes L_{\zeta_s}) = V_G(L_{\zeta_1}) \cap \cdots \cap V_G(L_{\zeta_s})$$
$$= V_G\langle\zeta_1\rangle \cap \cdots \cap V_G\langle\zeta_s\rangle = V_G\langle\zeta_1, \ldots, \zeta_s\rangle = V. \qquad\square$$

REMARK. In order to express V as $V_G\langle\zeta_1\rangle \cap \cdots \cap V_G\langle\zeta_s\rangle$ it is only necessary for some power of every element of $I(V)$ to be in the ideal generated by $\zeta_1, \ldots, \zeta_s$.

LEMMA 5.9.3. *If $\zeta_1 \in H^r(G,k)$ and $\zeta_2 \in H^s(G,k)$ then there is a short exact sequence*

$$0 \to \Omega^r L_{\zeta_2} \to L_{\zeta_1\zeta_2} \oplus (\text{projective}) \to L_{\zeta_1} \to 0.$$

PROOF. We tensor the sequence $0 \to L_{\zeta_2} \to \Omega^s k \xrightarrow{\hat\zeta_2} k \to 0$ with $\Omega^r k$ to obtain a sequence

$$0 \to \Omega^r L_{\zeta_2} \oplus (\text{projective}) \to \Omega^{r+s}k \oplus (\text{projective}) \xrightarrow{\Omega^r(\hat\zeta_2)} \Omega^r k \to 0.$$

Since $\widehat{\zeta_1\zeta_2} = \hat{\zeta}_1 \circ \Omega^r(\hat{\zeta})$ (cf. Section 2.6 of Volume I) we have a diagram

$$
\begin{array}{ccccccccc}
 & & 0 & & 0 & & & \\
 & & \downarrow & & \downarrow & & & \\
0 \longrightarrow \Omega^r L_{\zeta_2} \oplus \text{(projective)} & \longrightarrow & L_{\zeta_1\zeta_2} \oplus \text{(projective)} & \longrightarrow & L_{\zeta_1} & \longrightarrow & 0 \\
\| & & \downarrow & & \downarrow & & \\
0 \longrightarrow \Omega^r L_{\zeta_2} \oplus \text{(projective)} & \longrightarrow & \Omega^{r+s}k \oplus \text{(projective)} & \xrightarrow{\Omega^r(\hat{\zeta_2})} & L_{\zeta_1} & \longrightarrow & 0 \\
 & & \downarrow{\widehat{\zeta_1\zeta_2}} & & \downarrow{\hat{\zeta}_1} & & \\
 & & k & = & k & & \\
 & & \downarrow & & \downarrow & & \\
 & & 0 & & 0 & &
\end{array}
$$

The lemma now follows from the fact that projective modules are injective (Proposition 3.1.2 of Volume I). □

The following lemma gives us another way of viewing the modules L_ζ.

LEMMA 5.9.4. *Under the isomorphism*

$$H^n(G,k) \cong \text{Hom}_{kG}(\Omega^n k, k) \cong \text{Ext}^1_{kG}(\Omega^{n-1}k, k)$$

an element $\zeta \in H^n(G,k)$ *corresponds to an extension of the form*

$$0 \to k \to \Omega^{-1}L_\zeta \to \Omega^{n-1}k \to 0.$$

PROOF. Let P_{n-1} be the $(n-1)$st projective module in the minimal resolution of k as a kG-module, so that $\Omega^n k$ is a submodule of P_{n-1}. Then the lemma follows by applying the Snake Lemma, 2.3.9 of Volume I, to the diagram

$$
\begin{array}{ccccccccc}
0 & \longrightarrow & L_\zeta & \longrightarrow & \Omega^n k & \xrightarrow{\hat{\zeta}} & k & \longrightarrow & 0 \\
 & & \downarrow & & \downarrow & & \downarrow & & \\
0 & \longrightarrow & P_{n-1} & \longrightarrow & P_{n-1} & \longrightarrow & 0 & \longrightarrow & 0
\end{array}
$$

(cf. Exercise (v) at the end of Section 2.6 of Volume I). □

PROPOSITION 5.9.5. *Suppose M is a finitely generated kG-module. Then an element $\zeta \in H^n(G,k)$ is sent to zero under the map*

$$H^*(G,k) = \text{Ext}^*_{kG}(k,k) \xrightarrow{\otimes M} \text{Ext}^*_{kG}(M,M)$$

if and only if

$$\Omega^{-1}L_\zeta \otimes M \cong M \oplus \Omega^{n-1}M \oplus \text{(projective)}.$$

PROOF. By the lemma, the image of ζ under the map

$$
\begin{array}{ccc}
\operatorname{Ext}_{kG}^n(k,k) & \xrightarrow{\ \otimes M\ } & \operatorname{Ext}_{kG}^n(M,M) \\
\Big\downarrow{\scriptstyle\cong} & & \Big\downarrow{\scriptstyle\cong} \\
\operatorname{Ext}_{kG}^1(\Omega^{n-1}k,k) & \xrightarrow{\ \otimes M\ } & \operatorname{Ext}_{kG}^1(\Omega^{n-1}M,M)
\end{array}
$$

is represented by the sequence

$$0 \to M \to \Omega^{-1}L_\zeta \otimes M \to \Omega^{n-1}k \otimes M \to 0.$$

This image is zero if and only if this sequence splits, which by Lemma 2.6.2 of Volume I happens if and only if

$$\Omega^{-1}L_\zeta \otimes M \cong M \oplus (\Omega^{n-1}k \otimes M).$$

The proposition now follows from the fact that

$$\Omega^{n-1}k \otimes M \cong \Omega^{n-1}M \oplus (\text{projective}). \qquad \square$$

REMARK. Note that $\zeta \in H^n(G,k)$ is sent to zero under the map

$$H^*(G,k) = \operatorname{Ext}_{kG}^*(k,k) \xrightarrow{\ \otimes M\ } \operatorname{Ext}_{kG}^*(M,M)$$

if and only if cup product with ζ is identically zero on $\operatorname{Ext}_{kG}^*(M,M)$. So if this happens, we say that ζ **annihilates** $\operatorname{Ext}_{kG}^*(M,M)$.

If $\zeta \in H^n(G,k)$ vanishes on $V_G(M)$, then by the definition of $V_G(M)$, some power of ζ annihilates $\operatorname{Ext}_{kG}^*(M,M)$. In particular, it follows from Proposition 5.9.1 that some power of ζ annihilates $\operatorname{Ext}_{kG}^*(L_\zeta, L_\zeta)$. The following proposition gives us sharper information.

PROPOSITION 5.9.6. *If* $\zeta \in H^n(G,k)$ *then*
(i) ζ^2 *annihilates* $\operatorname{Ext}_{kG}^*(L_\zeta, L_\zeta)$.
(ii) *If the characteristic of the field k is odd and n is even then ζ annihilates* $\operatorname{Ext}_{kG}^*(L_\zeta, L_\zeta)$.

PROOF. (i) Since $\operatorname{Ext}_{kG}^r(\Omega^n k, L_\zeta) \cong \operatorname{Ext}_{kG}^{n+r}(k, L_\zeta)$, the short exact sequence

$$0 \to L_\zeta \to \Omega^n k \xrightarrow{\ \hat\zeta\ } k \to 0$$

gives rise to a long exact sequence

$$\cdots \to \operatorname{Ext}_{kG}^r(k, L_\zeta) \xrightarrow{\ \zeta\ } \operatorname{Ext}_{kG}^{n+r}(k, L_\zeta) \to \operatorname{Ext}_{kG}^r(L_\zeta, L_\zeta) \to \operatorname{Ext}_{kG}^{r+1}(k, L_\zeta)$$
$$\xrightarrow{\ \zeta\ } \operatorname{Ext}_{kG}^{n+r+1}(k, L_\zeta) \to \cdots$$

and hence to a short exact sequence

$$0 \to \operatorname{Ext}_{kG}^{n+r}(k, L_\zeta)/\operatorname{Im}(\zeta) \to \operatorname{Ext}_{kG}^r(L_\zeta, L_\zeta)$$
$$\to \operatorname{Ker}(\zeta \text{ on } \operatorname{Ext}_{kG}^{r+1}(k, L_\zeta)) \to 0.$$

Since ζ annihilates the left and right terms, ζ^2 annihilates the middle.

(ii) If $\mathbf{C}$ is a chain complex of kG-modules and $T = \langle \sigma \mid \sigma^2 = 1 \rangle$ is a cyclic group of order two then T acts on $\mathbf{C} \otimes \mathbf{C}$ via

$$\sigma(x \otimes y) = (-1)^{\deg(x)\deg(y)} y \otimes x.$$

If the characteristic of k is odd, then $\mathbf{C} \otimes \mathbf{C}$ decomposes into eigenspaces of T, which are again chain complexes of kG-modules. We define the **symmetric square** $S^2(\mathbf{C})$ to be the $+1$ eigenspace of this action, and **exterior square** $\Lambda^2(\mathbf{C})$ to be the -1 eigenspace. Taking eigenspaces on both sides of the Künneth formula

$$H(\mathbf{C} \otimes \mathbf{C}) \cong H(\mathbf{C}) \otimes H(\mathbf{C})$$

we obtain

$$H(S^2(\mathbf{C})) = S^2(H(\mathbf{C})), \quad H(\Lambda^2(\mathbf{C})) = \Lambda^2(H(\mathbf{C}))$$

Taking the exterior square of the complex

$$\mathbf{C}: \qquad 0 \to L_\zeta \to \Omega^n k \to 0$$

(whose homology is k in degree zero) we obtain the complex

$$\Lambda^2(\mathbf{C}): \qquad 0 \to S^2(L_\zeta) \to \Omega^n k \otimes L_\zeta \to \Lambda^2(\Omega^n k) \to 0$$

which is exact since the homology of $\mathbf{C} \otimes \mathbf{C}$ is k in degree zero, and is hence entirely in $S^2(\mathbf{C})$. (Note that the left-hand module in $\Lambda^2(\mathbf{C})$ is $S^2(L_\zeta)$, since L_ζ is in degree one so that σ acts via $x \otimes y \mapsto -y \otimes x$.)

Now

$$\Lambda^2(\Omega^n k) \oplus S^2(\Omega^n k) \cong \Omega^{2n} k \oplus (\text{projective}).$$

Since n is even and p is odd, $d = \dim \Omega^n k \equiv 1 \bmod p$, so that $\dim \Lambda^2(\Omega^n k) = d(d-1)/2 \equiv 0 \bmod p$ and $\dim S^2(\Omega^n k) = d(d+1)/2 \equiv 1 \bmod p$. Since projective modules have dimension divisible by p and $\Omega^{2n} k$ is indecomposable, the Krull–Schmidt theorem implies that we have $S^2(\Omega^n k) \cong \Omega^{2n} k \oplus (\text{projective})$, and $\Lambda^2(\Omega^n k)$ is projective. Thus the above exact sequence $\Lambda^2(\mathbf{C})$ shows that

$$S^2(L_\zeta) \oplus (\text{projective}) \cong \Omega^n k \otimes L_\zeta \cong \Omega^n(L_\zeta) \oplus (\text{projective})$$

and hence

$$S^2(L_\zeta) \cong \Omega^n(L_\zeta) \oplus (\text{projective}).$$

Similarly, if we take the symmetric square of the complex

$$\mathbf{C}': \qquad 0 \to \Omega^n k \xrightarrow{\hat{\zeta}} k \to 0$$

(whose homology is L_ζ in degree one) we obtain a complex

$$S^2(\mathbf{C}'): \qquad 0 \to \Lambda^2(\Omega^n k) \to \Omega^n k \xrightarrow{\hat{\zeta}} k \to 0$$

whose homology is $\Lambda^2(L_\zeta)$ in degree two, and hence an exact sequence

$$0 \to \Lambda^2(L_\zeta) \to \Lambda^2(\Omega^n k) \to L_\zeta \to 0.$$

Again using the fact that $\Lambda^2(\Omega^n k)$ is projective, we see that

$$\Lambda^2(L_\zeta) \cong \Omega(L_\zeta) \oplus (\text{projective}).$$

Putting these together, we have

$$L_\zeta \otimes L_\zeta \cong \Lambda^2(L_\zeta) \oplus S^2(L_\zeta) \cong \Omega(L_\zeta) \oplus \Omega^n(L_\zeta) \oplus \text{(projective)}$$

so that

$$\Omega^{-1}(L_\zeta) \otimes L_\zeta \cong L_\zeta \oplus \Omega^{n-1}(L_\zeta) \oplus \text{(projective)},$$

and hence by Proposition 5.9.5, ζ annihilates $\text{Ext}^*_{kG}(L_\zeta, L_\zeta)$. $\square$

REMARK. Let $G = \langle x, y \mid x^8 = y^2 = yxyx^{-3} = 1 \rangle$ be the semidihedral group of order 16, and k be a field of characteristic two. Let ζ be the element of $H^1(G, k)$ corresponding to the quaternion subgroup $\langle x^2, xy \rangle$ of order 8. Then for every $n > 1$, ζ^n does not annihilate $\text{Ext}^*_{kG}(L_{\zeta^n}, L_{\zeta^n})$.

EXERCISE. [32] Suppose that $G \cong \mathbb{Z}/p \times \mathbb{Z}/p$ with p odd, and $0 \neq \zeta \in H^2(G, \mathbb{F}_p)$ is in the image of the Bockstein map. Show that $L_\zeta \cong \Omega^2(L_\zeta)$.

Show that there is a subgroup $H \cong \mathbb{Z}/p$ and a map $L_\zeta \to \mathbb{F}_p$ which splits on restriction to H. Show that there is no such map $\Omega(L_\zeta) \to \mathbb{F}_p$, and therefore $L_\zeta \not\cong \Omega(L_\zeta)$.

Use Propositions 5.9.5 and 5.9.6 to show that $L_\zeta \otimes L_\zeta \cong L_\zeta \otimes \Omega(L_\zeta)$. Deduce that the element $[L_\zeta] - [\Omega(L_\zeta)]$ of the representation ring $a(G)$ is a non-zero nilpotent element (cf. Section 5.8 of Volume I).

5.10. Periodic modules

DEFINITION 5.10.1. A kG-module M is **periodic** if $\Omega^n M \cong M$ for some value of n. The minimal value of n for which this is true is called the **period** of M.

PROPOSITION 5.10.2 (Carlson [61]). Suppose M is a finitely generated indecomposable periodic kG-module of period n, with k algebraically closed. Then the nilpotent elements in $\text{Ext}^*_{kG}(M, M)$ form an ideal, and the ring $\text{Ext}^*_{kG}(M, M)$ is a direct sum of this ideal and a polynomial subring $k[x]$ in one variable x of degree n.

PROOF. Since M is indecomposable, $\text{End}_{kG}(M)$ is a local ring, so that modulo its radical it is isomorphic to k. Let $x \in \text{Ext}^n_{kG}(M, M)$ correspond to an isomorphism $\hat{x} : \Omega^n M \to M$. If $\zeta \in \text{Ext}^r_{kG}(M, M)$ is any element of positive degree represented by a map $\hat{\zeta} : \Omega^r M \to M$, then $\hat{\zeta}^n : \Omega^{rn} M \to M$ is either an isomorphism or nilpotent. If it is an isomorphism then it differs from a multiple of $\hat{x}^r$ by a nilpotent element. $\square$

COROLLARY 5.10.3. If M is a finitely generated indecomposable periodic kG-module and k is algebraically closed, then $V_G(M)$ is a single line through the origin in V_G.

PROOF. It follows from the proposition that $\tilde{V}_G(M)$ is an affine line $\mathbb{A}^1(k)$. Since $\tilde{V}_G(M)$ is a finite cover of $V_G(M)$, the latter is also a single line through the origin. $\square$

Conversely, we have the following:

THEOREM 5.10.4 (Eisenbud [100]). *If M is a finitely generated kG-module of complexity one, then M is a direct sum of periodic modules and projective modules.*

PROOF. (Carlson [69]) By Proposition 5.7.2, if M has complexity one then $V_G(M)$ is a finite union of lines through the origin in V_G. So we may choose a homogeneous element $\zeta \in H^n(G, k)$ such that $V_G\langle\zeta\rangle \cap V_G(M) = \{0\}$. Tensor the sequence

$$0 \to L_\zeta \to \Omega^n k \to k \to 0$$

with M to obtain a sequence

$$0 \to L_\zeta \otimes M \to \Omega^n k \otimes M \to M \to 0.$$

By Proposition 5.9.1 and Theorem 5.7.11 we have

$$V_G(L_\zeta \otimes M) = V_G(L_\zeta) \cap V_G(M) = V_G\langle\zeta\rangle \cap V_G(M) = \{0\}$$

and so by Proposition 5.7.2, $L_\zeta \otimes M$ is projective, and hence also injective (Proposition 3.1.2 of Volume I). Since $\Omega^n k \otimes M \cong \Omega^n M \oplus$ (projective), we have

$$M \cong \Omega^n M \oplus \text{(projective)}. \qquad \square$$

DEFINITION 5.10.5. *If M is periodic, an element $\zeta \in H^n(G, k)$ with $V_G\langle\zeta\rangle \cap V_G(M) = \{0\}$ is said to* **generate the periodicity** *of M.*

The proof of the above theorem shows that if $\zeta \in H^n(G, k)$ generates the periodicity of a periodic module M then the period of M divides n.

COROLLARY 5.10.6. *If $H^*(G, k)$ is finitely generated as a module over a subring generated by elements $x_1, \ldots, x_s$ in degrees $n_1, \ldots, n_s$ and M is an indecomposable finitely generated periodic kG-module, then the period of M divides one of the n_i.*

PROOF. The condition on the x_i ensures that the intersection of the hyperplanes they define is the origin. Now $V_G(M)$ is a single line through the origin by Corollary 5.10.3, so one of the x_i generates the periodicity of M. $\qquad \square$

COROLLARY 5.10.7. *If M is an indecomposable finitely generated periodic kE-module with E elementary abelian, then M has period one or two. If k has characteristic two, then M has period one.*

PROOF. This follows from the structure of $H^*(E, k)$ given in Corollary 3.5.7 of Volume I. $\qquad \square$

5.11. Andrews' theorem

In the last section, we described a method for obtaining upper bounds on the period of a periodic kG-module. In general, it is much harder to get lower bounds. However, a method of Andrews [14], which we now describe, enables us to determine the exact period of a periodic inflated module, in terms of its variety as a module for the quotient. We begin with a technical lemma.

LEMMA 5.11.1. *Suppose N is a normal p-subgroup of G, and M is a kG-module. Denote by $\tilde{N}$ the sum $\sum_{g \in N} g$ of the elements of N, as an element of the group algebra $kN \subseteq kG$. Then M has a non-zero projective summand as a kG-module if and only if $\tilde{N}M$ has a non-zero projective summand as a $k(G/N)$-module.*

PROOF. If M has a non-zero projective summand P as a kG-module then $\tilde{N}P$ is a non-zero projective summand of $\tilde{N}M$ as a $k(G/N)$-module. Conversely, suppose that $\tilde{N}M$ has a non-zero projective summand Q as a $k(G/N)$-module. Denote by π the projection of $\tilde{N}M$ onto Q, and write π as a transfer $\pi = \mathrm{Tr}_{1,G/N}(\alpha) = \mathrm{Tr}_{N,G}(\alpha)$, $\alpha \in \mathrm{End}_k(\tilde{N}M)$. Since N is normal in G, the transfer map $\mathrm{Tr}_{1,N}$ maps elements of $\mathrm{End}_k(M)$ to elements preserving the submodule $\tilde{N}M$, and gives a surjective map from $\mathrm{End}_k(M)$ onto $\mathrm{End}_k(\tilde{N}M)$. So there is an element $\beta \in \mathrm{End}_k(M)$ with the property that $\mathrm{Tr}_{1,N}(\beta)$ acts as α on $\tilde{N}M$. Thus $\mathrm{Tr}_{1,G}(\beta)$ acts as π on $\tilde{N}M$, and so $\mathrm{Tr}_{1,G}(\beta)$ is a non-nilpotent kG-module endomorphism of M. So we may choose $\phi \in \mathrm{End}_{kG}(M)$ in such a way that $\phi \circ \mathrm{Tr}_{1,G}(\beta)$ is a projection onto a kG-module summand of M. Since $\phi \circ \mathrm{Tr}_{1,G}(\beta) = \mathrm{Tr}_{1,G}(\phi \circ \beta)$, this summand is projective. $\square$

Now suppose we have a group extension

$$1 \to N \to G \to \bar{G} \to 1$$

with N cyclic of order p. Let $\varepsilon = 1$ if $p = 2$ and $\varepsilon = 2$ if p is odd. If M is a $k\bar{G}$-module, regarded as a kG-module by inflation, then $M \downarrow_N$ is a module with trivial action, and so $M \downarrow_N$ is periodic with period ε. So if M is periodic as a kG-module, then its period is certainly divisible by ε.

Let $\tilde{N}$ denote the sum $\sum_{g \in N} g$ of the elements of N, as an element of the group algebra kN. Then for any $r > 0$, $\tilde{N}\Omega_G^{\varepsilon r}(k)$ is a kG-module with N acting trivially, so we may regard it as a $k\bar{G}$-module. We set

$$V_r = V_{\bar{G}}(\tilde{N}\Omega_G^{\varepsilon r}(k)) \subseteq V_{\bar{G}}.$$

THEOREM 5.11.2 (Andrews). *If M is a $k\bar{G}$-module regarded as a kG-module by inflation, then the following are equivalent:*

(i) $\dim_k \Omega_G^{\varepsilon r}(M) = \dim_k M$

(ii) $\tilde{N}\Omega_G^{\varepsilon r}(M) = 0$

(iii) $\tilde{N}\Omega_G^{\varepsilon r}(k) \otimes M$ *is a projective $k\bar{G}$-module*

(iv) $\Omega_G^{\varepsilon r}(M) \cong M$

(v) $V_{\bar{G}}(\tilde{N}\Omega_G^{\varepsilon r}(k) \otimes M) = \{0\}$

(vi) $V_r \cap V_{\bar{G}}(M) = \{0\}$.

PROOF. By Schanuel's lemma we have

$$\Omega_G^{\varepsilon r}(M) \downarrow_N \cong \Omega_G^{\varepsilon r}(M \downarrow_N) \oplus Q \cong M \downarrow_N \oplus Q$$

where Q is a projective kN-module with $\tilde{N}\Omega_G^{\varepsilon r}(M) = \tilde{N}Q = Q^N$. Since N is a p-group, $Q = 0$ if and only if $Q^N = 0$, and so (i) is equivalent to (ii).

Since $\Omega_G^{\varepsilon r}(M)$ has no projective summands as a kG-module, the lemma implies that $\tilde{N}\Omega_G^{\varepsilon r}(M)$ has no projective summands as a $k\bar{G}$-module. Applying $\tilde{N}$ to the isomorphism

$$\Omega_G^{\varepsilon r}(k) \otimes M \cong \Omega_G^{\varepsilon r}(M) \oplus (\text{projective } kG\text{-module})$$

we obtain

$$\tilde{N}\Omega_G^{\varepsilon r}(k) \otimes M \cong \tilde{N}\Omega_G^{\varepsilon r}(M) \oplus (\text{projective } k\bar{G}\text{-module})$$

and so (ii) is equivalent to (iii).

Tensoring the short exact sequence

$$0 \to \tilde{N}\Omega_G^{\varepsilon r}(k) \to \Omega_G^{\varepsilon r}(k)^N \to k \to 0$$

with M, we see that if (iii) holds then

$$\Omega_G^{\varepsilon r}(k)^N \otimes M \cong M \oplus (\text{projective } k\bar{G}\text{-module}).$$

Since (iii) also implies that $Q = 0$ so that N acts trivially on $\Omega_G^{\varepsilon r}(M)$, we have

$$\Omega_G^{\varepsilon r}(k)^N \otimes M \cong (\Omega_G^{\varepsilon r}(k) \otimes M)^N \cong \Omega_G^{\varepsilon r}(M) \oplus (\text{projective } k\bar{G}\text{-module})$$

and hence (iii) implies (iv). It is easy to see that (iv) implies (i).

The equivalence of (iii) and (v) and the equivalence of (v) and (vi) are given by Proposition 5.7.2 and Theorem 5.7.11 respectively. $\square$

The varieties V_r are completely determined in [**37**].

5.12. The variety of an indecomposable kG-module is connected

This section is devoted to a generalisation of Corollary 5.10.3 to arbitrary (not necessarily periodic) finitely generated modules.

THEOREM 5.12.1 (Carlson [**70**]). *If M is a finitely generated kG-module and $V_G(M) = V_1 \cup V_2$ with $V_1 \cap V_2 = \{0\}$ then $M \cong M_1 \oplus M_2$ with $V_G(M_1) = V_1$ and $V_G(M_2) = V_2$.*

PROOF. We prove this theorem by induction on $\dim V_1 + \dim V_2$. The case where either V_1 or V_2 is $\{0\}$ is clear, so we shall assume both are non-zero. Thus we may choose homogeneous elements ζ_1 of degree r and ζ_2 of degree s such that $V_1 \subseteq V_G\langle\zeta_1\rangle$, $\dim(V_2 \cap V_G\langle\zeta_1\rangle) = \dim V_2 - 1$, $V_2 \subseteq V_G\langle\zeta_2\rangle$ and $\dim(V_1 \cap V_G\langle\zeta_2\rangle) = \dim V_1 - 1$. Since $V_G\langle\zeta_1\zeta_2\rangle = V_G\langle\zeta_1\rangle \cup V_G\langle\zeta_2\rangle \supseteq V_1 \cup V_2 = V_G(M)$, some power of $\zeta_1\zeta_2$ lies in $I_G(M)$. So by replacing ζ_1 and ζ_2 by suitable powers if necessary we may assume that $\zeta_1\zeta_2$ annihilates $\text{Ext}_{kG}^*(M,M)$. Thus by Proposition 5.9.5 we have

$$L_{\zeta_1\zeta_2} \otimes M \cong \Omega M \oplus \Omega^{r+s}M \oplus (\text{projective}).$$

Also, by Lemma 5.9.3 there is a short exact sequence

$$0 \to \Omega^r L_{\zeta_2} \to L_{\zeta_1\zeta_2} \oplus (\text{projective}) \to L_{\zeta_1} \to 0$$

so that tensoring with M we obtain a sequence of the form

$$0 \to \Omega^r L_{\zeta_2} \otimes M \to \Omega^{r+s} M \oplus \Omega M \oplus (\text{projective}) \to L_{\zeta_1} \otimes M \to 0.$$

By Proposition 5.7.3, Proposition 5.9.1 and Theorem 5.7.11, $V_G(\Omega^r L_{\zeta_2} \otimes M) = V_G\langle\zeta_2\rangle \cap V_G(M) = (V_1 \cap V_G\langle\zeta_2\rangle) \cup V_2$, so that by the inductive hypothesis $\Omega^r L_{\zeta_2} \otimes M \cong N_1 \oplus N_2$ with $V_G(N_1) = V_1 \cap V_G\langle\zeta_2\rangle$ and $V_G(N_2) = V_2$. Similarly $V_G(L_{\zeta_1} \otimes M) = V_G\langle\zeta_1\rangle \cap V_G(M) = V_1 \cup (V_2 \cap V_G\langle\zeta_1\rangle)$ so that by the inductive hypothesis $L_{\zeta_1} \otimes M \cong N_1' \oplus N_2'$ with $V_G(N_1') = V_1$ and $V_G(N_2') = V_2 \cap V_G\langle\zeta_1\rangle$.

Since $V_G(N_1') \cap V_G(N_2) = V_1 \cap V_2 = \{0\}$, it follows from Corollary 5.7.12 that we have $\text{Ext}^1_{kG}(N_1', N_2) = 0$. Similarly $V_G(N_2') \cap V_G(N_1) \subseteq V_1 \cap V_2 = \{0\}$ so that we have $\text{Ext}^1_{kG}(N_2', N_1) = 0$. It is easy to see that this forces the above sequence to decompose as a direct sum of two sequence of the form $0 \to N_1 \to N_1'' \to N_1' \to 0$ and $0 \to N_2 \to N_2'' \to N_2' \to 0$ with $V_G(N_1'') \subseteq V_1$ and $V_G(N_2'') \subseteq V_2$. By the Krull–Schmidt theorem and the invariance of varieties under Ω, the decomposition

$$\Omega^{r+s} M \oplus \Omega M \oplus (\text{projective}) \cong N_1'' \oplus N_2''$$

forces M to decompose as $M_1 \oplus M_2$ with $V_G(M_1) \subseteq V_1$ and $V_G(M_2) \subseteq V_2$. Since $V_G(M) = V_1 \cup V_2$ this forces $V_G(M_1) = V_1$ and $V_G(M_2) = V_2$. $\square$

COROLLARY 5.12.2. *If M is a finitely generated indecomposable kG-module, then the projective variety $\bar{V}_G(M)$ is connected in the Zariski topology.* $\square$

5.13. Example: dihedral 2-groups

In this section, we go through the example of the dihedral 2-groups in detail. Recall that the classification of the indecomposable modules was described in Section 4.11 of Volume I. We shall continue with the notation introduced there, so that

$$G = D_{4q} = \langle x, y \mid x^2 = y^2 = 1, (xy)^q = (yx)^q \rangle$$

(q a power of two). The kG-modules are given by:

(i) The modules $M(C)$ of the first kind corresponding to words C which alternate in $a^{\pm 1}$ and $b^{\pm 1}$ and which do not contain $(ab)^q$, $(ba)^q$ or their inverses.

(ii) The modules $M(C, \phi)$ of the second kind corresponding to words C which are not non-trivial powers, and no power of which contains $(ab)^q$, $(ba)^q$ or their inverses.

(iii) The projective indecomposable module $M((ab)^q(ba)^{-q}, \text{id}_k)$ of the second kind.

We shall assume that k is algebraically closed, so that we may apply the theory of varieties for modules.

We first deal with the case $q = 1$, since this is quite different from $q > 1$, and a lot easier. In this case, we have $H^*(G, k) = k[x_1, x_2]$ with $\deg(x_1) = \deg(x_2) = 1$ so that $V_G = \mathbb{A}^2(k)$ and $\bar{V}_G = \mathbb{P}^1(k)$. The indecomposable modules in this case (cf. Section 4.3 of Volume I) are the modules $\Omega^n(k)$

for $n \in \mathbb{Z}$, for which we have $V_G(\Omega^n(k)) = V_G$ by Proposition 5.7.3, and the modules L_{ζ^m} for $m \geq 1$ and $0 \neq \zeta \in H^1(G, k)$, for which we have $V_G(L_{\zeta^m}) = V_G\langle \zeta^m \rangle = V_G\langle \zeta \rangle$ by Proposition 5.9.1. Note that $\bar{V}_G\langle \zeta \rangle$ is the point in $\mathbb{P}^1(k)$ parametrising the module, as in the remark after Theorem 4.3.3 of Volume I.

In case $q > 1$, we may find $H^*(G, k)$ by repeated application of the spectral sequence of a central extension, and we find that

$$H^*(G, k) = k[x_1, x_2, z]/(x_1 x_2)$$

with $\deg(x_1) = \deg(x_2) = 1$ and $\deg(z) = 2$. We choose the labelling in such a way that the generator x_1 corresponds to the subgroup $\langle xyx, y \rangle$ of index two, while x_2 corresponds to $\langle x, yxy \rangle$. Thus

$$\bar{V}_G = \operatorname{proj} k[x_1, x_2, z]/(x_1 x_2) = \mathbb{P}_a^1 \cup \mathbb{P}_b^1$$

where $\mathbb{P}_a^1$ and $\mathbb{P}_b^1$ are projective lines over k intersecting in the common point at infinity $\mathbb{P}_a^1 \cap \mathbb{P}_b^1 = \{\infty_a = \infty = \infty_b\}$. We choose the notation so that $\mathbb{P}_a^1 = \operatorname{proj} k[x_1, z]$ and $\mathbb{P}_b^1 = \operatorname{proj} k[x_2, z]$, and so that $\lambda x_1^2 + \mu x_2^2 + z = 0$ is the equation of the pair of points $\{\lambda_a, \mu_b\} \subseteq (\mathbb{P}_a^1 \cup \mathbb{P}_b^1) \setminus \{\infty\}$.

THEOREM 5.13.1. *The varieties for the kG-modules, $G = D_{4q}$, $q \geq 1$, are given as follows:*

(i) $\bar{V}_G(M(C)) = \mathbb{P}_a^1 \cup \mathbb{P}_b^1$ *if* $C \sim a^{\pm 1} \dots b^{\pm 1}$

           $\mathbb{P}_b^1$ *if* $C \sim a^{\pm 1} \dots a^{\pm 1}$ *but*

                       $C \not\sim (ab)^{q-1} a(b^{-1}(ab)^{q-1} a)^r$, $r \geq 0$

           $\mathbb{P}_a^1$ *if* $C \sim b^{\pm 1} \dots b^{\pm 1}$ *but*

                       $C \not\sim (ba)^{q-1} b(a^{-1}(ba)^{q-1} b)^r$, $r \geq 0$

           $\{0_b\}$ *if* $C \sim (ab)^{q-1} a(b^{-1}(ab)^{q-1} a)^r$, $r \geq 0$

           $\{0_a\}$ *if* $C \sim (ba)^{q-1} b(a^{-1}(ba)^{q-1} b)^r$, $r \geq 0$

(ii) $\bar{V}_G(M(C, \begin{pmatrix} \lambda & 1 & & 0 \\ & \ddots & \ddots & \\ & & \lambda & 1 \\ 0 & & & \lambda \end{pmatrix})) = \{\infty\}$ *unless* $C \sim a^{-1} b(ab)^{q-1}$,

                                     $b^{-1} a(ba)^{q-1}$ *or* $(ab)^q (ba)^{-q}$

                          $\{\lambda_a\}$ *if* $C \sim a^{-1} b(ab)^{q-1}$

                          $\{\lambda_b\}$ *if* $C \sim b^{-1} a(ba)^{q-1}$

(iii) $\bar{V}_G(M((ab)^q (ba)^{-q}, \operatorname{id}_k) = \emptyset$.

PROOF. (i) The first case follows from Corollary 5.8.5. For the remaining cases, we need to calculate the dimension of $V_G(M(C))$ (which is of course one greater than the dimension of $\bar{V}_G(M(C))$). By Proposition 5.7.2, $\dim V_G(M(C)) = 1$ if and only if $M(C)$ has complexity one, which by Theorem 5.10.4 happens if and only if $M(C)$ is periodic. In this case, by Corollary 5.10.3, $\bar{V}_G(M(C))$ is a single point in $\bar{V}_G$, and by Corollary 5.10.6, $M(C)$

has period two. Recall from Section 4.17 of Volume I that $\Omega^2 M(C) \cong M(L_q R_q C)$, and that $C \sim L_q R_q C$ if and only if $C \sim (R_q)^r(ab)^{q-1}a = (ab)^{q-1}a(b^{-1}(ab)^{q-1}a)^r$ or $C \sim (R_q)^r(ba)^{q-1}b = (ba)^{q-1}b(a^{-1}(ba)^{q-1}b)^r$ for some $r \geq 0$.

Now if $C \sim a^{\pm 1} \ldots a^{\pm 1}$ then it follows from the explicit description of the actions of x and y that $M(C)\downarrow_{\langle x \rangle}$ is free, while $M(C)\downarrow_{\langle y \rangle}$ is not free. Thus $\bar{V}_G(M(C))$ does not contain 0_a but does contain 0_b. By Theorem 5.12.1, $\bar{V}_G(M(C))$ is connected, so if $M(C)$ is periodic then $\bar{V}_G(M(C)) = \{0_b\}$ and if $M(C)$ is not periodic then $\bar{V}_G(M(C)) = \mathbb{P}^1_b$. Similarly for $C \sim b^{\pm 1} \ldots b^{\pm 1}$, if $M(C)$ is periodic then $\bar{V}_G(M(C)) = \{0_a\}$ and if $M(C)$ is not periodic then $\bar{V}_G(M(C)) = \mathbb{P}^1_a$.

(ii) According to Section 4.17 of Volume I, these modules $M(C, \phi)$ are all periodic, so that by Corollary 5.10.3, $\bar{V}_G(M(C, \phi))$ is a single point in $\bar{V}_G$. If $\bar{V}_G(M(C, \phi)) \neq \{\infty\}$ then either $\bar{V}_G(M(C, \phi)) \cap \mathbb{P}^1_a = \emptyset$ or $\bar{V}_G(M(C, \phi)) \cap \mathbb{P}^1_b = \emptyset$.

If $\bar{V}_G(M(C, \phi)) \cap \mathbb{P}^1_a = \emptyset$ then by Corollary 5.7.9 we have $\bar{V}_H(M(C, \phi)) = \emptyset$, where H is the subgroup $\langle xyx, y \rangle$ of index two in G. Thus by Proposition 5.7.2, $M(C, \phi)\downarrow_H$ is projective, and so the rank of the matrix $\tilde{H} = \sum_{h \in H} h$ (which is a generator for $\mathrm{Soc}(kH)$) on $M(C, \phi)$ is equal to $\dim_k M(C, \phi)/|H|$. Setting $X = 1 + x$ and $Y = 1 + y$ in kG, so that $X^2 = Y^2 = 0$, we have (cf. Lemma 4.11.1 of Volume I)

$$\tilde{H} = ((1 + xyx)(1 + y))^{q/2} = (Y + XY + YX + XYX)Y)^{q/2}$$

$$= (YXY + XYXY)^{q/2} = (YX)^{q-1}Y + (XY)^q.$$

Since $M(C, \phi)$ is a non-projective indecomposable kG-module,

$$\tilde{G} = \sum_{g \in G} g = (XY)^q$$

acts as zero on $M(C, \phi)$, and so $\tilde{H}$ acts in the same way as $(YX)^{q-1}Y$. Looking at the way in which X and Y act on $M(C, \phi)$ (Section 4.11 of Volume I), we see that the only way $(YX)^{q-1}Y$ can act with rank equal to $\dim_k M(C, \phi)/2q$ is to have $C \sim (a^{-1}b(ab)^{q-1})^r$ for some $r \geq 1$ (recall that $(ba)^q$ must not appear in any power of C). Since C is not allowed to be a power of a word of smaller length, we have $r = 1$, and $C \sim a^{-1}b(ab)^{q-1}$. Similarly, if $\bar{V}_G(M(C, \phi)) \cap \mathbb{P}^1_b = \emptyset$ then $C \sim b^{-1}a(ba)^{q-1}$.

To complete the proof, we must determine the varieties

$$\bar{V}_G(M(a^{-1}b(ab)^{q-1}, \phi)) \quad \text{and} \quad \bar{V}_G(M(b^{-1}a(ba)^{q-1}, \phi)).$$

Let $\zeta = (\lambda x_1^2 + \mu x_2^2 + z)^r \in H^{2r}(G, k)$. We claim that

$$L_\zeta \cong M\left(a^{-1}b(ab)^{q-1}, \begin{pmatrix} \lambda & 1 & & 0 \\ & \ddots & \ddots & \\ & & \lambda & 1 \\ 0 & & & \lambda \end{pmatrix}\right) \oplus M\left(b^{-1}a(ba)^{q-1}, \begin{pmatrix} \mu & 1 & & 0 \\ & \ddots & \ddots & \\ & & \mu & 1 \\ 0 & & & \mu \end{pmatrix}\right),$$

where the matrices on the right-hand side are $r \times r$ matrices. Given this claim, it follows from Proposition 5.9.1 that the variety of this direct sum is

$\{\lambda_a, \mu_b\}$, so that by applying Proposition 5.7.5 and comparing for different values of λ and μ, we see that the second and third cases of part (ii) are proved.

To prove the claim, we argue as follows. We have

$$\Omega^{2r}(k) \cong M((b^{-1}a(ba)^{q-1})^{-r}(a^{-1}b(ab)^{q-1})^r),$$

a module of dimension $4qr + 1$. According to the schema described in Section 4.11 of Volume I, we have an ordered basis $z_0, \ldots, z_{4qr}$ of $\Omega^{2r}(k)$ corresponding to this word. With respect to this basis, the element $x_1^{2(r-s)}z^s \in H^{2r}(G, k) \cong \mathrm{Hom}_{kG}(\Omega^{2r}(k), k)$ corresponds to the homomorphism sending $z_{2q(2r-s)}$ to 1 and all other z_i to zero, while $x_2^{2(r-s)}z^s \in H^{2r}(G, k)$ corresponds to the homomorphism sending z_{2qs} to 1 and all other z_i to zero. We shall show that

$$L_\zeta \cap \langle z_{2qr}, \ldots, z_{4qr} \rangle \cong M\left(a^{-1}b(ab)^{q-1}, \begin{pmatrix} \lambda & 1 & & 0 \\ & \ddots & \ddots & \\ & & \lambda & 1 \\ 0 & & & \lambda \end{pmatrix}\right)$$

$$L_\zeta \cap \langle z_0, \ldots, z_{2qr} \rangle \cong M\left(b^{-1}a(ba)^{q-1}, \begin{pmatrix} \mu & 1 & & 0 \\ & \ddots & \ddots & \\ & & \mu & 1 \\ 0 & & & \mu \end{pmatrix}\right).$$

We concentrate on the first of these isomorphism, as the second follows by symmetry. Let

$$V_0 = L_\zeta \cap \langle z_{2qr}, z_{2q(r+1)}, \ldots, z_{4qr} \rangle$$
$$V_i = \langle z_{2qr+i}, z_{2q(r+1)+i}, \ldots, z_{2q(2r-1)+i} \rangle \qquad 1 \leq i \leq 2q - 1.$$

and take as basis for V_0 the preimages of $z_{2q(r+1)}, \ldots, z_{4qr}$ under the projection onto these basis vectors of $\Omega^{2r}(k)$. In terms of these bases, we see that $L_\zeta \cap \langle z_{2qr}, \ldots, z_{4qr} \rangle$ has been given as the schema for $M(a^{-1}b(ab)^{q-1}, \phi)$ as described in Section 4.11 of Volume I, where

$$\phi = \begin{pmatrix} r\lambda & \binom{r}{2}\lambda^2 & \cdots & \lambda^r \\ 1 & 0 & \cdots & 0 \\ 0 & 1 & \cdots & 0 \\ \vdots & & & \vdots \\ 0 & 0 & \cdots & 1 & 0 \end{pmatrix}.$$

Some elementary linear algebra shows that this matrix is conjugate to the matrix $\begin{pmatrix} \lambda & 1 & & 0 \\ & \ddots & \ddots & \\ & & \lambda & 1 \\ 0 & & & \lambda \end{pmatrix}$. Thus we have produced two submodules of L_ζ of the appropriate isomorphism types, which intersect in $\{0\}$ and span L_ζ. This completes the identification of L_ζ.

(iii) This follows from Proposition 5.7.2, since $M((ab)^q(ba)^{-q}, \mathrm{id}_k)$ is projective. $\qquad \square$

5.14. Multiple complexes

The main theorem of this section is a generalisation of Theorem 5.10.4 to arbitrary (not necessarily periodic) finitely generated modules. Namely, we shall show that in a suitable sense, a module of complexity c possesses a c-fold periodic projective resolution. The hard part of the proof is involved in checking that certain modules are projective. We shall give two proofs of this, one involving the machinery of varieties and the other using a hypercohomology argument. The latter is more opaque, but works in greater generality.

We begin by recalling from Section 2.6 of Volume I the basic construction we shall be using. Suppose $\zeta \in H^n(G, R) \cong \mathrm{Ext}^n_{RG}(R, R)$. We choose a cocycle $\hat{\zeta} : \tilde{\Omega}^n R \to R$ representing ζ, where $\tilde{\Omega}^n R$ is the nth kernel in a projective resolution $\mathbf{P}$ of R as an RG-module. By making $\mathbf{P}$ large enough, we may assume $\hat{\zeta}$ is surjective. We denote its kernel by L_ζ, and form the pushout diagram

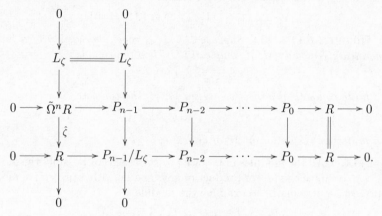

The bottom row of this diagram is an n-fold extension representing the element $\zeta \in \mathrm{Ext}^n_{RG}(R, R)$. We denote by $\mathbf{C}_\zeta$ the chain complex

$$0 \to P_{n-1}/L_\zeta \to P_{n-2} \to \cdots \to P_0 \to 0$$

formed by truncating the bottom row of this diagram. Thus we have

$$H_i(\mathbf{C}_\zeta) \cong \begin{cases} R & \text{if } i = 0, n-1 \\ 0 & \text{otherwise.} \end{cases}$$

We write $\tilde{\zeta}$ for the generator of degree $n-1$, and 1 for the generator of degree zero.

The complex $\mathbf{C}_\zeta$ should be thought of as a sort of algebraic analogue of a sphere with G-action, with ζ being the transgression of the fundamental class of the sphere.

We also write $\mathbf{C}^{(\infty)}_\zeta$ for the chain complex

$$\cdots \to P_1 \to P_0 \to P_{n-1}/L_\zeta \to P_{n-2} \to \cdots \to P_1 \to P_0 \to 0$$

obtained by splicing together infinitely many copies of $\mathbf{C}_\zeta$ in positive degree. It is an exact complex except in degree zero, where the homology is R.

DEFINITION 5.14.1. *We say that homogeneous elements* $\zeta_1, \ldots, \zeta_c$ *of the ring* $H^{\cdot}(G, k)$ **cover the variety** *of a finitely generated* kG-*module* M *of complexity* c *if*

$$V_G\langle \zeta_1 \rangle \cap \cdots \cap V_G\langle \zeta_c \rangle \cap V_G(M) = \{0\}.$$

Note that by Proposition 5.7.2, $\dim V_G(M) = c$, so that by the Noether Normalisation Lemma 5.4.5 there exist homogeneous elements $\zeta_1, \ldots, \zeta_c$ generating a polynomial subalgebra

$$k[\zeta_1, \ldots, \zeta_c] \subseteq H^{\cdot}(G, k)/I_G(M)$$

over which $H^{\cdot}(G, k)/I_G(M)$ is finitely generated as a module. Thus the corresponding map $V_G(M) \to \mathbb{A}^c(k)$ is finite, so that the preimage of the origin is a finite set. But this map is also homogeneous, and so the preimage of the origin is the origin. Therefore for such a choice of $\zeta_1, \ldots, \zeta_c$ we have

$$V_G\langle \zeta_1 \rangle \cap \cdots \cap V_G\langle \zeta_c \rangle \cap V_G(M) = \{0\}.$$

THEOREM 5.14.2. [34] *Suppose* $\zeta_1, \ldots, \zeta_c$ *cover the variety of a finitely generated* kG-*module* M *of complexity* c. *Then the complex*

$$\mathbf{C}_{\zeta_1} \otimes \cdots \otimes \mathbf{C}_{\zeta_c} \otimes M$$

is a finite complex of projective kG-*modules. The complex*

$$\mathbf{C}_{\zeta_1}^{(\infty)} \otimes \cdots \otimes \mathbf{C}_{\zeta_c}^{(\infty)} \otimes M$$

is a projective resolution of M *of growth* c.

PROOF. All the modules in $\mathbf{C}_{\zeta_i}$ are projective except possibly the module P_{n_i-1}/L_{ζ_i}. Since the tensor product of any module with a projective module is projective, it remains to examine the module

$$P_{n_1-1}/L_{\zeta_1} \otimes \cdots \otimes P_{n_c-1}/L_{\zeta_c} \otimes M.$$

By Proposition 5.7.3, Theorem 5.7.11 and Proposition 5.9.1, the variety of this module is

$$V_G\langle \zeta_1 \rangle \cap \cdots \cap V_G\langle \zeta_c \rangle \cap V_G(M)$$

which is $\{0\}$ by the choice of $\zeta_1, \ldots, \zeta_c$. It thus follows from Proposition 5.7.2 that this module is also projective.

It now follows by Proposition 3.4.4 that $\mathbf{C}_{\zeta_1}^{(\infty)} \otimes \cdots \otimes \mathbf{C}_{\zeta_c}^{(\infty)} \otimes M$ is a projective resolution of M. Since each $\mathbf{C}_{\zeta_i}^{(\infty)}$ is a periodic complex, the rate of growth of the tensor product is exactly c. □

REMARK. The above theorem shows that a finitely generated module always has a projective resolution which may be expressed as a tensor product of periodic complexes, and of the same polynomial rate of growth as the minimal resolution. However, in general the minimal resolution may not be written as a tensor product in this way. But it appears in all the examples calculated that the minimal resolution of a module of complexity c can always be written as the total complex of a c-fold complex in which the rows, columns, etc. are all eventually periodic. For example, if $G = GL_3(\mathbb{F}_2)$ and $\operatorname{char}(k) = 2$, then there are three simple modules in the principal block, namely the trivial module k, the natural three dimensional simple module M and its dual N. The minimal resolution of the trivial module is the total complex of a double complex of the form given as follows.

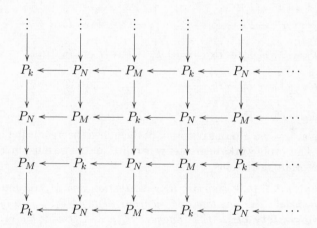

As another example, if G is the alternating group A_6 and $\operatorname{char}(k) = 3$, then there are four simple modules in the principal block, namely the trivial module k, two distinct three dimensional modules L and M, and a four dimensional module N. The minimal resolution of the trivial module is the total complex of a double complex of the form given as follows.

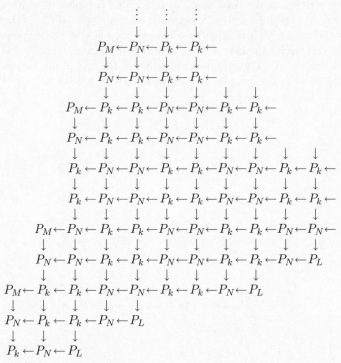

We now give an alternative proof of the above theorem using hypercohomology. This proof works in greater generality, and so we must first generalise the appropriate definitions.

DEFINITION 5.14.3. *Suppose R is Noetherian and M is a finitely generated RG-module. We say that homogeneous elements $\zeta_1, \ldots, \zeta_c$ of $H^*(G, R)$* **cover the variety** *of M if the images of $\zeta_1, \ldots, \zeta_c$ under the map*

$$H^*(G, R) = \text{Ext}^*_{RG}(R, R) \xrightarrow{\otimes M} \text{Ext}^*_{RG}(M, M)$$

generate a subring over which $\text{Ext}^*_{RG}(M, M)$ is finitely generated as a module.

REMARKS. In case $R = k$ is a field, this agrees with the previous definition of covering the variety of M. By Corollary 4.2.4 to Evens' theorem, a set of homogeneous elements covering the variety of M always exists. We shall see below that under some moderately weak hypotheses, M has a projective resolution which is a tensor product of c periodic complexes, and which therefore has growth exactly c. Thus a minimal such c may be thought of as the complexity of M. Note that in this generality there is no unique minimal resolution, so that our previous definition over a field does not work.

We begin with a lemma. Recall from Proposition 3.4.3 that if **D** is a chain complex of RG-modules, bounded above, then we have a hypercohomology spectral sequence

$$\text{Ext}^p_{RG}(H_q(\mathbf{C}_\zeta), \mathbf{D}) \Rightarrow \text{Ext}^{p+q}_{RG}(\mathbf{C}_\zeta, \mathbf{D}).$$

Since $H_q(\mathbf{C}_\zeta)$ is only non-zero for $q = n - 1$ ($n = \deg(\zeta)$) and $q = 0$, the only possible non-zero differential in the above spectral sequence is d_n.

LEMMA 5.14.4. *Suppose* $\mathbf{D}$ *is a chain complex of RG-modules, bounded above. In the two row spectral sequence*

$$E^{pq}_2(\zeta) = \text{Ext}^p_{RG}(H_q(\mathbf{C}_\zeta), \mathbf{D}) \Rightarrow \text{Ext}^{p+q}_{RG}(\mathbf{C}_\zeta, \mathbf{D})$$

the differential d_n *is given by*

$$d_n(\alpha.\tilde{\zeta}) = \alpha.\zeta \in E^{p+n,0}_n(\zeta) = H^{p+n}(G, \mathbf{D}).$$

PROOF. We first remark that since $H_q(\mathbf{C}_\zeta) = R$ if $q = 0$ or $n - 1$ and is zero elsewhere, we have

$$E^{pq}_2(\zeta) = \begin{cases} H^p(G, \mathbf{D}) & \text{if } q = 0 \text{ or } n - 1 \\ 0 & \text{otherwise.} \end{cases}$$

The differential d_n may be obtained as follows. We have a short exact sequence of chain complexes

$$0 \to \mathbf{C}_\zeta \to \mathbf{C}^{(\infty)}_\zeta \to \mathbf{C}^{(\infty)}_\zeta[n] \to 0.$$

If $\mathbf{P}$ is a projective resolution of R as an RG-module, we have a short exact sequence

$$0 \to \mathbf{P} \otimes \mathbf{C}_\zeta \to \mathbf{P} \otimes \mathbf{C}^{(\infty)}_\zeta \to \mathbf{P} \otimes \mathbf{C}^{(\infty)}_\zeta[n] \to 0$$

and hence a short exact sequence of cochain complexes

$$0 \to \text{Hom}_{RG}(\mathbf{P} \otimes \mathbf{C}^{(\infty)}_\zeta, \mathbf{D})[-n] \to \text{Hom}_{RG}(\mathbf{P} \otimes \mathbf{C}^{(\infty)}_\zeta, \mathbf{D})$$
$$\to \text{Hom}_{RG}(\mathbf{P} \otimes \mathbf{C}_\zeta, \mathbf{D}) \to 0.$$

Now $\mathbf{P} \otimes \mathbf{C}^{(\infty)}_\zeta$ is again a projective resolution of R as an RG-module, so the long exact sequence in cohomology of this short exact sequence of cochain complexes is

$$\cdots \to \text{Ext}^{r-1}_{RG}(\mathbf{C}_\zeta, \mathbf{D}) \to H^{r-n}(G, \mathbf{D}) \xrightarrow{d_n} H^r(G, \mathbf{D}) \to \text{Ext}^r_{RG}(\mathbf{C}_\zeta, \mathbf{D}) \to \cdots$$

This is the long exact sequence associated to the above two-row spectral sequence, and so the marked homomorphism is d_n.

To identify this map as multiplication by ζ, we argue as follows. In general, multiplication by ζ may be thought of in the following way. A cocycle representing the element ζ lifts to a map of projective resolutions $\mathbf{P} \to \mathbf{P}[n]$. If $\alpha \in H^{r-n}(G, \mathbf{D})$ is represented by a chain map $\mathbf{P} \to \mathbf{D}[r - n]$ then the product $\alpha.\zeta$ is represented by shifting and composing

$$\mathbf{P} \to \mathbf{P}[n] \to \mathbf{D}[r].$$

So all we need check is that the map of resolutions

$$\mathbf{P} \otimes \mathbf{C}^{(\infty)}_\zeta \to \mathbf{P} \otimes \mathbf{C}^{(\infty)}_\zeta[n]$$

given above represents the element ζ in cohomology. But this follows from the diagram

$$\cdots \to (\mathbf{P} \otimes \mathbf{C}_\zeta^{(\infty)})_n \to (\mathbf{P} \otimes \mathbf{C}_\zeta^{(\infty)})_{n-1} \to \cdots \to (\mathbf{P} \otimes \mathbf{C}_\zeta^{(\infty)})_0 \to R \to 0$$

$$\cdots \to (\mathbf{P} \otimes \mathbf{C}_\zeta^{(\infty)})_0$$

$$0 \longrightarrow R \longrightarrow P_{n-1}/L_\zeta \longrightarrow \cdots \longrightarrow P_0 \longrightarrow R \to 0$$

since the bottom row is an exact sequence representing ζ.   □

If $\zeta_1, \ldots, \zeta_c \in H^*(G, R)$ are homogeneous elements, then Corollary 2.7.2 of the Künneth Theorem in Volume I shows that $H_*(\bigotimes_{i=1}^c \mathbf{C}_{\zeta_i})$ has a basis consisting of elements of the form $x_1 \otimes \cdots \otimes x_c$ where each x_i is equal to either 1 or $\tilde\zeta_i$. We also write $\tilde\zeta_i$ for the element $1 \otimes \cdots \otimes \tilde\zeta_i \otimes \cdots \otimes 1$ of $H_*(\bigotimes_{i=1}^c \mathbf{C}_{\zeta_i})$.

THEOREM 5.14.5. [35] *Suppose R is a Noetherian ring with the property that for each prime p dividing $|G|$, either p is invertible in R or R/pR is Artinian, and M is an RG-lattice (i.e., a finitely generated RG-module which is projective as an R-module). If $\zeta_1, \ldots, \zeta_c \in H^*(G, R)$ cover the variety of M, then the complex*

$$\mathbf{C}_{\zeta_1} \otimes \cdots \otimes \mathbf{C}_{\zeta_c} \otimes M$$

is a finite complex of projective RG-modules. The complex

$$\mathbf{C}_{\zeta_1}^{(\infty)} \otimes \cdots \otimes \mathbf{C}_{\zeta_c}^{(\infty)} \otimes M$$

is a projective resolution of M.

PROOF. As in the proof of Theorem 5.14.2, it suffices to show that the module

$$P_{n_1-1}/L_{\zeta_1} \otimes \cdots \otimes P_{n_c-1}/L_{\zeta_c} \otimes M$$

is projective. We first show that it is sufficient to prove that $\mathbf{C} \otimes M$ has a finite projective resolution, where $\mathbf{C} = \mathbf{C}_{\zeta_1} \otimes \cdots \otimes \mathbf{C}_{\zeta_c}$. For if $\mathbf{P} \to \mathbf{C} \otimes M$ is a finite projective resolution, then by enlarging $\mathbf{P}$ if necessary, we may assume this map is surjective. The kernel is therefore a finite exact sequence of modules, all except possibly one of which is projective. Moreover, as R-modules, all the modules involved are projective, and so by Corollary 3.6.5 of Volume I, $P_{n_1-1}/L_{\zeta_1} \otimes \cdots \otimes P_{n_c-1}/L_{\zeta_c} \otimes M$ is projective as an RG-module.

By Corollary 4.2.7, in order to prove that $\mathbf{C} \otimes M$ has a finite projective resolution, it suffices to prove that $\mathrm{Ext}^n_{RG}(\mathbf{C} \otimes M, \mathbf{C} \otimes M) = 0$ for all sufficiently large n. We start by showing that $\mathrm{Ext}^n_{RG}(\mathbf{C} \otimes M, M) = 0$ for all sufficiently large n. Since M is an RG-lattice, $\mathrm{Ext}^n_{RG}(\mathbf{C} \otimes M, M) \cong \mathrm{Ext}^n_{RG}(\mathbf{C}, \mathrm{Hom}_R(M, M))$, so we have a spectral sequence

$$E_2^{pq} = \mathrm{Ext}^p_{RG}(H_q(\mathbf{C}), \mathrm{Hom}_R(M, M)) \Rightarrow \mathrm{Ext}^{p+q}_{RG}(\mathbf{C} \otimes M, M).$$

The map $\mathbf{C} \to \mathbf{C}_{\zeta_i}$ given by the augmentation $\mathbf{C}_{\zeta_j} \to R$ for $j \neq i$ gives rise to a map of spectral sequences to the above spectral sequence from

$$E_2^{pq}(\zeta_i) = \mathrm{Ext}_{RG}^p(H_q(\mathbf{C}_{\zeta_i}), \mathrm{Hom}_R(M, M)) \Rightarrow \mathrm{Ext}_{RG}^{p+q}(\mathbf{C}_{\zeta_i} \otimes M, M).$$

The latter spectral sequence was examined in the lemma, where it was shown that the differential d_{n_i} is given by $d_{n_i}(\alpha.\tilde{\zeta_i}) = \alpha.\zeta_i$. It follows that the same formula holds in the original spectral sequence E_j^{pq}.

Now E_2^{**} is finitely generated as a module over E_2^{0*} (which is a free R-module dual to $H_*(\mathbf{C})$). Since the differentials are E_2^{*0}-module homomorphisms, it follows that E_∞^{**} is also finitely generated as a module over E_2^{*0}. But the elements $\zeta_1, \ldots, \zeta_c$ in the kernel of $E_2^{*0} \to E_\infty^{*0}$ act as zero on E_∞^{**}, and so E_∞^{**} is finitely generated as a module over

$$E_2^{*0}/(\zeta_1, \ldots, \zeta_c) = \mathrm{Ext}_{RG}^*(M, M)/(\zeta_1, \ldots, \zeta_c).$$

Since $\mathrm{Ext}_{RG}^*(M, M)$ is finitely generated as a module over the subring generated by the images of $\zeta_1, \ldots, \zeta_c$, the quotient by the images of multiplication by these elements is a finitely generated R-module. It follows that E_∞^{**} is a finitely generated R-module, and hence so is $\mathrm{Ext}_{RG}^*(\mathbf{C} \otimes M, M)$.

Finally, the spectral sequence

$$\mathrm{Ext}_{RG}^p(\mathbf{C} \otimes M, H_{-q}(\mathbf{C}) \otimes M) \Rightarrow \mathrm{Ext}_{RG}^{p+q}(\mathbf{C} \otimes M, \mathbf{C} \otimes M)$$

has as its E_2 term a finite number of non-zero rows, each isomorphic to $\mathrm{Ext}_{RG}^*(\mathbf{C} \otimes M, M)$. So the E_2 term is a finitely generated R-module, and hence so are the E_∞ term and $\mathrm{Ext}_{RG}^*(\mathbf{C} \otimes M, \mathbf{C} \otimes M)$. □

COROLLARY 5.14.6. *Suppose $\zeta_1, \ldots, \zeta_c \in H^*(G, \mathbb{Z})$ are homogeneous elements generating a subring over which $H^*(G, \mathbb{Z})$ is finitely generated as a module. Then for any (commutative) coefficient ring R, $\mathbf{C} = R \otimes_{\mathbb{Z}} \mathbf{C}_{\zeta_1} \otimes \cdots \otimes \mathbf{C}_{\zeta_r}$ is a finite complex of projective RG-modules with $H_0(\mathbf{C}) \cong R$, and $H_i(\mathbf{C})$ is a direct sum of copies of R for all i.* □

REMARK. The existence of a finite complex $\mathbf{C}$ of projective RG-modules with $H_i(\mathbf{C})$ a direct sum of copies of R for all i, as in the theorem, can also be proved topologically as follows. A faithful complex representation of G gives rise to an embedding into a compact unitary group $G \hookrightarrow U(n)$. If we choose a cellular division of $U(n)$ such that the image of a cell under the action of G is always a cell, then the complex of cellular chains with values in R is a chain complex of free RG-modules. Since $U(n)$ is connected, every element of G acts in a way which may be deformed to the identity, and hence trivially on homology.

5.15. Gaps in group cohomology

In this and the next section, we give some examples of theorems which may be proved using the theory of multiple complexes developed in the last section.

Our first theorem is best stated and proved in terms of Tate cohomology, so we begin with a quick review of this subject.

Let G be a finite group and R be a commutative ring of coefficients. If

$$\cdots \to P_2 \to P_1 \to P_0$$

is a projective resolution of R as an RG-module then

$$\operatorname{Hom}_R(P_0, R) \to \operatorname{Hom}_R(P_1, R) \to \operatorname{Hom}_R(P_2, R) \to \cdots$$

is an exact sequence of modules which are injective relative to the trivial subgroup, or equivalently projective (see Proposition 3.6.4 of Volume I). Splicing these sequences together, we obtain a doubly infinite sequence

$$\mathbf{P}_{\pm}: \qquad \cdots \to P_2 \to P_1 \to P_0 \to P_{-1} \to P_{-2} \to \cdots$$

where $P_{-n-1} = \operatorname{Hom}_R(P_n, R)$. This is called a **complete resolution** of R. The cohomology groups of the complex $\operatorname{Hom}_{RG}(\mathbf{P}_{\pm}, M)$ are the Tate cohomology groups $\hat{H}^*(G, M)$. Thus for $n > 0$ we have $\hat{H}^*(G, M) \cong H^*(G, M)$. Since $\operatorname{Hom}_{RG}(P_{-n-1}, M) \cong P_n \otimes_{RG} M$, for $n > 0$ we have $\hat{H}^{-n-1}(G, M) \cong H_n(G, M)$. The group $\hat{H}^0(G, M)$ is the quotient of $H^0(G, M) = M^G$ by the image of $\operatorname{Tr}_{1,G}$. In the notation of Definition 3.6.2 of Volume I we have $\hat{H}^0(G, M) = M^{G,1}$. The group $\hat{H}^{-1}(G, M)$ is the dually defined submodule of $H_0(G, M) = R \otimes_{RG} M = M/[G, M]$.

It is easy to see that a short exact sequence of RG-modules $0 \to M_0 \to M_1 \to M_2 \to 0$ gives rise to a doubly infinite long exact sequence in Tate cohomology extending the usual long exact sequence (Proposition 2.5.3 (ii) of Volume I) in ordinary cohomology.

THEOREM 5.15.1. *Suppose G is a finite group and R is a commutative ring of coefficients. Let $\mathbf{C}$ be the finite complex of projective RG-modules described in Corollary 5.14.6 (or Theorem 5.14.5 with $M = R$). Then there is a spectral sequence whose E_2 term is given by*

$$E_2^{pq} = \operatorname{Hom}_R(H_q(\mathbf{C}), \hat{H}^p(G, M)),$$

and which converges to zero.

There is also a spectral sequence whose E_2 term is given by

$$E_2^{pq} = \operatorname{Hom}_R(H_q(\mathbf{C}), H^p(G, M)),$$

and which converges to $H^*(\operatorname{Hom}_{RG}(\mathbf{C}, M))$.

PROOF. Let $\mathbf{P}_{\pm}$ be a complete resolution of R as an RG-module, and let E_0^{pq} be the double complex $\operatorname{Hom}_{RG}(\mathbf{P}_{\pm} \otimes \mathbf{C}, M)$, with horizontal differential d_1 coming from $\mathbf{P}_{\pm}$ and vertical differential d_0 coming from $\mathbf{C}$. If we look at the spectral sequence in which d_1 is performed first, we see that since C_i is projective, each row is a split exact sequence of projectives, and so E_1 is zero. Thus the cohomology of the total complex is zero, and so the spectral sequence in which d_0 is performed first also converges to zero. In the latter spectral sequence we have

$$E_1^{pq} \cong \operatorname{Hom}_{RG}(\mathbf{P}_{\pm} \otimes H_q(\mathbf{C}), M) \cong \operatorname{Hom}_R(H_q(\mathbf{C}), \operatorname{Hom}_{RG}(\mathbf{P}_{\pm}, M))$$

since $H_q(\mathbf{C})$ is a direct sum of copies of R with trivial G-action. Now the horizontal differential is the one induced by the boundary map on $\mathbf{P}_\pm$, and so E_2 is as given.

If instead of a complete resolution $\mathbf{P}_\pm$ we use a projective resolution $\mathbf{P}$ of R as an RG-module, then the spectral sequence in which d_1 is performed first has $E_1 = \operatorname{Hom}_{RG}(\mathbf{C}, M)$ concentrated along the vertical axis and $E_2 = E_\infty = H^*(\operatorname{Hom}_{RG}(\mathbf{C}, M))$. The spectral sequence in which d_0 is performed first has

$$E_1 \cong \operatorname{Hom}_R(H_q(\mathbf{C}), \operatorname{Hom}_{RG}(\mathbf{P}, M))$$

and E_2 as given. $\qquad\square$

THEOREM 5.15.2. [38] *Given a finite group G, there exists a positive integer r such that for any commutative ring R of coefficients and any RG-module M, if $\hat{H}^n(G, M) = 0$ for $r + 1$ consecutive values of n, then $\hat{H}^n(G, M) = 0$ for all n, positive and negative.*

PROOF. We first prove this in the case where M is projective as an R-module. Let $\mathbf{C}$ be the finite complex of projective RG-modules described in Corollary 5.14.6, and let r be its length, so that $\mathbf{C}$ has the form

$$0 \to C_r \to \cdots \to C_1 \to C_0 \to 0.$$

Then $H_r(\mathbf{C}) \cong H_0(\mathbf{C}) \cong R$, and so the top and bottom rows of the spectral sequence described in the above theorem have the form

$$E_2^{nr} \cong E_2^{n0} \cong \hat{H}^n(G, M).$$

Suppose that $\hat{H}^n(G, M)$ is not always zero, but is zero for $r + 1$ consecutive values of n, say $m, m + 1, \ldots, m + r$, with either $\hat{H}^{m-1}(G, M)$ or $\hat{H}^{m+r+1}(G, M)$ non-zero. Thus either $E_2^{m-1,r} \neq 0$ or $E_2^{m+r+1,0} \neq 0$. However, $E_2^{pq} = 0$ for $m \leq p \leq m + r$. The differential d_k on the E_k page takes E_k^{pq} to $E_k^{p+k,q-k-1}$, and $d_k = 0$ for $k > r + 1$. So whichever of $E_2^{m-1,r}$ or $E_2^{m+r+1,0}$ is non-zero, this group can never be killed at any stage in the spectral sequence. This contradicts the fact that the spectral sequence converges to zero.

We now deal with the general case where M is not necessarily projective as an R-module. In this case, we first remark that since $\hat{H}^n(G, M)$ does not change when the coefficient ring R is replaced by $\mathbb{Z}$ (cf. the remark after Definition 2.4.4 in Volume I), we may as well assume R is $\mathbb{Z}$. Note at this stage that we have not assumed that M is finitely generated. Now if

$$0 \to M' \to F \to M \to 0$$

is a short exact sequence with F a free $\mathbb{Z}G$-module, then M' is $\mathbb{Z}$-free. Since $\hat{H}^n(G, F) = 0$, we have $\hat{H}^n(G, M) \cong \hat{H}^{n+1}(G, M')$ and so the theorem for M follows from the theorem for M'. $\qquad\square$

5.16. Isomorphisms in group cohomology

As an application of the theorem of the last section, we present a modified version due to Serge Bouc (private communication) of an algebraic proof by Benson and Evens [**39**] that if a homomorphism of finite groups induces an isomorphism in integral cohomology then it is a group isomorphism. This was first proved by Evens (unpublished, but announced in [**101**]) by embedding G into a compact unitary group as in the remark after Corollary 5.14.6; Jackowski [**356**] independently found a similar proof. A related but much harder theorem of Mislin [**194**] gives a group theoretic characterisation of the homomorphisms of finite groups which induce an isomorphism in mod p cohomology.

THEOREM 5.16.1. *Let $\phi : G' \to G$ be a homomorphism of finite groups. If*

$$\phi^* : H^*(G, \mathbb{Z}) \to H^*(G', \mathbb{Z})$$

is an isomorphism, then ϕ is an isomorphism.

PROOF. We begin by proving that if ϕ is surjective, with kernel N, and ϕ^* is an epimorphism, then ϕ is an isomorphism. In this case, ϕ^* is the inflation map, which is the horizontal edge homomorphism for the Lyndon–Hochschild–Serre spectral sequence for the extension

$$1 \to N \to G' \to G \to 1.$$

If this is an epimorphism, it follows that the E_∞ page of this spectral sequence is concentrated along the horizontal axis, and so the vertical edge homomorphism is zero in positive degrees. Thus the restriction map from $H^*(G', \mathbb{Z})$ to $H^*(N, \mathbb{Z})$ is zero in positive degrees. Theorem 4.1.3 shows that if $g \neq 1$ is an element of N then the restriction map from $H^*(G', \mathbb{Z})$ to $H^*(\langle g \rangle, \mathbb{Z})$ is non-zero in some positive degree, so it follows from this that $N = 1$, and ϕ is an isomorphism.

Since a general homomorphism can be factored as the composite of a surjective and an injective group homomorphism, it follows from what we have just proved, that ϕ is injective. So we regard G' as a subgroup of G via ϕ. Consider the map $\mathbb{Z} \to \mathbb{Z}_{G'} \uparrow^G$ given by sending 1 to $\sum_{g \in G/G'} g \otimes 1$. Write M for the cokernel, so that we have a short exact sequence of $\mathbb{Z}G$-modules

$$0 \to \mathbb{Z} \to \mathbb{Z}_{G'} \uparrow^G \to M \to 0.$$

According to the exercise at the end of Section 2.8 of Volume I, the corresponding map in cohomology

$$H^*(G, \mathbb{Z}) \to H^*(G, \mathbb{Z}_{G'} \uparrow^G) \cong H^*(G', \mathbb{Z})$$

is the restriction map. So the long exact sequence in cohomology (Proposition 2.5.3(ii) of Volume I) shows that $H^n(G, M) = 0$ for all $n > 0$. It now follows from Theorem 5.15.2 that $\hat{H}^n(G, M) = 0$ for all $n \in \mathbb{Z}$, so that $\hat{H}^n(G, \mathbb{Z}) \cong \hat{H}^n(G', \mathbb{Z})$ for all $n \in \mathbb{Z}$. But $\hat{H}^0(G, \mathbb{Z}) \cong \mathbb{Z}/|G|$, so $|G| = |G'|$ and ϕ is an isomorphism. $\square$

5.17. Poincaré duality

In this section, taken from [**35**], we show that if k is a field, and $H^*(G, k)$ is finitely generated as a module over the subring generated by homogeneous elements $\zeta_1, \ldots, \zeta_r$ (so that these elements cover the variety of the trivial module) then the complex of projective modules $\bigotimes_{i=1}^r \mathbf{C}_{\zeta_i}$ described in Theorem 5.14.2 (with $M = k$) is homotopy equivalent to its dual, suitably shifted in degree. We begin with a general lemma about homotopy equivalences.

LEMMA 5.17.1. *Suppose* $\mathbf{C}$ *and* $\mathbf{D}$ *are finite chain complexes of finitely generated projective* kG*-modules, and* $f : \mathbf{C} \to \mathbf{D}$ *is a chain map. Then the following are equivalent.*

(i) f *induces an isomorphism in homology* $f_* : H_*(\mathbf{C}) \xrightarrow{\cong} H_*(\mathbf{D})$.

(ii) f *is a homotopy equivalence.*

(iii) *There exist decompositions* $\mathbf{C} = \mathbf{C}' \oplus \mathbf{P}'$, $\mathbf{D} = \mathbf{D}' \oplus \mathbf{Q}'$, *where* $\mathbf{P}'$ *and* $\mathbf{Q}'$ *are exact sequences of projective modules, and the restriction of* f *to* $\mathbf{C}'$ *is an isomorphism* $f : \mathbf{C}' \xrightarrow{\cong} \mathbf{D}'$.

PROOF. It is clear that (iii) $\Rightarrow$ (ii) $\Rightarrow$ (i), so we shall prove that (i) $\Rightarrow$ (iii). Suppose $f : \mathbf{C} \to \mathbf{D}$ induces an isomorphism in homology. By adding an exact sequence of projective modules $\mathbf{Q}$ to $\mathbf{C}$ we may make f surjective, and still a homology isomorphism. Denote by $\mathbf{P}$ the kernel of $f : \mathbf{C} \oplus \mathbf{Q} \twoheadrightarrow \mathbf{D}$. The long exact sequence in homology shows that $\mathbf{P}$ is an exact sequence of projectives. Since projective kG-modules are also injective, the sequence

$$0 \to \mathbf{P} \to \mathbf{C} \oplus \mathbf{Q} \to \mathbf{D} \to 0$$

splits, and so $\mathbf{C} \oplus \mathbf{Q} \cong \mathbf{D} \oplus \mathbf{P}$. The result now follows from the Krull–Schmidt theorem for finite chain complexes of finitely generated kG-modules. □

Now if $\zeta \in H^n(G, k)$ is represented by a cocycle $\hat{\zeta} : \Omega^n k \to k$, we may dualise to obtain a map $\hat{\zeta}^* : k \to \Omega^{-n} k$. Applying Ω^n to this map, we obtain a map $\Omega^n(\hat{\zeta}^*) : \Omega^n k \to k$. The relationship between this map and the original map $\hat{\zeta}$ is given in the following proposition.

PROPOSITION 5.17.2. *If* $\zeta \in H^n(G, k)$ *then*

$$\Omega^n(\hat{\zeta}^*) = (-1)^{n(n+1)/2} \hat{\zeta}.$$

There is a map of chain complexes

$$
\begin{array}{ccccccccc}
0 & \to & P_{n-1}/L_\zeta & \to & P_{n-2} & \to & \cdots & \to & P_1 & \longrightarrow & P_0 & \longrightarrow & 0 \\
& & \downarrow & & \downarrow & & & & \downarrow & & \downarrow & & \\
0 & \longrightarrow & P_0^* & \longrightarrow & P_1^* & \to & \cdots & \to & P_{n-2}^* & \to & (P_{n-1}/L_\zeta)^* & \longrightarrow & 0
\end{array}
$$

inducing an isomorphism on homology.

PROOF. We build a commutative diagram

$$
\begin{array}{ccccccccc}
0 & \longrightarrow & \Omega^n k & \longrightarrow & P_{n-1} & \longrightarrow \cdots \longrightarrow & P_0 & \longrightarrow & k & \longrightarrow 0 \\
& & \downarrow{\scriptstyle\hat\zeta^*\otimes 1} & & \downarrow{\scriptstyle\hat\zeta^*\otimes 1} & & \downarrow{\scriptstyle\hat\zeta^*\otimes 1} & & \downarrow{\scriptstyle\hat\zeta^*} & \\
0 \to & \Omega^{-n}k \otimes \Omega^n k & \to \Omega^{-n}k \otimes P_{n-1} & \to \cdots \to & \Omega^{-n}k \otimes P_0 & \to \Omega^{-n}k \to 0 \\
& \downarrow{\scriptstyle\gamma} & & \downarrow & & \downarrow & & \| & \\
0 & \longrightarrow & k & \longrightarrow & P_0^* & \longrightarrow \cdots \longrightarrow & P_{n-1}^* & \longrightarrow & \Omega^{-n}k \to 0.
\end{array}
$$

In this diagram, the bottom set of vertical arrows has been filled in using the fact that both rows are exact sequences of projective modules, except at the ends, and γ is obtained by restricting the previous map. Now modulo maps which factor through a projective module, there is only one dimension of maps from $\Omega^{-n}k \otimes \Omega^n k$ to k. It follows that γ is some multiple $\lambda_n.$ev of the evaluation map (regarding $\Omega^{-n}k$ as the dual of $\Omega^n k$). Now the composite of the left-hand vertical maps in the above diagram is $\Omega^n(\hat\zeta^*)$, and so we have

$$\Omega^n(\hat\zeta^*) = \lambda_n.\mathrm{ev} \circ (\hat\zeta^* \otimes 1) = \lambda_n.\hat\zeta.$$

(There are no maps from $\Omega^n k$ to k which factor through a projective module).

It remains to determine the constants λ_n. It is clear that λ_n only depends on n, and not on G or ζ. Since the map $\zeta \mapsto \lambda_n.\zeta$ is an anti-automorphism of $H^*(G,k)$ (duality reverses Yoneda composition) we have

$$\lambda_m\xi.\lambda_n\zeta = \lambda_{m+n}\zeta.\xi,$$

so that

$$\lambda_m\lambda_n = (-1)^{mn}\lambda_{m+n}.$$

It thus remains to determine λ_1. Since the transpose inverse of the matrix $\left(\begin{smallmatrix}1 & \alpha \\ 0 & 1\end{smallmatrix}\right)$ is $\left(\begin{smallmatrix}1 & 0 \\ -\alpha & 1\end{smallmatrix}\right)$, we see that $\lambda_1 = -1$, and so by induction $\lambda_n = (-1)^{n(n+1)/2}$.

Now the left-hand map, $\gamma \circ (\hat\zeta^* \otimes 1) = (-1)^{n(n+1)/2}\hat\zeta$, in the above diagram has kernel L_ζ, while the right-hand map $\hat\zeta^*$ has cokernel L_ζ^*, and so by passing to the appropriate quotient on the top row and subcomplex on the bottom row, we obtain a map of complexes

$$
\begin{array}{ccccccccc}
0 \to & k & \to P_{n-1}/L_\zeta & \to P_{n-2} & \to \cdots \longrightarrow & P_1 & \longrightarrow & P_0 & \longrightarrow k \to 0 \\
{\scriptstyle (-1)^{n(n-1)/2}}\downarrow & \downarrow & & \downarrow & & \downarrow & & \downarrow & \| \\
0 \to & k & \longrightarrow P_0^* & \longrightarrow P_1^* & \to \cdots \to & P_{n-2}^* & \to & (P_{n-1}/L_\zeta)^* & \to k \to 0
\end{array}
$$

and hence the required homology equivalence. $\square$

THEOREM 5.17.3. *Suppose that $H^*(G,k)$ is finitely generated as a module over the subring generated by $\zeta_1, \ldots, \zeta_r$ in degrees $n_1, \ldots, n_r$ with each $n_i \geq 2$. Then the complex*

$$\mathbf{C} = \bigotimes_{i=1}^{r} \mathbf{C}_{\zeta_i}$$

is a direct sum of a complex $\mathbf{C}'$ *satisfying Poincaré duality in formal dimension* $s = \sum_{i=1}^{r}(n_i - 1)$,

$$\mathbf{Hom}_k(\mathbf{C}', k)[s] \cong \mathbf{C}'$$

and an exact complex $\mathbf{P}'$ *of projective modules.*

PROOF. By the proposition, we have homology equivalences

$$\mathbf{Hom}_k(\mathbf{C}_{\zeta_i}, k)[n_i - 1] \to \mathbf{C}_{\zeta_i}.$$

Putting these together, we obtain a homology equivalence

$$\mathbf{Hom}_k(\mathbf{C}, k)[s] \cong \mathbf{C}$$

By Theorem 5.14.2, $\mathbf{C}$ is a finite complex of projective modules, and so the theorem now follows by applying Lemma 5.17.1. $\qquad\square$

COROLLARY 5.17.4. *If* $\mathbf{C}$ *and* s *are as in the theorem, then the Poincaré series* $f(t) = \sum_{i=0}^{s} t^i \dim_k \mathrm{Hom}_{kG}(C_i, k)$ *satisfies the functional equation* $t^s f(1/t) = f(t)$. $\qquad\square$

5.18. Cohen–Macaulay cohomology rings

Recall from Section 5.4 that a finitely generated graded k-algebra is Cohen–Macaulay if there is a polynomial subring over which it is finitely generated and free as a module.

If E is an elementary abelian p-group then $H^*(E, k)$ is Cohen–Macaulay. Quillen's calculations (see Section 2.9) show that the cohomology rings of general linear groups at primes other than the natural one are Cohen–Macaulay, and his work on extraspecial 2-groups (see Section 5.5) shows that these also have Cohen–Macaulay cohomology rings. If G is a semidihedral 2-group then $H^*(G, k)$ is not Cohen–Macaulay, see Evens and Priddy [109]. Also, if G is a split metacyclic p-group with p odd, then usually $H^*(G, k)$ is not Cohen–Macaulay, see Diethelm [91]. Note that if G has maximal elementary abelian p-subgroups of different ranks, then by Quillen's Stratification Theorem 5.6.3 the variety is not equidimensional and so $H^*(G, k)$ is not Cohen–Macaulay.

In this section, we shall use the Poincaré duality of the last section to show that the Poincaré series satisfies a certain functional equation in the Cohen–Macaulay case.

THEOREM 5.18.1. [35] *Suppose* G *is a finite group and* k *is a field of characteristic* p. *If* $H^*(G, k)$ *is Cohen–Macaulay, then the Poincaré series*

$$p(t) = \sum_{r \geq 0} t^r \dim_k H^r(G, k)$$

is a rational function of t *satisfying the functional equation*

$$p(1/t) = (-t)^{r_p(G)} p(t),$$

where $r_p(G)$ *is the* p-*rank of* G.

PROOF. The fact that $p(t)$ is a rational function of t follows from the Proposition 5.3.1.

Let $\zeta_1, \dots, \zeta_r$ be elements of degrees $n_1, \dots, n_r$ (with each $n_i \geq 2$) generate a polynomial subring over which $H^*(G, k)$ is a finitely generated free module. Thus by Theorem 5.3.8 we have $r = r_p(G)$. Let $\mathbf{C}$ be the complex described in Theorem 5.14.5 (with $M = k$). According to Theorem 5.15.1, there is a spectral sequence whose E_2 term is given by

$$E_2^{pq} = \mathrm{Hom}_k(H_q(\mathbf{C}), H^p(G, k)) = H^*(G, k) \otimes \Lambda^*(\tilde{\zeta}_1, \dots, \tilde{\zeta}_r),$$

where $\Lambda^*(\tilde{\zeta}_1, \dots, \tilde{\zeta}_r)$ is an exterior algebra on generators $\tilde{\zeta}_i$ in degree $n_i - 1$, and converging to $H^*(\mathrm{Hom}_{kG}(\mathbf{C}, k))$.

Arguing as in the proof of Theorem 5.14.5, for each ζ_i we have a homomorphism of spectral sequences

$$E_2^{pq}(\zeta_i) = \mathrm{Hom}_k(H_q(\mathbf{C}_{\zeta_i}), H^p(G, k)) \to E_2^{pq}.$$

By Lemma 5.14.4 we have $d_{n_i}(\alpha.\tilde{\zeta}_i) = \alpha.\zeta_i$. Since the ζ_i form a regular sequence in $H^*(G, k)$, one can check that this determines all the differentials (see the remarks below), and the E_∞ page is $H^*(G, k)/(\zeta_1, \dots, \zeta_r)$ concentrated along the bottom row. We thus have

$$H^*(G, k)/(\zeta_1, \dots, \zeta_r) \cong H^*(\mathrm{Hom}_{kG}(\mathbf{C}, k)).$$

Thus if $f(t)$ is the Poincaré series of $H^*(\mathrm{Hom}_{kG}(\mathbf{C}, k))$, we have

$$p(t) = f(t)/\prod_{i=1}^{r}(1 - t^{n_i}).$$

Since by Corollary 5.17.4 $f(t)$ satisfies $t^s f(1/t) = f(t)$, where $s = \sum_{i=1}^{r}(n_i - 1)$, we have

$$p(1/t) = f(1/t)/\prod_{i=1}^{r}(1 - t^{-n_i}) = t^{-s} f(t).\prod_{i=1}^{r} t^{n_i}/(t^{n_i} - 1) = (-t)^r p(t). \qquad \square$$

REMARKS. Strictly speaking, in the above proof, we have assumed that the spectral sequence has a multiplicative structure, and we have not shown this to be the case. One can get round this by using more homomorphisms of spectral sequences, see Section 9 of [**35**] for further details.

Alternatively, one can try to impose a multiplicative structure. For this purpose, one really needs a comultiplication on $\mathbf{C}$. In Carlson [**75**], it is shown that as long as ζ annihilates $\mathrm{Ext}_{kG}^*(L_\zeta, L_\zeta)$ (cf. Propositions 5.9.5 and 5.9.6; in particular this always happens if p is odd), and as long as we use the dual $\mathrm{Hom}_k(\mathbf{C}_{\zeta_i}, k)[n_i - 1]$ instead of $\mathbf{C}_{\zeta_i}$ (which we may do with impunity according to the last section), there is a comultiplication on each $\mathbf{C}_{\zeta_i}$ with the desired properties. For $p = 2$ the situation is less clear.

Even if $H^*(G, k)$ is not Cohen–Macaulay, the spectral sequence described in the above theorem still gives very strong restrictions on the possible shape of $H^*(G, k)$. In [**35**], all the differentials in this spectral sequence are described in terms of matric Massey products (see also [**31**]).

Many theorems in the cohomology of finite groups generalise in an obvious way to compact Lie groups. The following corollary is, of course, false in general for compact Lie groups.

COROLLARY 5.18.2. *Suppose G is a finite group and k is a field of characteristic p. If $H^*(G, k)$ is a polynomial ring, then $p = 2$, the generators are in degree one, and $G/O_{2'}(G)$ is an elementary abelian 2-group.*

PROOF. Since a polynomial ring is Cohen–Macaulay, the functional equation of the theorem must be satisfied. This can only happen if the generators are in degree one. It follows that $p = 2$, since otherwise degree one elements square to zero.

Since $H^1(G, k) \cong \mathrm{Hom}(G, k)$, it follows that G has an elementary abelian quotient G/N with the property that the map $G \to G/N$ induces an isomorphism $H^*(G/N, k) \to H^*(G, k)$. Since this is the horizontal edge homomorphism for the Lyndon–Hochschild–Serre spectral sequence of the group extension, we deduce that the vertical edge homomorphism, namely restriction from $H^*(G, k)$ to $H^*(N, k)$ is zero in positive degrees. By Theorem 4.1.3, this implies that N has no elements of order two, so that $N = O_{2'}(G)$. □

RECENT PROGRESS: Since the first edition of this book was published, the theory of varieties for modules has been extended to infinitely generated modules for finite groups [310, 311] using Rickard's theory of idempotent modules in the stable category [394]. The main difference from the finitely generated case is that instead of a single subvariety, a module has associated to it a collection of subvarieties of V_G. This rests on a version of Dade's lemma for infinitely generated modules in [311] in which one has to restrict not just to cyclic shifted subgroups defined over the original field, but also to "generic" cyclic shifted subgroups defined over the function fields of subvarieties.

This work has had a number of applications. It was used in [312] to classify the thick subcategories of the stable category of finitely generated modules for a p-group. In [305] it was used to prove most of the conjectures made in [38] on the vanishing of cohomology, although it has later been shown by Carlson (see section 13 of [328]) that this can be proved without appealing to infinitely generated modules. In [307, 308] the theory of varieties of modules is extended to a large class of infinite groups. The point here is that even if a module starts off finitely generated, if we restrict to a subgroup of infinite index it need not end up that way.

The **nucleus** is defined to be the subvariety Y_G of V_G given by the union of the images of the restriction maps $\mathrm{res}^*_{G,H} : V_H \to V_G$ as H runs over the subgroups of G whose centraliser is not p-nilpotent. This concept was introduced in [38], and its importance has become more apparent in [305] where it is seen to control the modules in the principal block having no cohomology, and in [312] where it plays a role in the classification of thick subcategories of the stable category of finitely generated modules.

CHAPTER 6

Group actions and the Steinberg module

The material in this Chapter is based on a lecture given by Peter Webb at the 1986 Arcata conference on Representation Theory of Finite Groups. I would like to thank him for supplying me with an early copy of the published version of this talk [**282**].

6.1. G-simplicial complexes

We shall be interested in group actions on topological spaces. Since we shall be interested in homotopical properties of group actions, and every topological space has the same weak homotopy type as its simplicial complex of singular chains, it is no real restriction to limit our attention to simplicial complexes with G acting *simplicially*, i.e., in such a way that the image of a simplex under a group element is always a simplex. Mostly our attention will be focused on the action of G on various finite simplicial complexes arising in a natural way from the subgroup structure of G. We shall be interested in representation theoretic invariants of these actions, and we shall concentrate on a definition for an arbitrary finite group of a *generalised Steinberg module*, which will be a virtual projective module which agrees for a Chevalley group (up to sign) with the usual Steinberg module.

DEFINITION 6.1.1. *Suppose G acts simplicially on a simplicial complex* Δ. *We say that Δ is a G-**simplicial complex** if whenever an element of G stabilises a simplex of Δ setwise, then it stabilises it pointwise.*

Note that if the action does not satisfy this last condition, we can take a *barycentric subdivision* and then it will.

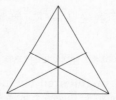

We'll define barycentric subdivision formally below, but informally you should think of it as being obtained inductively from the original simplicial complex by inserting a new vertex in the middle of each simplex, and joining it to the previously subdivided simplices. In particular, the underlying topological space is unchanged by this process.

If Δ is a G-simplicial complex, and k is a commutative ring of coefficients (we are mostly interested in the case where k is a field), we obtain a chain complex as follows. First we need to say a word or two about orientations.

DEFINITION 6.1.2. *An* **oriented simplex** *is a simplex together with a total ordering on the set of vertices, where two orderings are regarded as giving the same orientation if and only if they differ by an even permutation. The* **opposite orientation** *of a simplex is obtained from the given orientation by applying an odd permutation.*

DEFINITION 6.1.3. *We define the nth* **chain group** *of Δ,*

$$C_n(\Delta) = C_n(\Delta; k)$$

to be the k-module with the oriented simplices as generators, and relators saying that each oriented simplex is equal to minus the same simplex with the opposite orientation.

It follows from the fact that the setwise stabiliser of a simplex stabilises it pointwise, that $C_n(\Delta)$ is a permutation module. Although the permutation basis depends on a choice of orientations consistent over G-orbits, the isomorphism type of the permutation representation is well defined, and gives us a well defined element $c_n(\Delta)$ of the Burnside ring $b(G)$ (see Chapter 5 of Volume I).

BOUNDARY MAPS AND HOMOLOGY. We define boundary maps

$$\partial_n : C_n(\Delta) \to C_{n-1}(\Delta)$$

as follows. If $(x_0, \dots, x_n)$ is an oriented simplex, we define

$$\partial(x_0, \dots, x_n) = \sum_{i=0}^{n} (-1)^i (x_0, \dots, x_{i-1}, x_{i+1}, \dots, x_n).$$

It is easy to check that $\partial_n \circ \partial_{n+1} = 0$, and that the boundary maps are kG-module homomorphisms, and $C_n(\Delta)$ is a chain complex of kG-modules. Thus the homology groups

$$H_n(\Delta) = H_n(\Delta; k) = \mathrm{Ker}(\partial_n)/\mathrm{Im}(\partial_{n+1})$$

are also kG-modules.

It is also often convenient to work with reduced homology.

DEFINITION 6.1.4. *The* **augmented chain complex** *of Δ is the chain complex with an extra copy of k inserted in degree -1, where the boundary homomorphism*

$$\partial_0 = \varepsilon : C_0(\Delta; k) \to k$$

is taken to be the **augmentation map** *sending each zero-simplex to the identity element of k.*

The **reduced homology groups** $\tilde{H}_n(\Delta; k)$ *are defined to be the homology groups of the augmented chain complex, namely*

$$\tilde{H}_n(\Delta; k) = \begin{cases} H_n(\Delta; k) & n \neq 0 \\ \text{Ker}(H_0(\Delta; k) \to k) & n = 0 \end{cases}$$

6.2. *G*-posets

DEFINITION 6.2.1. *A* **G-poset** *is a partially ordered set together with a G-action which preserves the partial order (i.e., $a < b$ implies $ga < gb$).*

If X is a G-poset, then we form a G-simplicial complex $|X|$, called the **simplicial realisation** of X, as follows. The n-simplices are the totally ordered subsets

$$x_0 < x_1 < \cdots < x_n.$$

This can be viewed as the topological realisation of a simplicial set (see Section 1.8) in which there are also degenerate simplices where equalities are allowed:

$$x_0 \leq x_1 \leq \cdots \leq x_n.$$

The face and degeneracy maps are given by omitting and repeating elements in the obvious way.

The **boundary** of such a simplex is

$$\sum_{i=0}^{n}(-1)^i(x_0 < \cdots < x_{i-1} < x_{i+1} < \cdots < x_n).$$

Conversely, given a G-simplicial complex Δ, we form a G-poset $S(\Delta)$ whose elements are the simplices of Δ, and where $a < b$ if and only if the vertices of the simplex a are a subset of the vertices of b.

Any invariant defined for simplicial complexes or G-simplicial complexes is defined for posets or G-posets by passing to the simplicial realisation. So for example we define $H_n(X; k) = H_n(|X|, k)$ and $\pi_n(X, x_0) = \pi_n(|X|, x_0)$.

DEFINITION 6.2.2. *The* **barycentric subdivision** $sd(\Delta)$ *is defined as the simplicial realisation of the poset of simplices, $|S(\Delta)|$.*

EXERCISE. Show that the underlying topological space of $sd(\Delta)$ is homeomorphic to that of Δ by a homeomorphism which preserves the G-action.

EXAMPLES. The following are examples of G-posets defined using the subgroup structure of G.

1. $\mathcal{S}_p(G) = \{\text{non-trivial } p\text{-subgroups of } G\}$
2. $\mathcal{A}_p(G) = \{\text{non-trivial elementary abelian } p\text{-subgroups of } G\}$
3. $\mathcal{B}_p(G) = \{\text{non-trivial } p\text{-subgroups } P \text{ of } G \text{ satisfying } P = O_p(N_G(P))\}$

We demand that the subgroups be non-trivial, since otherwise the corresponding simplicial complex would be a cone, and hence contractible (see the next section but one for further details). In fact, these complexes turn out to be homotopy equivalent in a way which preserves the G-action, as we shall see in the next few sections.

We discuss coverings of posets (cf. Definition 1.6.12) via invertible local coefficient systems (or invertible systems for short). This is a special case of the concept of a local coefficient system, discussed in detail in the next chapter.

DEFINITION 6.2.3. *An* **invertible system** *of sets (abelian groups, etc.) $\mathcal{F}$ on a poset X is a functor from X, regarded as a category, to the category of sets (abelian groups, etc.) with the property that every map $\mathcal{F}(x) \to \mathcal{F}(x')$ associated to an inequality $x \leq x'$ is an isomorphism. Invertible systems on X form a category, in which the morphisms are the natural transformations.*

Given an invertible system of sets $\mathcal{F}$ on X, we obtain a poset $\tilde{X}$ whose elements are pairs (x, u) with $x \in X$, $u \in \mathcal{F}(x)$, and with $(x, u) \leq (x', u')$ if and only if $x \leq x'$, and u' is the image of u under the map $\mathcal{F}(x) \to \mathcal{F}(x')$ corresponding to this inequality. There is an obvious map of posets $\tilde{X} \to X$ whose simplicial realisation is a covering map $|\tilde{X}| \to |X|$.

Conversely, given a covering space $E \to |X|$, we obtain an invertible system $\mathcal{F}$ by setting $\mathcal{F}(x)$ equal to the inverse image in E of the point in $|X|$ corresponding to x.

It is easy to check that the above describes an equivalence of categories between covering spaces of $|X|$ and invertible systems on X. In particular, if X is simply connected (i.e., X is connected and $\pi_1(X, x_0) = \{1\}$), then every invertible system is isomorphic to a constant functor with values in some set. By the same token, there is an invertible system $\mathcal{F}$ on X such that the corresponding poset $\tilde{X}$ is the **universal cover** of the poset X. By Theorem 1.6.13, for this $\mathcal{F}$ we have $\mathcal{F}(x) \cong \pi_1(X, x)$.

6.3. The Lefschetz Invariant

Suppose Δ is a finite G-simplicial complex. We wish to associate to Δ various representation theoretic invariants, and hence get information about representations out of information about group actions, and vice-versa.

First we investigate what happens in characteristic zero.

PROPOSITION 6.3.1. *Suppose k is a field of characteristic zero. Then we have*

$$\sum_i (-1)^i H_i(\Delta, k) = \sum_i (-1)^i C_i(\Delta, k)$$

as elements of the representation ring $a(G)$ (see Chapter 5 of Volume I).

PROOF. Since we are in characteristic zero, all sequences of G-modules split. The chain complex

$$\ldots \to C_{n+1}(\Delta, k) \xrightarrow{\partial_{n+1}} C_n(\Delta, k) \xrightarrow{\partial_n} C_{n-1}(\Delta, k) \to \ldots$$

gives us short exact sequences

$$0 \to \text{Im}(\partial_{n+1}) \to \text{Ker}(\partial_n) \to H_n(\Delta, k) \to 0$$
$$0 \to \text{Ker}(\partial_n) \to C_n(\Delta, k) \to \text{Im}(\partial_n) \to 0.$$

So in $a(G)$ we have

$$H_n(\Delta, k) = \text{Ker}(\partial_n) - \text{Im}(\partial_{n+1}), \qquad C_n(\Delta, k) = \text{Ker}(\partial_n) + \text{Im}(\partial_n),$$

and these expressions clearly have the same alternating sum. $\square$

In particular the **Euler characteristic**

$$\chi(\Delta) = \sum_i (-1)^i \dim C_i(\Delta, k)$$

only depends on the homology of Δ, and is hence a homotopy invariant. The **reduced Euler characteristic** is the corresponding concept for the augmented chain complex, namely $\tilde{\chi}(\Delta) = \chi(\Delta) - 1$.

If the characteristic of k divides the order of G, the above proposition is no longer true, but we can still make the following definition.

DEFINITION 6.3.2. *The* **Lefschetz module** $L_G(\Delta, k)$ *is defined to be the virtual module*

$$L_G(\Delta, k) = \sum_i (-1)^i C_i(\Delta, k) \in a(G).$$

The **reduced Lefschetz module** *is*

$$\tilde{L}_G(\Delta, k) = \sum_i (-1)^i \tilde{C}_i(\Delta, k) = L_G(\Delta, k) - 1 \in a(G).$$

EXAMPLE. If char $k = p$, we call $\tilde{L}_G(\mathcal{S}_p(G), k)$ the **generalised Steinberg module** for G at the prime p. We shall see later that it is a virtual projective module (i.e., it lies in $a(G, 1)$, the linear span of the projective modules in $a(G)$). In the case where G is a Chevalley group in characteristic p, it turns out to be equal to plus or minus the usual Steinberg module. We shall have more to say about this later.

The Lefschetz module has a precursor in the Burnside ring $b(G)$ (see Chapter 5 of Volume I), namely the Lefschetz invariant, defined as follows. Let $c_n(\Delta)$ denote the set of n-simplices in Δ, regarded as a permutation representation of G, and hence as an element of $b(G)$.

DEFINITION 6.3.3. *The* **Lefschetz invariant** *of Δ is defined to be*

$$\Lambda_G(\Delta) = \sum_i (-1)^i c_i(\Delta) \in b(G).$$

The **reduced Lefschetz invariant** *of Δ is $\tilde{\Lambda}_G(\Delta) = \Lambda_G(\Delta) - 1$.*

Under the natural map $b(G) \to a(kG)$ we have $\Lambda_G(\Delta) \mapsto L_G(\Delta, k)$. Recall from Section 5.4 of Volume I that we have a natural inclusion

$$b(G) \hookrightarrow \bigoplus_{H \leq_G G} \mathbb{Z}$$

$$X \mapsto |X^H|.$$

In particular

$$\Lambda_G(\Delta) \mapsto \sum_i (-1)^i . |(\Delta_{(i)})^H| = \chi(\Delta^H)$$

the Euler characteristic of the fixed point set Δ^H. So knowing $\Lambda_G(\Delta)$ is equivalent to knowing $\chi(\Delta^H)$ for all $H \leq G$.

EXERCISES. 1. Show that $\sum_i (-1)^i H_i(\Delta, k)$ and $\sum_i (-1)^i C_i(\Delta, k)$ have the same Brauer character, or equivalently the same image in the Grothendieck ring of kG-modules.

2. Show that if Δ and Δ' are finite G-simplicial complexes, then
(i) $\Lambda_G(\Delta \dot\cup \Delta') = \Lambda_G(\Delta) + \Lambda_G(\Delta')$,
(ii) $\Lambda_G(\Delta \times \Delta') = \Lambda_G(\Delta).\Lambda_G(\Delta')$.

6.4. Equivariant homotopy

DEFINITION 6.4.1. *Suppose that X and Y are G-simplicial complexes. We say $f : X \to Y$ is a G-map if $f(gx) = gf(x)$ for all $x \in X$ and $g \in G$. If $f, f' : X \to Y$ are G-maps, we say f and f' are G-homotopic (written $f \simeq_G f'$) if there is a G-map $F : X \times I \to Y$ (where I is the unit interval $[0,1]$ with trivial G-action) such that $F(x,0) = f(x)$ and $F(x,1) = f'(x)$.*

We say X and Y are G-homotopy equivalent (written $X \simeq_G Y$) if there are G-maps $f : X \to Y$ and $f' : Y \to X$ such that the composites are G-homotopic to the identity maps $f \circ f' \simeq_G \mathrm{id}_Y$ and $f' \circ f \simeq_G \mathrm{id}_X$.

Clearly, if $X \simeq_G Y$ then $X^H \simeq Y^H$ for all subgroups $H \leq G$. It is a remarkable fact that the converse is also true.

THEOREM 6.4.2. *Suppose that X and Y are two G-simplicial complexes and $f : X \to Y$ is a G-map with the property that for all subgroups $H \leq G$, f restricts to an ordinary homotopy equivalence $f^H : X^H \simeq Y^H$. Then f is a G-homotopy equivalence.*

PROOF. In fact this statement is true for a much wider class of spaces and maps. See for example Tammo tom Dieck's book [89, Proposition 8.2.4 and Remark 8.2.5] and also James and Segal [138]. We shall prove the theorem in the context of G-CW-complexes, where G is supposed to act cellularly and f is a cellular map. The proof is modelled on the proof of Whitehead's Theorem 1.5.8 but with G-action incorporated. Let M_f denote the mapping cylinder of f. For each subgroup H of G, we have a long exact sequence

$$\cdots \to \pi_{n+1}(M_f^H, X^H) \to \pi_n(X^H) \xrightarrow{f_*^h} \pi_n(Y^H) \to \pi_n(M_f^H, X^H) \to \cdots$$

and so since f_*^H is an isomorphism, we deduce that we have $\pi_n(M_f^H, X^H) = 0$ for all subgroups H of G and all $n \geq 0$.

Now suppose (K, L) is a relative G-CW-complex and $g : (K, L) \to (M_f, X)$ is a cellular G-map. By induction on skeleta, the restriction of g to the n-skeleton of (K, L) is G-homotopic relative to L to a cellular G-map sending $K^{(n)}$ into X. Namely, once the $(n-1)$-skeleton is in, for each orbit of G on the cells, say $G/H \times$ (cell), we choose a representative cell in this orbit, with stabiliser H. Since $\pi_n(M_f^H, X^H) = 0$, there is no obstruction to moving it (non-equivariantly) into X, and then we use the G-action to see how to move the rest of the orbit. We may fit these G-homotopies together by doing the first skeleton in the interval $[0, \frac{1}{2}]$, the second in $[\frac{1}{2}, \frac{3}{4}]$, and so on, to show that g is G-homotopic relative to L to a G-map sending K into X.

Applying the above statement to the identity map from (M_f, X) to itself, we see that the inclusion $X \subseteq M_f$ is a G-homotopy equivalence, and hence so is f. $\square$

REMARK. It is *not* true that if the restrictions of two maps f, $f' : X \to Y$ to fixed point sets of H are homotopic as maps from X^H to Y^H for all subgroups $H \leq G$ then $f \simeq_G f'$.

PROPOSITION 6.4.3. *Suppose that X and Y are two finite G-simplicial complexes. Then we have* (i) $\Leftrightarrow$ (ii) $\Rightarrow$ (iii) $\Rightarrow$ (iv) $\Rightarrow$ (v) *in the following list of statements:*

(i) $X \simeq_G Y$
(ii) $X^H \simeq Y^H$ *for all* $H \leq G$
(iii) $\chi(X^H) = \chi(Y^H)$ *for all* $H \leq G$
(iv) $\Lambda_G(X) = \Lambda_G(Y) \in b(G)$
(v) $L_G(X) = L_G(Y) \in a(kG)$.

PROOF. This is clear from the definitions and the above theorem. $\square$

We shall be applying this to several G-simplicial complexes arising from the subgroup structure of G, and we shall mostly be interested in properties of the corresponding Lefschetz modules.

Now recall that Conlon's Induction Theorem (5.6.8 of Volume I) implies that two elements of $b(G)$ have the same image in $a(kG)$ if their restrictions to all p-hypo-elementary subgroups (i.e., subgroups H with $H/O_p(H)$ cyclic) are equal. So by the same chain of reasoning as above, we have the following.

PROPOSITION 6.4.4. *Suppose k is a field of characteristic p. Suppose X and Y are two G-simplicial complexes such that for all p-hypo-elementary subgroups $H \leq G$, we have $\chi(X^H) = \chi(Y^H)$. Then $L_G(X, k) = L_G(Y, k) \in a(kG)$.* $\square$

We now turn to the equivariant homotopy theory of G-posets. If X and Y are G-posets then we have equivariant homeomorphisms

$$|X^{\mathrm{op}}| \cong |X|, \qquad |X \times Y| = |X| \times |Y|.$$

Here, the product on the right is taken in the category of compactly generated spaces, as in the remark after Definition 1.5.3. This statement can be seen by going through simplicial sets, where the behaviour with respect to products is clearer, and then using the fact that topological realisation commutes with products for simplicial sets (see Section 1.8).

If $f : X \to Y$ is a G-map of G-posets (i.e., order preserving G-map of the underlying sets) then f induces a map $|f| : |X| \to |Y|$. If two maps of posets (resp. G-posets) $f, f' : X \to Y$ induce homotopic (resp. G-homotopic) maps of simplicial complexes, then we say that f and f' are **homotopic**, (resp. G-**homotopic**) and write $f \simeq f'$ (resp. $f \simeq_G f'$). We say that a poset X is **contractible** if $|X|$ is contractible.

LEMMA 6.4.5. *If $f, f' : X \to Y$ are maps of posets such that for all $x \in X$, $f(x) \le f'(x)$, then $f \simeq f'$.*

PROOF. Denote by $(0 < 1)$ the poset with two elements and one inequality. Then f and f' define a map of posets

$$
\begin{aligned}
F \;:\; (0 < 1) \times X &\to Y \\
(0, x) &\mapsto f(x) \\
(1, x) &\mapsto f'(x)
\end{aligned}
$$

Since $|(0 < 1)|$ is the unit interval, $|F|$ is a homotopy from $|f|$ to $|f'|$. $\qquad \square$

DEFINITION 6.4.6. *We say a poset X is **conically contractible** if there exists a map of posets $f : X \to X$ and an element $x_0 \in X$ such that for all $x \in X$, $x \le f(x) \ge x_0$.*

By the above lemma, if X is conically contractible, then it is contractible.

EXAMPLE. If a simplicial complex Δ is a cone, with vertex x_0, then $S(\Delta)$ is conically contractible, since we may take for f the map sending each simplex to the simplex obtained by adjoining x_0 if it is not already a vertex.

6.5. Quillen's lemma

Our main tool for proving that complexes are homotopy equivalent is a lemma of Quillen, which we now discuss.

DEFINITION 6.5.1. *If $f : X \to Y$ is a map of posets and $y \in Y$, we define*

$$
\begin{aligned}
f/y &= \{x \in X \mid f(x) \le y\} \subseteq X \\
y \backslash f &= \{x \in X \mid f(x) \ge y\} \subseteq X
\end{aligned}
$$

LEMMA 6.5.2 (Quillen). *Suppose that $f : X \to Y$ is a map of posets with the property that f/y (resp. $y \backslash f$) is contractible for all $y \in Y$. Then the corresponding map $|f|$ of simplicial complexes is a homotopy equivalence.*

PROOF. We shall describe the proof given in Quillen [**217**], Section 7. Another proof may be found in Quillen [**214**], Theorem A.

It suffices to prove that $f_* : \pi_1(X, x_0) \to \pi_1(Y, f(x_0))$ is an isomorphism, so that f lifts to a map of universal covers $\tilde{f} : \tilde{X} \to \tilde{Y}$, and that

$$\tilde{f}_* : H_*(\tilde{X}) \to H_*(\tilde{Y})$$

is an isomorphism. For then by the Whitehead Theorems 1.4.3 and 1.5.8, $\tilde{f}$ is a homotopy equivalence. Then by Theorem 6.4.2, $\tilde{f}$ is a $\pi_1(X, x_0)$-homotopy equivalence and so f is a homotopy equivalence.

To show that $f_* : \pi_1(X, x_0) \to \pi_1(Y, f(x_0))$ is an isomorphism, by Theorem 1.6.3 and the discussion following Definition 6.2.3, it suffices to show that f induces an equivalence of categories between invertible systems on X and Y. If $\mathcal{F}$ is an invertible system on Y, then the pullback $f^*\mathcal{F}$ is the invertible system on X defined by $f^*\mathcal{F}(x) = \mathcal{F}(f(x))$. If $x \le x'$ then the map $f^*\mathcal{F}(x) \to f^*\mathcal{F}(x')$ is the map $\mathcal{F}(f(x)) \to \mathcal{F}(f(x'))$ associated to the inequality $f(x) \le f(x')$.

Conversely, if $\mathcal{F}'$ is an invertible system on X, then we define an invertible system on Y as follows. Given $y \in Y$, the poset f/y is contractible, and hence simply connected. So $\mathcal{F}'$ is isomorphic to a constant functor on f/y, and so one can identify the set associated to each point in f/y with the set $\varinjlim_{x \in f/y} \mathcal{F}(x)$. We set

$$f_!\mathcal{F}(y) = \varinjlim_{x \in f/y} \mathcal{F}(x).$$

It is easy to check that f^* and $f_!$ give an equivalence of categories.

Now the morphism $\tilde{f} : \tilde{X} \to \tilde{Y}$ satisfies the same hypothesis as f, so we now assume X and Y are simply connected, and we must show that $f_* : H_*(X) \to H_*(Y)$ is an isomorphism. We do this by examining the spectral sequence of the double complex (see Section 3.4) E^0_{**} defined as follows. We let E^0_{pq} be the vector space over k whose basis consists of pairs of chains of inequalities, $(y_0 < \cdots < y_p)$ in Y and $(x_0 < \cdots < x_q)$ in f/y_0. The horizontal differential d_1 is given by the boundary map (see Section 6.1) on $(y_0 < \cdots < y_p)$ and the identity map on $(x_0 < \cdots < x_q)$. The vertical differential d_0 is given by the identity map on $(y_0 < \cdots < y_p)$ and the boundary map on $(x_0 < \cdots < x_q)$, multiplied by $(-1)^p$ to make the squares anticommute rather than commute.

With respect to the horizontal differential, this double complex has, for each chain $(x_0 < \cdots < x_q)$ in X, the chain complex of the subposet of Y consisting of those y with $y \ge f(x_q)$. This poset has a minimal element, and is hence contractible. So if we do the horizontal differential first, we get $E^1 = C_*(X)$ concentrated on the vertical axis, and $E^\infty = E^2 = H_*(X)$.

On the other hand, if we do the vertical differential first, since f/y is contractible for each $y \in Y$, we have $E^1 = C_*(Y)$ concentrated on the horizontal axis, and $E^\infty = E^2 = H_*(Y)$. One can verify that the horizontal edge homomorphism is $f_* : H_*(X) \to H_*(Y)$, which is hence an isomorphism.

The proof with $y \backslash f$ contractible is given by replacing X and Y by the opposite posets. $\square$

EXAMPLE. If X is a G-poset, we can make a simplicial complex $\Delta(X)$ by allowing chains of not necessarily strict inequalities

$$x_0 \leq x_1 \leq \cdots \leq x_n.$$

This poset is in general much larger than $|X|$, and is usually infinite even when X is finite. However, the inclusion

$$\mathcal{S}(|X|) \hookrightarrow \mathcal{S}(\Delta(X))$$

is a G-homotopy equivalence since for each subgroup $H \leq G$ a non-strict H-invariant chain has a unique maximal strict subchain, so that the strict subchains of such a chain form a conically contractible poset.

CONJECTURE (WEBB). The quotient complex $\mathcal{S}_p/G$ is contractible.

(Warning: one might be tempted to think this follows immediately from Sylow's theorems, but the point is that $\mathcal{S}_p$ needs to be barycentrically subdivided before the quotient may be thought of as a simplicial complex.)

EXERCISES. Prove the following.

(i) The inclusion of a G-poset X into its barycentric subdivision $\mathcal{S}(|X|)$ is a G-homotopy equivalence. Of course, Quillen's lemma is not by any means the most elementary way of proving this!

(ii) The inclusion of a G-poset X into the G-poset of open subsets of $|X|$ is a G-homotopy equivalence. Thus in particular the homology groups of a G-simplicial complex are "G-topological" invariants.

6.6. Equivalences of subgroup complexes

QUILLEN'S EQUIVALENCE $\mathcal{A}_p(G) \simeq \mathcal{S}_p(G)$. Recall that we denote by $\mathcal{S}_p(G)$ the poset of non-trivial p-subgroups of G, and by $\mathcal{A}_p(G)$ the poset of non-trivial elementary abelian p-subgroups of G.

THEOREM 6.6.1 (Quillen, Thévenaz). *The inclusion of* $\mathcal{A}_p(G)$ *in* $\mathcal{S}_p(G)$ *is a* G-*homotopy equivalence.*

PROOF. Let $i : \mathcal{A}_p(G) \hookrightarrow \mathcal{S}_p(G)$ be the inclusion. By Theorem 6.4.2 and Quillen's Lemma 6.5.2 it suffices to show that for each subgroup $H \leq G$ and for each non-trivial p-subgroup $P \leq G$ normalised by H, the poset $(i/P)^H = \mathcal{A}_p(P)^H$ is contractible.

Denote by $P_0 = \Omega_1 Z(P)$ the subgroup of the centre of P generated by elements of order p, so that P_0 is a non-trivial central elementary abelian

p-subgroup of P, also normalised by H. Then for any non-trivial elementary abelian subgroup $A \leq P$ normalised by H, we have

$$A \leq \langle A, P_0 \rangle \geq P_0.$$

Since $\langle A, P_0 \rangle$ is elementary abelian, this means that $\mathcal{A}_p(P)^H$ is conically contractible. $\square$

COMMENT. The homotopy equivalence $\mathcal{A}_p(G) \simeq \mathcal{S}_p(G)$ is due to Quillen [217]. Thévenaz [263] showed that the homotopy equivalence is in fact G-equivariant.

Quillen conjectured that $\mathcal{S}_p(G)$ is contractible if and only if $O_p(G) \neq 1$. The "if" is clear since if $O_p(G) \neq 1$, then $\mathcal{S}_p(G)$ is conically contractible via

$$P \leq PO_p(G) \geq O_p(G).$$

He also proved his conjecture in the case where G is solvable. It should be remarked that if $\mathcal{S}_p(G)$ is G-contractible then $O_p(G) \neq 1$, since in particular G has to have a fixed point on $\mathcal{S}_p(G)$.

BOUC'S EQUIVALENCE $\mathcal{B}_p(G) \simeq \mathcal{S}_p(G)$. Recall that we denote by $\mathcal{B}_p(G)$ the poset of non-trivial p-subgroups which are equal to the O_p of their normaliser.

Bouc's observation is that if $P \in \mathcal{S}_p(G)^H, P \notin \mathcal{B}_p(G)^H$, then

$$\{Q \in \mathcal{S}_p(G)^H \mid Q > P\}$$

is contractible, since

$$Q \geq N_Q(P) \leq N_Q(P)O_pN_G(P) \geq O_pN_G(P).$$

To make use of this, we introduce the following notation:

DEFINITION 6.6.2. *If X is a poset and $x, y \in X$, we set*

$$[x, y]_X = \{z \in X \mid x \leq z \leq y\}$$
$$(x, y)_X = \{z \in X \mid x < z < y\}$$

etc. In particular,

$$[x, -)_X = \{z \in X \mid x \leq z\}$$
$$(x, -)_X = \{z \in X \mid x < z\}$$

If $B \subseteq X$, we set

$$e_X(B) = \{x \in X \mid (-, x]_X \cap B \text{ is contractible}\}$$
$$f_X(B) = \{x \in X \mid [x, -)_X \cap B \text{ is contractible}\}$$

LEMMA 6.6.3. *If $B \subseteq Y \subseteq e_X(B)$ (resp. $B \subseteq Y \subseteq f_X(B)$) then these inclusions are homotopy equivalences. In particular, if $e_X(B) = X$ then $Y \simeq X$.*

PROOF. Apply Quillen's Lemma 6.5.2 to the inclusions $B \hookrightarrow Y$ and $B \hookrightarrow e_X(B)$. $\square$

DEFINITION 6.6.4. *Let*

$$X^* = \{x \in X \mid (x, -)_X \text{ is not contractible}\}$$
$$X_* = \{x \in X \mid (-, x)_X \text{ is not contractible}\}$$

A poset X has **bounded height** *if there is an upper bound for the length of a chain*

$$x_0 < x_1 < \cdots < x_n.$$

This is the same as saying that the simplicial realisation of X is a finite dimensional complex.

PROPOSITION 6.6.5 (Bouc). *Suppose that X has bounded height. If B is a subset of X containing X^* (resp. containing X_*) then the inclusions $X^* \subseteq B \subseteq X$ (resp. $X_* \subseteq B \subseteq X$) are homotopy equivalences.*

PROOF. It suffices to treat the inclusion $X^* \subseteq B$, since $X^* \subseteq X$ is a special case of this. Denote by i the inclusion $X^* \hookrightarrow B$. We claim that for $b \in B$, the poset $b \backslash i = [b, -)_{X^*}$ is contractible, whereupon the proposition follows by applying Quillen's Lemma 6.5.2.

We prove this claim by induction on the height of X. If $b \in X^*$, the claim is clear, since $[b, -)_{X^*}$ has a unique minimal element, and is therefore conically contractible. Otherwise, we have $[b, -)_{X^*} = (b, -)_{X^*}$. What we know from the definition of X^* is that $(b, -)_X$ is contractible, so we need to show that the inclusion $(b, -)_X \hookrightarrow (b, -)_{X^*}$ is a homotopy equivalence. But this is a case of the proposition of strictly smaller height, and therefore the proposition is proved by induction.   □

We can now apply this proposition together with the observation made earlier to deduce Bouc's theorem.

THEOREM 6.6.6 (Bouc, Thévenaz). *The inclusion of $\mathcal{B}_p(G)$ in $\mathcal{S}_p(G)$ is a G-homotopy equivalence.*

PROOF. This follows by applying Theorem 6.4.2 and the above proposition, since we have $\mathcal{B}_p(G)^H \supseteq (\mathcal{S}_p(G)^H)^*$.   □

COMMENT. The statement that $\mathcal{B}_p(G) \simeq \mathcal{S}_p(G)$ is due to Bouc [49], and Thévenaz showed that the homotopy equivalence is in fact a G-homotopy equivalence.

EXERCISE. Let $\mathcal{Z}_p(G)$ be the poset of non-trivial elementary abelian p-subgroups $A \leq G$ such that $A = \Omega_1 O_p Z C_G(A)$. Prove that the inclusion is a homotopy equivalence $\mathcal{Z}_p(G) \simeq_G \mathcal{S}_p(G)$.

6.7. The generalised Steinberg module

DEFINITION 6.7.1. *We define the* **generalised Steinberg module** *for a finite group G at a prime p to be*

$$St_p(G) = \tilde{L}_G(|\mathcal{S}_p(G)|) \in a(G).$$

REMARKS. Since $\mathcal{S}_p(G) \simeq_G \mathcal{A}_p(G) \simeq_G \mathcal{B}_p(G)$, we could have used any of these complexes in the above definition.

We shall see in the next section what this has to do with the classical definition of Steinberg module for a Chevalley group.

THEOREM 6.7.2 (Quillen, Webb). *The generalised Steinberg module is a virtual projective module*

$$St_p(G) \in a(G, 1).$$

PROOF. Since $St_p(G)$ clearly lies in $a(G)$, it suffices to show that it lies in $A(G, 1) = \mathbb{C} \otimes_{\mathbb{Z}} a(G,)$. Recall from Section 5.6 of Volume I that we have a commutative diagram

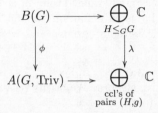

with H p-hypo-elementary and g an element of $H/O_p(H)$. Under the top map, $\tilde{\Lambda}_G(\Delta)$ goes to the element whose Hth coordinate is $\tilde{\chi}(\Delta^H)$. Under the bottom map, the ideal $A(G, 1)$ corresponds to those (H, b) for which $O_p(H) = 1$. Thus the theorem follows from the following lemma.

LEMMA 6.7.3. *Suppose that H is a p-hypo-elementary subgroup of G with $O_p(H) \neq 1$. Then $\mathcal{S}_p(G)^H$ is conically contractible, and in particular the reduced Euler characteristic $\tilde{\chi}(\mathcal{S}_p(G)^H) = 0$.*

PROOF. If $P \in \mathcal{S}_p(G)^H$ then $P \leq PO_p(H) \geq O_p(H)$.   □

COMMENT. The fact that the character of $St_p(G)$ is the character of a virtual projective module is due to Quillen [**217**]. The statement that this alternating sum is virtual projective in the representation ring $a(G)$ is due to Webb.

COROLLARY 6.7.4 (Ken Brown's theorem). *The Euler characteristic*

$$\chi(\mathcal{S}_p(G)) \equiv 1 \pmod{|G|_p}$$

PROOF. The dimension of any projective module is divisible by $|G|_p$ since on restriction to a Sylow p-subgroup it must be free. But the dimension of $St_p(G)$ is $\chi(\mathcal{S}_p(G)) - 1$.   □

REMARK. This statement may be thought of as analogous to the statement that the number of Sylow p-subgroups is congruent to 1 mod p. In fact, Thévenaz has proved a relative version of Ken Brown's theorem of which these two statements are the extreme examples.

EXAMPLE. It is not true that all the $\tilde{H}_i(\mathcal{S}_p(G), k)$ are necessarily projective. The following is an unpublished example of Alperin. Let H be a semidirect product of $\mathbb{Z}/q$ by $\mathbb{Z}/p$, where $p|(q-1)$ and the action is non-trivial. Let G be the wreath product of H with $\mathbb{Z}/p$, so that $|G| = p^{p+1}q^p$. Then every elementary abelian subgroup of G of order bigger than p^2 is contained in the normal subgroup H^p. So as long as $p > 3$, the pth homology

$$\tilde{H}_p(\mathcal{S}_p(G), k) \cong \tilde{H}_p(\mathcal{A}_p(G), k) \cong \tilde{H}_p(\mathcal{A}_p(H^p), k)$$

has dimension p^p. Since the dimension of a projective module is divisible by p^{p+1}, this module is not projective.

6.8. Chevalley groups: a crash course

In the case where G is a Chevalley group in characteristic p, we wish to compare $St_p(G)$ with the usual definition via the Tits building. We begin with a quick review of Chevalley groups, and we shall illustrate with the example of $GL_n(\mathbb{F}_q)$.

LINEAR ALGEBRAIC GROUPS. In Section 5.4, we gave a quick review of varieties and commutative algebra. We briefly recall the definition, and then define linear algebraic groups.

DEFINITION 6.8.1. *An* **affine variety** *is a set of points in an affine space, given by the simultaneous vanishing of a set of polynomials. A* **linear algebraic group** *over an algebraically closed field k of characteristic p is an affine variety V together with a compatible group structure. This means that the multiplication map $V \times V \to V$ and the inverse map $V \to V$ are morphisms of varieties (i.e., the coordinates of the map are given by polynomials).*

For example, $SL_n(k)$ can be regarded as the set of zeroes of the determinant function minus one. The group $GL_n(k)$ can be viewed as a linear algebraic group by taking coordinates $x_{11}, \ldots, x_{nn}, y$, and using the polynomial equation $y \det(x_{ij}) = 1$. It turns out that every linear algebraic group is isomorphic to a closed subgroup of $GL_n(k)$ for some n.

Other examples of linear algebraic groups are the orthogonal groups $O_n(k)$, the special orthogonal groups $SO_n(k)$, the symplectic groups $Sp_{2n}(k)$, and the unitriangular groups $\text{Uni}_n(k)$ consisting of all $n \times n$ upper triangular matrices with ones on the diagonal.

A good background reference for the theory of linear algebraic groups is Humphreys [**134**].

DEFINITION 6.8.2. *A linear algebraic group is called* **connected** *if it is connected in the Zariski topology (in other words, the underlying variety cannot be written as a disjoint union of two proper subvarieties each defined by the vanishing of some polynomials). It is called* **unipotent** *if it is isomorphic to a subgroup of $\text{Uni}_n(k)$ for some value of n. It is called* **reductive** *if it has no non-trivial normal closed connected unipotent subgroups. Thus all the examples given above, with the exception of $\text{Uni}_n(k)$ are reductive linear algebraic groups. Of these, $O_n(k)$ is not connected, but the rest are.*

An algebraic group is **defined**[1] **over** $k_0 \leq k$ if the equations defining it as a variety and the equations giving the multiplication and inversion can be written with coefficients in k_0. For example, $GL_n(k)$ is defined over the ground field $\mathbb{F}_p$.

CLASSIFICATION. The classification of reductive linear algebraic groups depends on the notion of a root system. A **torus** is a linear algebraic group isomorphic to a direct product of copies of the multiplicative group of the field. For example, the group of diagonal matrices of a given size is a torus. It turns out that in a reductive group G, the maximal subgroups isomorphic to a torus (the **maximal tori**) are all conjugate. The **Lie rank** of G is the dimension of a maximal torus. Now, the characters (i.e., one-dimensional representations) of a torus T form a free abelian group $\chi(T)$ under tensor product, whose rank is equal to the dimension of the torus. Inside $\chi(T)$, the **root system** of G is the set of non-zero characters of T occurring in the restriction to T of the action of G by (the derivative of) conjugation on its Lie algebra (i.e., the tangent space at the identity element).

For example, if $G = GL_n(k)$ then T can be taken to be the subgroup of diagonal matrices (commuting diagonalisable matrices are simultaneously diagonalisable), and the characters of T correspond to sequences $r_1, \ldots, r_n$ of integers, the corresponding character of T being given by

$$\operatorname{diag}(x_1, \ldots, x_n) \mapsto x_1^{r_1} \ldots x_n^{r_n}.$$

The Lie algebra of G is the Lie algebra $gl_n(k)$ of all $n \times n$ matrices. The eigenspaces of T with non-zero eigenvalues, in its conjugation action on $gl_n(k)$, are the matrices with a non-zero entry in only one (given) position, off the diagonal. If this entry is in the (ij)th place, then the corresponding character is the one with $r_i = +1$, $r_j = -1$, and all other $r_k = 0$. These roots lie in the hyperplane $\sum r_i = 0$ of $\chi(T)$.

It turns out that the root system of a connected reductive group determines the isomorphism type of the group modulo its largest normal abelian subgroup. The root system is an (abstract) root system in the sense of Humphreys [**133**, Chapter 3] where it is also explained how to classify root systems. Every root system is a direct sum of indecomposable root systems, and the indecomposable root systems are classified by their **Dynkin diagram**. A list of the Dynkin diagrams is given in the above reference on p. 58, or in Chapter 4 of Volume I. Their names are A_l, $l \geq 1$; B_l, $l \geq 2$; C_l, $l \geq 3$; D_l, $l \geq 4$; E_6, E_7, E_8, F_4 and G_2. The subscript refers to the (Lie) rank.

For example, the root system described above for $GL_n(k)$ is an indecomposable root system of type A_{n-1}. The root system for $SO_{2n+1}(k)$ is of type B_n, for $Sp_{2n}(k)$ it is of type C_n, and for $SO_{2n}(k)$ it is of type D_n. The remaining Dynkin diagrams correspond to the so-called "exceptional types" of connected reductive group.

---

[1] Actually, if k_0 is not perfect, the situation is more subtle, but that need not concern us here as we are primarily interested in the case where k_0 is finite.

THE FROBENIUS MORPHISM. If a linear algebraic group G is defined over the finite field of $q = p^e$ elements, then the **Frobenius morphism** F with respect to this finite field is the morphism of algebraic groups given by raising the coordinates to the qth power. If G is connected and reductive then the **Chevalley group** $G(\mathbb{F}_q)$ is defined to be the group G^F of fixed points of F on G. Note that there are also Chevalley groups over infinite fields or rings, defined in a related fashion. We shall not be concerned with these here.

For example, if $G = GL_n(k)$, then $G^F = GL_n(\mathbb{F}_q)$, since F acts by raising the matrix entries to the qth power, and $\mathbb{F}_q$ is exactly the set of solutions of the equation $x^q = x$.

PARABOLIC SUBGROUPS. If H is a closed subgroup of a linear algebraic group G, then there is a theorem of Chevalley (see Humphreys [**134**, p. 80]) which says that there is a finite dimensional representation of G with the property that H is exactly the stabiliser of some one-dimensional subspace. Taking the orbit of the corresponding point in projective space gives the coset space G/H the structure of a quasiprojective variety.

DEFINITION 6.8.3. *A* **parabolic subgroup** *of a reductive group G is a proper subgroup P with the property that the orbit space G/P is a projective variety.*

A **Borel subgroup** *B of G is a maximal connected solvable closed subgroup.*

It turns out that a subgroup is parabolic if and only if it contains a Borel subgroup, and that the Borel subgroups are all conjugate. The parabolic subgroups containing a given Borel subgroup are in one-one correspondence with the proper subsets of the set of vertices of the Dynkin diagram. If G is defined over $\mathbb{F}_q$ then there is always an F-stable Borel subgroup, and the parabolic subgroups containing it are also F-stable. If P is an F-stable parabolic then we call P^F a parabolic subgroup of G^F.

For example, in the case $G = GL_n(k)$, the Borel subgroups are the conjugates of the group of upper triangular matrices, i.e., the stabilisers of complete flags. A **flag** in the vector space V on which G acts is a chain of proper subspaces

$$0 < V_1 < \cdots < V_s < V.$$

A flag is **complete** if there is one subspace of each possible dimension. The parabolic subgroups containing a given Borel subgroup are the stabilisers of subflags of the corresponding complete flag. Thus with respect to a suitable basis the parabolics look something like this:

$$\begin{pmatrix} * & * & * \\ \hline 0 & * & * \\ \hline 0 & 0 & * \end{pmatrix}$$

In the case of $Sp_{2n}(k)$ and $SO_n(k)$ the corresponding concepts are the flags of isotropic subspaces. In the case of $SO_{2n}(k)$, the fact that there are two types

of maximal isotropic subspaces is reflected in the fact that the D_n diagram
has a bifurcation at one end.

The **unipotent radical** of a parabolic subgroup is the largest normal
unipotent subgroup. In the above example, the unipotent radical has the
form:

$$\left(\begin{array}{c|c|c} I & * & * \\ \hline 0 & I & * \\ \hline 0 & 0 & I \end{array} \right)$$

If P is F-stable, then so is its unipotent radical U, and we call U^F the
unipotent radical of P^F. In fact these are related by $U^F = O_p(P^F)$.

THE POSET $\mathcal{B}_p(G)$.

THEOREM 6.8.4 (Borel, Tits). *If G^F is a Chevalley group then a p-sub-*
group U is equal to $O_p N_{G^F}(U)$ if and only if $N_{G^F}(U)$ is parabolic and U is
its unipotent radical.

PROOF. This was first proved in A. Borel and J. Tits [45]. A simpler
proof (but one which only applies to the finite Chevalley groups, which is no
inconvenience to us here!) may be found in N. Burgogne and C. Williamson [57]. We shall give a proof for the particular case where $G^F = GL_n(\mathbb{F}_q)$.
Suppose U is a p-subgroup of G^F. Denote by V_1 the fixed space of U (which
is automatically non-trivial since we are in characteristic p), and inductively
define V_i to be the subspace containing V_{i-1} with the property that V_i/V_{i-1}
is the fixed space of U on V/V_{i-1}. It is clear by induction that V_i is U-
invariant, so that the definition makes sense at the next stage. Now let H
denote the set of all elements of G^F which stabilise the V_i and act trivially
on the V_i/V_{i-1}. Then H is a p-group (since it is contained in $\mathrm{Uni}_n(\mathbb{F}_q)$
with respect to a suitable basis) and it contains U and is normalised by
$N = N_{G^F}(U)$ since it is defined in an invariant way in terms of U. If $U \neq H$
then $U < N_H(U) \leq O_p(N)$. Thus $U = H$ is the unipotent radical of the
parabolic subgroup N corresponding to the flag $0 < V_1 < \cdots < V_s < V$. □

Thus the poset $\mathcal{B}_p(G^F)$ is the opposite poset to the poset of parabolic
subgroups ordered by inclusion.

THE TITS BUILDING. The Tits building of a Chevalley group G^F is
defined to be the simplicial complex whose simplices are indexed by the parabolic subgroups of G^F, and where the inclusion relation between simplices
is the opposite of the inclusion relation between parabolic subgroups. Thus
the vertices (0-simplices) correspond to the maximal parabolics, the edges (1-
simplices) to "sub-maximal" parabolics, and so on. The simplices of maximal
dimension have dimension $(r-1)$, where r is the Lie rank, and correspond to
the Borel subgroups. Every simplex is contained in one of maximal dimension. It is clear that $\mathcal{B}_p(G^F)$ is the poset of simplices of the Tits building,
and so $|\mathcal{B}_p(G^F)|$ is the barycentric subdivision of the Tits building, and is
hence homeomorphic to it.

For example, in the case $G^F = GL_n(\mathbb{F}_q)$, the parabolic subgroups correspond to the chains of proper subspaces (i.e., subspaces which are non-zero and not equal to the whole space) of the space V^F over $\mathbb{F}_q$ on which G^F acts, and so the Tits building is the simplicial realisation of the poset $Subsp(V^F)$ of proper subspaces of V^F. Similarly in the symplectic and orthogonal cases, but with proper subspaces replaced by proper isotropic subspaces. In general, it is possible to realise the building as the simplicial realisation of a poset, but this involves making a choice of an orientation for each of the edges of the Dynkin diagram.

THEOREM 6.8.5 (Solomon [**248**], Tits). *The Tits building of G^F is homotopy equivalent to a bouquet of spheres of the same dimension (i.e., take some spheres, and identify a point in each sphere, to give a single base-point, like a bunch of balloons; this construction is called the* **wedge product***). The dimension of the spheres is $(r - 1)$, and the number of spheres is q to the power of half the number of roots in the corresponding root system.*

PROOF. We shall give Quillen's proof [**215**] of this fact, in the case where $G^F = GL_n(\mathbb{F}_q)$. The proof goes by induction on n. For $n = 2$ the building is clearly a collection of disjoint points, which is a bouquet of 0-spheres (a 0-sphere is just two disjoint points), so suppose $n \geq 3$. Choose a line (one-dimensional subspace) L in V^F, and denote by $\mathcal{H}$ the set of hyperplanes complementary to L.

We claim that the poset $Y = Subsp(V^F) \setminus \mathcal{H}$ is contractible. To show this, consider the quotient map $q : Y \to \underbrace{Subsp}(V^F/L)$, where $\underbrace{Subsp}$ denotes the poset of all proper subspaces, together with the zero subspace. This poset has a unique minimal element, and is hence contractible. Thus by Quillen's Lemma 6.5.2, it suffices to show that q/W is contractible for each element W of $\underbrace{Subsp}(V^F/L)$. But this is clear, since this sub-poset has a unique maximal element, namely the preimage in V^F of W.

It follows that the building is homotopy equivalent to the complex obtained by contracting the subcomplex corresponding to Y down to a point. The latter complex may be described as follows. For $H \in \mathcal{H}$, let $\mathrm{Link}(H)$ denote the subcomplex of the building consisting of those simplices not containing H, but which are contained in simplices containing H. Then the contracted complex is the wedge product of the unreduced suspensions of the complexes $\mathrm{Link}(H)$ (the unreduced suspension of a space is formed by taking a direct product with the unit interval, and identifying each end of the resulting "cylinder" to a point). But $\mathrm{Link}(H)$ is just the complex of proper subspaces of H.

The number and dimension of the spheres is easy to calculate from this argument.   □

REMARKS. (i) With a bit of work, this proof may be made to generalise to the remaining classes of Chevalley groups. It simply needs to be expressed in the language of parabolic subgroups. This is left as a (difficult) exercise for the interested reader.

(ii) Of course, there is nothing special about the finiteness of the field for this argument. In fact it even works for non-commutative division rings.

THE STEINBERG MODULE. It follows from the above theorem that the building has only one non-trivial reduced homology group, and this occurs in degree $(r - 1)$. This reduced homology group is what is usually referred to as the Steinberg module of a Chevalley group. The preceding discussion makes it clear that it differs from our definition of $St_p(G^F)$ only by a sign $(-1)^{r-1}$.

REMARK. It has been shown by Lusztig that the ideal of virtual projective modules in the representation ring of a Chevalley group is the principal ideal generated by the Steinberg module. Note, however, that for a general group, this ideal is not necessarily principal.

6.9. Steinberg module inversion and Alperin's conjecture

We defined the generalised Steinberg module in terms of an alternating sum of permutation modules. In this section, we develop an inversion formula by the means of which we can invert this relationship and express the trivial module as a sum of modules induced from generalised Steinberg modules.

PROPOSITION 6.9.1 (Bouc, Thévenaz). *If X is a finite G-poset, then*

$$\Lambda_G(X) = - \sum_{x \in X/G} \mathrm{ind}_{\mathrm{Stab}_G(x),G} \tilde{\Lambda}_{\mathrm{Stab}_G(x)}((x, -)_X)$$

PROOF. We sort the simplices $x_0 < \cdots < x_n$ of $|X|$ into clumps according to the G-orbit of x_0. The contribution to $\Lambda_G(X)$ from such a clump will be given as follows. Choose an element x in the orbit. Then we get the contribution from those simplices $x_0 < \cdots < x_n$ with $x_0 = x$, regarded as an element of the Burnside ring $b(\mathrm{Stab}_G(x))$, and then induced up to G. It is easy to see that this part of the chain complex of X is isomorphic to the chain complex of $(x, -)$ but with the dimension increased by one. The shift of dimension accounts for the sign change.   □

We now wish to apply this to the complex $\mathcal{S}_p(G) \cup \{1\}$ of *all* p-subgroups of G. If P is a p-subgroup of G, then $\mathrm{Stab}_G(P) = N_G(P)$, and the complex $(P, -)$ is identified by the following proposition.

PROPOSITION 6.9.2 (Quillen, Webb). *If P is a p-subgroup of G, then there is a $N_G(P)$-homotopy equivalence*

$$(P, -) \simeq_{N_G(P)} \mathcal{S}_p(N_G(P)/P)$$

where the right-hand side is regarded as a $N_G(P)$-module via the quotient map $N_G(P) \to N_G(P)/P$.

In particular,

$$\tilde{\Lambda}_{N_G(P)}((P, -)) = \tilde{\Lambda}_{N_G(P)}(\mathcal{S}_p(N_G(P)/P))$$

$$\tilde{L}_{N_G(P)}((P, -)) = \inf_{N_G(P)/P, N_G(P)}(St_p(N_G(P)/P))$$

in $b(G)$ and $a(kG)$ respectively.

PROOF. By Theorem 6.4.2, we only need show that

$$(P, -)^H \simeq \mathcal{S}_p(N_G(P)/P)^H$$

for all $H \leq N_G(P)$. Let

$$i : \mathcal{S}_p(N_G(P)/P)^H \hookrightarrow (P, -)^H$$

denote the inclusion and

$$r : (P, -)^H \to \mathcal{S}_p(N_G(P)/P)^H$$
$$Q \mapsto N_Q(P)/P.$$

Then it is easy to check that $r \circ i = \mathrm{id}$ while $i \circ r(Q) \leq Q$, so that i and r are homotopy inverses. $\square$

COMMENT. The homotopy equivalence is due to Quillen, and the fact that the homotopy equivalence is equivariant is due to Webb.

Putting these together, we obtain the following theorem.

THEOREM 6.9.3. *In the Burnside ring $b(G)$, we have the identity*

$$1 = - \sum_{\substack{\text{ccls of} \\ p\text{-subgroups} \\ 1 \leq P \leq G}} \mathrm{ind}_{N_G(P), G}\, \tilde{\Lambda}_{N_G(P)}(\mathcal{S}_p(N_G(P)/P)).$$

and in the representation ring $a(kG)$, the corresponding identity is the **Steinberg module inversion formula**

$$1 = - \sum_{\substack{\text{ccls of} \\ p\text{-subgroups} \\ 1 \leq P \leq G}} \mathrm{ind}_{N_G(P), G}\, \inf_{N_G(P)/P, N_G(P)} St_p(N_G(P)/P) \qquad \square$$

REMARKS. (i) If $P \neq O_p N_G(P)$ then $\mathcal{S}_p(N_G(P)/P)$ is contractible, and so we have $St_p(N_G(P)/P) = 0$. Thus the Steinberg module inversion formula only has contributions from subgroups $P = O_p N_G(P)$.

(ii) We used Conlon's induction theorem to show that $St_p(G)$ is a virtual projective. Conversely the above formula shows that these statements are equivalent, since in $A(G)$, any virtual projective can be expressed as a $\mathbb{C}$-linear combination of modules induced from cyclic p'-subgroups.

LENGTH FUNCTIONS. For the rest of this section, we assume k is algebraically closed.

DEFINITION 6.9.4. *A* **length function** *on finite groups is an integer valued function on isomorphism classes of finite groups.*

The following are the length functions we shall be interested in:

$l(G)$ = number of isomorphism classes of irreducible kG-modules

= number of conjugacy classes of p'-elements

$f_0(G)$ = number of isomorphism classes of projective irreducibles

("blocks of defect zero").

If α is a length function, then α gives rise to a linear map, also denoted by α, defined as follows.

$$\alpha : b(G) \to \mathbb{Z}$$

$$(G/H) \mapsto \alpha(H).$$

PROPOSITION 6.9.5. *Suppose that α and β are length functions, and that α satisfies $\alpha(G) = \alpha(G/O_p(G))$. Then the following are equivalent:*

(i) $\displaystyle \alpha(G) = \sum_{\substack{\text{ccls of} \\ p\text{-subgroups} \\ 1 \le P \le G}} \beta(N_G(P)/P)$

(ii) $\beta(G) = -\alpha(\tilde{\Lambda}(\mathcal{S}_p(G)))$.

PROOF. Apply α to the Steinberg module inversion formula to obtain

$$\alpha(G) = - \sum_{\substack{\text{ccls of} \\ p\text{-subgroups} \\ 1 \le P \le G}} \alpha(\tilde{\Lambda}_{N_G(P)}(\mathcal{S}_p(N_G(P)/P))).$$

Substituting (ii) in this yields (i), while substituting (i), and (ii) for all strictly smaller groups (by induction) yields (ii). $\qquad\square$

ALPERIN'S CONJECTURE. Alperin has conjectured that for all finite groups G,

$$l(G) = \sum_{\substack{\text{ccls of} \\ p\text{-subgroups} \\ 1 \le P \le G}} f_0(N_G(P)/P)$$

where $l(G)$ and $f_0(G)$ are as above. If you know about Green correspondence, the mnemonic for this statement is "the number of simple modules is equal to the number of simple Green correspondents". Geoff Robinson's reformulation of this conjecture is obtained by applying the above proposition. This reformulation states that

$$f_0(G) = -l(\tilde{\Lambda}_G(\mathcal{S}_p(G))).$$

The treatment of this reformulation using length functions is due to Webb.

RECENT PROGRESS: Since the first edition of this book was published, Peter Symonds [404] had proved the conjecture of Webb appearing in Section 6.5. A lot of work has been done on Alperin's conjecture, but it still appears to be elusive.

CHAPTER 7

Local coefficients on subgroup complexes

7.1. Local coefficients

In this Chapter we present a version of the theory of Mark Ronan and Steve Smith [**222, 223, 224, 241**] in which representation theoretic information about a group is obtained in a natural way from constructions involving local coefficients on G-simplicial complexes. The results are most comprehensible for the case where G is a Chevalley group over a field of the same characteristic, mostly because of a theorem of Steve Smith on irreducibility of fixed point sets. Note that in some of their papers, Ronan and Smith use the word *sheaf* or *presheaf* to describe what we refer to as a local coefficient system. We refrain from using the word sheaf because the usual gluing condition for sheaves plays no rôle. I hope this does not cause too much confusion.

All the material of this Chapter is due to Ronan and Smith unless otherwise explicitly stated; any reformulation that has occurred, and all mistakes that have crept in, are my responsibility.

DEFINITION 7.1.1. *A G-**equivariant local coefficient system** (or just an equivariant coefficient system) $\mathcal{F}$ of vector spaces (abelian groups, sets, etc.) on a G-poset X assigns to each $x \in X$ a vector space (abelian group, set, etc.), to each inequality $x \le y$ a **restriction map***

$$\mathrm{res}_{y,x} : \mathcal{F}(y) \to \mathcal{F}(x)$$

satisfying

(i) $\mathrm{res}_{x,x} = \mathrm{id} : \mathcal{F}(x) \to \mathcal{F}(x)$

(ii) *If $x \le y \le z$ then $\mathrm{res}_{y,x} \circ \mathrm{res}_{z,y} = \mathrm{res}_{z,x}$*

and for each $g \in G$, $x \in X$ a map $g_* : \mathcal{F}(x) \to \mathcal{F}(gx)$ in such a way that these maps constitute a group action. In other words, $g_* h_* = (gh)_* : \mathcal{F}(x) \to \mathcal{F}(ghx)$, and $1_* = \mathrm{id} : \mathcal{F}(x) \to \mathcal{F}(x)$.

If you like, $\mathcal{F}$ is an "equivariant functor" on X considered as a "G-category". Note that the G-action makes $\mathcal{F}(x)$ into a representation of $\mathrm{Stab}_G(x)$.

DEFINITION 7.1.2. *A G-**equivariant local coefficient system** on a G-simplicial complex Δ is the same as an equivariant local coefficient system on the poset of simplices $S(\Delta)$. In other words, to each simplex σ we assign* *a vector space $\mathcal{F}(\sigma)$, and to each inclusion $\tau \subseteq \sigma$ a restriction map $\mathcal{F}(\sigma) \to \mathcal{F}(\tau)$ (satisfying transitivity etc. as before) together with a G-action.*

Conversely, given an equivariant coefficient system on a G-poset X, we obtain an equivariant coefficient system on $|X|$ via

$$(x_0 < \cdots < x_n) \mapsto \mathcal{F}(x_n)$$

with the obvious restriction maps and G-action. Thus on the barycentric subdivision of a G-complex, we assign to each simplex the value on the smallest simplex of the original complex which contains it (thinking of the simplicial complex and its barycentric subdivision as having the same underlying topological space via the natural homeomorphism).

EXAMPLE. If M is a kG-module and X is any G-poset or G-simplicial complex, then we have the **constant coefficient system** κ_M which assigns M to each element (resp. simplex), and where all the restriction maps are the identity. The G-action is the same as it is on M.

The coefficient systems we shall be most interested in are the **fixed point coefficient systems** of kG-modules. If M is a kG-module, then the fixed point coefficient system $\mathcal{F}_M$ on the G-poset $\mathcal{B}_p(G)$ (or any other G-poset of subgroups of G; see Section 6.2) is defined by

$$\mathcal{F}_M(P) = M^P,$$

the space of fixed points of the p-subgroup P on M. The G-action is given via the G-action on M, and the restriction maps are inclusions. Note that if $m \in M^P$ then $gm \in gPg^{-1}$.

REMARK. If G is a Chevalley group, $\mathcal{B}_p(G)$ is the building of G, and then this is the fixed point sheaf in the sense of Ronan and Smith (see for example [**223, 224**]).

DEFINITION 7.1.3. *A **morphism** $\mathcal{F} \to \mathcal{G}$ of G-equivariant coefficient systems over X consists of maps of vector spaces $\mathcal{F}(x) \to \mathcal{G}(x)$, for each element (resp. simplex) x of X, in such a way that the following diagram commutes whenever $x < y$*

$$
\begin{array}{ccc}
\mathcal{F}(y) & \longrightarrow & \mathcal{G}(y) \\
\downarrow & & \downarrow \\
\mathcal{F}(x) & \longrightarrow & \mathcal{G}(x)
\end{array}
$$

and also commuting with the G-action. We write $\mathrm{Hom}_{X,G}(\mathcal{F},\mathcal{G})$ for the vector space of morphisms from $\mathcal{F}$ to $\mathcal{G}$.

*The coefficient system $\mathcal{F}$ is a **sub-coefficient system** of $\mathcal{G}$ if $\mathcal{F}(x) \subseteq \mathcal{G}(x)$ for all x. The **quotient coefficient system** is $(\mathcal{F}/\mathcal{G})(x) = \mathcal{F}(x)/\mathcal{G}(x)$.*

*There are also obvious notions of **direct sum, exact sequence, indecomposability**, etc.*

EXAMPLES. (i) The fixed point coefficient system $\mathcal{F}_M$ is a sub-coefficient system of the constant coefficient system κ_M, and so we may form the quotient coefficient system $\kappa_M/\mathcal{F}_M$.

(ii) Let $G = GL_3(\mathbb{F}_2)$, the group of three by three matrices over $\mathbb{F}_2$. The permutation representation on the seven points of the projective plane over $\mathbb{F}_2$ decomposes as a direct sum of a trivial one dimensional module and an indecomposable module X of dimension six. The module X is a non-split extension

$$0 \to V \to X \to V^* \to 0$$

where V is the natural three dimensional module and V^* is its dual. It happens that taking the fixed point space of any non-trivial p-subgroup in the above sequence yields a split short exact sequence of modules for the normaliser. However, the splittings cannot be done in a way consistent with the restriction maps, and so the short exact sequence of fixed point coefficient systems is not split. In general, for a short exact sequence of modules, we only get a long exact sequence of fixed point coefficient systems, since taking fixed points is left exact but not right exact.

DEFINITION 7.1.4. *A coefficient system $\mathcal{F}$ is* **non-degenerate** *if $\mathcal{F}(x) \neq 0$ for all x.*

For example, the fixed point coefficient systems discussed above are non-degenerate, since a p-group always has a fixed point on any module in characteristic p.

7.2. Constructions on coefficient systems

Suppose $\mathcal{F}$ is a G-equivariant coefficient system on a G-poset or G-simplicial complex X.

DEFINITION 7.2.1. *The* **inverse limit** *(or* **projective limit***) of $\mathcal{F}$, written $\varprojlim \mathcal{F}$, is a vector space V together with maps $V \to \mathcal{F}(x)$ for each $x \in X$, commuting with restriction maps, and universal with respect to this property. Universal in this context means that given another W and maps as above, there is a unique map of vector spaces $W \to V$ whose composite with each given map $V \to \mathcal{F}(X)$ is the given map $W \to \mathcal{F}(x)$.*

Dually, the **direct limit** *(also known as the* **inductive limit** *or* **colimit** *of $\mathcal{F}$), written $\varinjlim \mathcal{F}$, is a vector space V together with maps $\mathcal{F}(x) \to V$ for each $x \in X$, commuting with restriction maps, and universal with respect to this property. This time, universal means that given another W and maps as above, there is a unique map of vector spaces $V \to W$ whose composite with each given map $\mathcal{F}(X) \to V$ is the given map $\mathcal{F}(x) \to W$.*

We shall also write H^0 for $\varprojlim$ and H_0 for $\varinjlim$, for reasons which I hope will become apparent.

The usual methods may be used to show that these concepts exist and are unique up to canonical isomorphism. It is helpful to have a concrete model for these concepts. The space $\varprojlim \mathcal{F}$ may be constructed as the subspace of $\prod_{x \in X} \mathcal{F}(x)$ consisting of elements whose coordinates are related by the restriction maps whenever the corresponding elements of the poset (resp. simplices) are comparable.

Dually, the space $\varinjlim \mathcal{F}$ may be constructed as the quotient of $\bigoplus_{x \in X} \mathcal{F}(x)$ by the subspace generated by elements with all except two coordinates zero, and the remaining two related by a restriction map.

It is not hard to prove the above universal properties from these constructions. This is left as a (boring) exercise for the reader. Note that the universal properties ensure that the spaces $\varprojlim \mathcal{F}$ and $\varinjlim \mathcal{F}$ come equipped with a G-action, so that they are kG-modules.

EXAMPLES. If X is the poset with trivial group action

(where an arrow denotes an inequality in which the head of the arrow is less than the tail) then $\varinjlim \mathcal{F}$ is the same as the **pushout** of the vector spaces at the vertices. Dually, if X is the poset with trivial group action

then $\varprojlim \mathcal{F}$ is the same as the **pullback** of the vector spaces at the vertices.

EXERCISE. If $0 \to \mathcal{F} \to \mathcal{G} \to \mathcal{G}/\mathcal{F} \to 0$ is a short exact sequence of coefficient systems, show that
  (i) $0 \to H^0\mathcal{F} \to H^0\mathcal{G} \to H^0(\mathcal{G}/\mathcal{F})$ and
  (ii) $H_0\mathcal{F} \to H_0\mathcal{G} \to H_0(\mathcal{G}/\mathcal{F}) \to 0$
are exact (in other words, $\varprojlim$ is left exact while $\varinjlim$ is right exact; the side of the exactness can be remembered by the direction in which the arrow points).

If $f : X \to Y$ is a map of G-posets, there are various ways of passing from a G-equivariant coefficient system on X to a G-equivariant coefficient system on Y and vice-versa.

DEFINITION 7.2.2. *If $\mathcal{F}$ is a coefficient system on X and $\mathcal{G}$ is a coefficient system on Y, we define*
  (i) $f^{-1}\mathcal{G}(x) = \mathcal{G}(f(x))$. *For example, if $i : X \hookrightarrow Y$ is an inclusion, then $i^{-1}\mathcal{G}$ is the* **restriction** *of $\mathcal{G}$ to X.*
  (ii) $f_*\mathcal{F}(y) = \varprojlim_{x \in y\backslash f} \mathcal{F}(x)$ *(so that in particular f_* is left exact). The notations $y\backslash f$ and f/y is explained in Section 6.5.*
  (iii) $f_!\mathcal{F}(y) = \varinjlim_{x \in f/y} \mathcal{F}(x)$ *(so that in particular $f_!$ is right exact).*

LEMMA 7.2.3. *If $f : X \to Y$ is a map of G-posets, then there are natural isomorphisms*
  (i) $\operatorname{Hom}_{X,G}(\mathcal{F}, f^{-1}\mathcal{G}) \cong \operatorname{Hom}_{Y,G}(f_!\mathcal{F}, \mathcal{G})$
  (ii) $\operatorname{Hom}_{X,G}(f^{-1}\mathcal{G}, \mathcal{F}) \cong \operatorname{Hom}_{Y,G}(\mathcal{G}, f_*\mathcal{F})$

PROOF. (i) Giving a map $\mathcal{F} \to f^{-1}\mathcal{G}$ is the same as giving a consistent set of maps $\mathcal{F}(x) \to \mathcal{G}(f(x))$ for all $x \in X$. By the universal property of direct limits, this is the same as giving a consistent set of maps

$$\varinjlim_{x \in f/y} \mathcal{F}(x) \to \mathcal{G}(y),$$

i.e., a map $f_!\mathcal{F} \to \mathcal{G}$. Note that if $y = f(x_0)$ then

$$\varinjlim_{x \in f/y} \mathcal{F}(x) = \mathcal{F}(x_0).$$

The proof of (ii) is dual.   □

Note that in particular if we take $\mathcal{F} = f^{-1}\mathcal{G}$ in the above lemma, the maps corresponding to the identity map under the above natural isomorphisms are naturally defined maps $f_! f^{-1}\mathcal{G} \to \mathcal{G}$ and $\mathcal{G} \to f_* f^{-1}\mathcal{G}$.

EXAMPLE: FROBENIUS RECIPROCITY. If we take Y to be the poset consisting of a single point in the above lemma, we obtain the following. G-equivariant coefficient systems on (pt) are the same thing as kG-modules. If we denote by f the unique map from X to (pt), then $f^{-1}(M_{(pt)}) = \kappa_M$, the constant coefficient system on X with values in M.

PROPOSITION 7.2.4. *If $\mathcal{F}$ is a G-equivariant coefficient system on X, and M is a kG-module, then there are natural isomorphisms*
  (i) $\operatorname{Hom}_{X,G}(\mathcal{F}, \kappa_M) \cong \operatorname{Hom}_{Y,G}(H_0(\mathcal{F}), M)$
  (ii) $\operatorname{Hom}_{X,G}(\kappa_M, \mathcal{F}) \cong \operatorname{Hom}_{Y,G}(M, H^0(\mathcal{F}))$.   □

Taking $\mathcal{F} = \mathcal{F}_M$ we obtain a natural map $H_0(\mathcal{F}_M) \to M$, corresponding to the inclusion $\mathcal{F}_M \hookrightarrow \kappa_M$. We shall be investigating this map in the section after next.

DEFINITION 7.2.5. *If $i : X \hookrightarrow Y$ is an inclusion and $\mathcal{F}$ is a G-equivariant coefficient system on X, then $i_!\mathcal{F}$ is the* **universal extension** *of $\mathcal{F}$ to Y.*

The reason for making this definition is the following characterisation.

PROPOSITION 7.2.6. *Let $i : X \hookrightarrow Y$ be an inclusion, and $\mathcal{G}$ a coefficient system on Y. Then the following are equivalent.*
  (i) *The natural map $i_! i^{-1}\mathcal{G} \to \mathcal{G}$ is surjective.*
  (ii) *For each $y \in Y$, $y \notin X$, $\mathcal{G}(y)$ is generated by the $\operatorname{res}_{x,y}\mathcal{G}(x)$ for $x \geq y$, $x \in X$.*   □

In case these equivalent conditions hold, we say that $\mathcal{G}$ is **generated** by its restriction to X. Thus the universal extension $i_!\mathcal{F}$ has the property that *whenever the restriction of $\mathcal{G}$ to X is isomorphic to $\mathcal{F}$, and $\mathcal{G}$ is generated by this restriction, then $\mathcal{G}$ is a quotient of $i_!\mathcal{F}$.*

7.3. Chain complexes and homology of coefficient systems

A coefficient system $\mathcal{F}$ on a G-simplicial complex Δ gives rise to a chain complex as follows (cf. Section 6.1):

DEFINITION 7.3.1. *We define the nth chain group of Δ with coefficients in $\mathcal{F}$, $C_n(\Delta; \mathcal{F})$ to be the k-module generated by the $\mathcal{F}(\sigma)$ as σ ranges over the oriented n-simplices of Δ, with relators saying that each element of $\mathcal{F}(\sigma)$ is equal to minus the corresponding element of $\mathcal{F}(\sigma^{\mathrm{op}})$, where σ^{op} denotes the same simplex with the opposite orientation.*

The vector space $C_n(\Delta; \mathcal{F})$ inherits a G-module structure from the action of G on $\mathcal{F}$.

Just as the chain groups $C_n(\Delta; k)$ of Section 6.1 are permutation modules, the modules $C_n(\Delta; \mathcal{F})$ are sums of modules induced from stabilisers of n-simplices. The case of $C_n(\Delta; k)$ is simply the special case of the constant system with coefficients in the trivial kG-module k.

The boundary map

$$\delta_n : C_n(\Delta; \mathcal{F}) \to C_{n-1}(\Delta; \mathcal{F})$$

is defined analogously to the case of trivial coefficients, but using the alternating sum of the restriction maps to faces of a simplex. If $u \in \mathcal{F}(\sigma)$ with $\sigma = (x_0, \dots, x_n)$ then

$$\delta(u) = \sum_{i=0}^{n} (-1)^i \mathrm{res}_{\sigma,\sigma_i}(u)$$

where $\sigma_i = (x_0, \dots, x_{i-1}, x_{i+1}, \dots, x_n)$.

It is again easy to check that $\delta_n \circ \delta_{n+1} = 0$, and that the boundary maps are kG-module homomorphisms, so that the homology groups

$$H_n(\Delta; \mathcal{F}) = \mathrm{Ker}(\delta_n)/\mathrm{Im}(\delta_{n+1})$$

are also kG-modules.

EXERCISE. Show that $H_0(\Delta; \mathcal{F})$ is the same as the kG-module $H_0(\mathcal{F})$ introduced in the last section. For this reason, we may use the notation $H_i(\mathcal{F})$ for $H_i(\Delta; \mathcal{F})$ without ambiguity.

EXAMPLES. (i) (Solomon, Tits) If $\Delta = |\mathcal{B}_p(G)|$ with G a Chevalley group in characteristic p, then

$$H_i(\kappa_M) = \begin{cases} M & i = 0 \\ M \otimes St & i = \mathrm{rank}(G) - 1 \\ 0 & \text{otherwise.} \end{cases}$$

(ii) (Lusztig [169]) If $G = SL_n(\mathbb{F}_q)$ and M is the natural n-dimensional module, then

$$H_i(\mathcal{F}_M) = \begin{cases} M & i = 0 \\ 0 & \text{otherwise.} \end{cases}$$

(iii) If $G = Sp_4(\mathbb{F}_q)$ and M is the natural 4-dimensional module, then

(a) if $p \neq 2$ then $H_0(\mathcal{F}_M) = M$;

(b) whereas if $p = 2$ then $\dim H_0(\mathcal{F}_M) = 5$, and there is a non-split short exact sequence

$$0 \to k \to H_0(\mathcal{F}_M) \to M \to 0.$$

In fact $H_0(\mathcal{F}_M)$ is the natural orthogonal module for $Sp_4(\mathbb{F}_q) \cong O_5(\mathbb{F}_q)$ in this latter case. We shall have more to say about this example in the next section.

LONG EXACT SEQUENCES. Recall from Section 2.3 of Volume I that a short exact sequence of chain complexes $0 \to \mathbf{C}' \to \mathbf{C} \to \mathbf{C}'' \to 0$ gives rise to a long exact sequence of homology groups

$$\cdots \to H_i(\mathbf{C}') \to H_i(\mathbf{C}) \to H_i(\mathbf{C}'') \to H_{i-1}(\mathbf{C}') \to$$
$$\cdots \to H_0(\mathbf{C}) \to H_0(\mathbf{C}'') \to 0.$$

Thus if $0 \to \mathcal{F} \to \mathcal{G} \to \mathcal{G}/\mathcal{F} \to 0$ is a short exact sequence of coefficient systems, then there is a long exact sequence of homology groups

$$\cdots \to H_i(\mathcal{F}) \to H_i(\mathcal{G}) \to H_i(\mathcal{G}/\mathcal{F}) \to H_{i-1}(\mathcal{F}) \to \cdots.$$

If you know about left derived functors, you will recognise that this is saying that the H_i are the left derived functors of $\varinjlim$. In a similar way, one can construct right derived functors H^i of $\varprojlim$ using the Čech approach to cohomology, via the covering given by the simplices. We shall not be using H^i in this context.

EXAMPLE. From the short exact sequence

$$0 \to \mathcal{F}_M \to \kappa_M \to \kappa_M/\mathcal{F}_M \to 0$$

we obtain a long exact sequence which ends with

$$\cdots \to H_0(\mathcal{F}_M) \to H_0(\kappa_M) \to H_0(\kappa_M/\mathcal{F}_M) \to 0.$$

Since $H_0(\kappa_M) = M$, this says that the natural map $H_0(\mathcal{F}_M) \to M$ is surjective if and only if $H_0(\kappa_M/\mathcal{F}_M) = 0$. Under these conditions we say that M is **generated by fixed points**. This is always the case, for example, if M is irreducible.

7.4. Symplectic and orthogonal groups

In this section we examine some examples, and attempt to convince the reader that the information contained in the module $H_0(\mathcal{F}_M)$ is just the information recoverable from **local information** about the structure of M. This idea will be made more formal in the ensuing sections. It should also become apparent that finding $H_0(\mathcal{F})$ for a coefficient system $\mathcal{F}$ is equivalent to the **embedding problem**: given an abstract incidence structure of points, lines, planes etc. defined by the subgroup structure of the group, when is there a projective space in which they embed? From this point of view, the statement that the natural module M for $GL_n(k)$ is equal to $H_0(\mathcal{F}_M)$ says that we can recover projective space from the incidence structure of its points and lines. This can be thought of as a weak form of the fundamental theorem of projective geometry. Our first example shows that the corresponding theorem for symplectic geometry is only true in odd characteristic; in characteristic two, we end up constructing an orthogonal space of one larger dimension.

THE SYMPLECTIC AND ODD DIMENSIONAL ORTHOGONAL GROUPS. We now examine the $2n$-dimensional natural module M for the symplectic group $Sp_{2n}(\mathbb{F}_q)$. As mentioned in the section on Chevalley groups (Section 6.8), this is the group associated to the B_n Dynkin diagram over the field $\mathbb{F}_q$.

Recall that an **isotropic subspace** of a symplectic space is a subspace on which the symplectic form vanishes identically. The parabolic subgroups of the symplectic group are just the stabilisers of flags of isotropic subspaces (**isotropic flags**). Thus the Borel subgroups are the stabilisers of complete isotropic flags $0 < V_1 < \cdots < V_n$ with $\dim(V_i) = i$. The maximal parabolics (which are as usual labelled by the vertices of the Dynkin diagram) correspond to isotropic subspaces according to the following diagram.

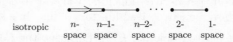

| isotropic | n- | n–1- | n–2- | 2- | 1- |
| | space | space | space | space | space |

Thus the building of G has as vertices the isotropic subspaces, as edges the isotropic flags of length 2, etc. up to $(n-1)$-simplices corresponding to the complete isotropic flags. In other words, this is just the simplicial realisation of the poset of non-zero isotropic subspaces of the symplectic space M.

The coefficient system $\mathcal{F}_M$ assigns to each simplex the smallest subspace in the corresponding flag, since this is the fixed space of the O_p of the corresponding parabolic subgroup. Thus this coefficient system is the one obtained by the usual process from the coefficient system on the poset of non-zero isotropic subspaces which assigns to each element of the poset the corresponding isotropic subspace. We call this the **tautological coefficient system**. So $\mathcal{F}_M$ is the simplicial realisation of the tautological coefficient system.

What is $H_0(\mathcal{F}_M)$? Since two isotropic 1-spaces are contained in an isotropic subspace if and only if they are contained in an isotropic 2-space, the general construction of limits shows that $H_0(\mathcal{F}_M)$ is the quotient of the direct sum of all the isotropic 1-spaces by the relators coming from the isotropic 2-spaces. If v is an isotropic vector in M, we shall write $\hat{v}$ for the corresponding element of the corresponding 1-space in $H_0(\mathcal{F}_M)$. In these terms, $H_0(\mathcal{F}_M)$ is generated by the $\hat{v}$ for isotropic points $v \in M$ subject to the relations $\hat{v} = \hat{w} + \hat{x}$ whenever $v = w + x$ and v, w and x lie in an isotropic 2-space. The surjection $H_0(\mathcal{F}_M) \to M$ takes $\hat{v}$ to v.

THEOREM 7.4.1. (i) *If k has odd characteristic then $H_0(\mathcal{F}_M) \cong M$.*

(ii) *If k has characteristic two then there is a short exact sequence*

$$0 \to k \to H_0(\mathcal{F}_M) \to M \to 0.$$

PROOF. (i) Choose isotropic subspaces $X = \langle x_1, \ldots, x_n \rangle$ and $Y = \langle y_1, \ldots, y_n \rangle$ of dimension n with $(x_i, y_j) = \delta_{ij}$. Then $\hat{X}$ and $\hat{Y}$ are n-dimensional isotropic subspaces of $H_0(\mathcal{F}_M)$, and we must show that each $\hat{z} \in H_0(\mathcal{F}_M)$ is in the linear span of $\hat{X}$ and $\hat{Y}$. Let $z = x + y$ with $x \in X$ and $y \in Y$. If $(x, y) = 0$, we are done, so assume $(x, y) = a \neq 0$. Choose $x' \in X$ with $(x', y) = 0$. Since k has odd characteristic, we may define

$$x_0 = \tfrac{1}{2}(x + x') \qquad x = x_0 + x_0'$$
$$x_0' = \tfrac{1}{2}(x - x') \qquad x' = x_0 - x_0'$$

Choose y_0 with $(x_0, y_0) = 0$ and $(x_0', y_0) = a/2$, and let $y_0' = y - y_0$. Then

$$(x_0 + y_0, x_0' + y_0') = (x_0, y_0') + (y_0, x_0') = a/2 - a/2 = 0$$

and

$$z = (x_0 + y_0) + (x_0' + y_0')$$

so that

$$\langle \hat{z} \rangle \subseteq \langle \widehat{(x_0 + y_0)}, \widehat{(x_0' + y_0')} \rangle \subseteq \langle \hat{x}_0, \hat{y}_0, \hat{x}_0', \hat{y}_0' \rangle.$$

(ii) In order to understand the situation in characteristic two, it is necessary to say something about orthogonal forms in characteristic two. An orthogonal form in characteristic two consists of the following data. First of all we are given a symmetric bilinear form $b(x, y) = b(y, x)$. Then we are also given a **quadratic form** $q(x)$ which is a map satisfying

$$q(x + y) = q(x) + q(y) + b(x, y).$$

In particular this implies that $b(x, x) = 0$, so that $b(x, y)$ is a symplectic form. In odd dimension, this symplectic form necessarily has a non-zero radical, and this radical is invariant under the action of the orthogonal group. If the orthogonal form is non-degenerate, the radical can only have one dimension, since the quadratic form is (semi)linear on this radical. The quotient of the $(2n + 1)$-dimensional orthogonal space by the radical of the symplectic form is thus a non-degenerate $(2n)$-dimensional symplectic space. Given a (necessarily isotropic) vector in the symplectic space, among its preimages is exactly one isotropic vector in the orthogonal space. Thus there is a surjective map from $H_0(\mathcal{F}_M)$ onto an orthogonal space of one larger dimension. Case (ii) of the theorem will therefore follow from the next theorem, which is independent of characteristic.

THEOREM 7.4.2. *Let V be a non-degenerate orthogonal $(2n + 1)$-dimensional space $(n \geq 1)$, and let $\mathcal{F}_V$ be the tautological coefficient system on the poset of non-trivial isotropic subspaces of V. Then the natural map $H_0(\mathcal{F}_V) \to V$ is an isomorphism.*

PROOF. Choose a basis $x_1, \ldots ,x_n, y_1, \ldots ,y_n, t$ with $X = \langle x_1, \ldots , x_n \rangle$ and $Y = \langle y_1, \ldots , y_n \rangle$ isotropic, $(x_i, y_j) = \delta_{ij}$ and $q(t) = 0$.

$$\left(\begin{array}{cc|c} 0 & I & ? \\ I & 0 & ? \\ \hline ? & ? & 0 \end{array} \right)$$

(EXERCISE: Show this is always possible; you'll need to use the finiteness of the field.)

We must show that $H_0(\mathcal{F}_V) = \hat{X} \oplus \hat{Y} \oplus \langle \hat{t} \rangle$; i.e., if $x + y + \lambda t$ is singular then $\widehat{(x + y + \lambda t)} \in \hat{X} \oplus \hat{Y} \oplus \langle \hat{t} \rangle$.

CASE 1. If $\lambda = 0$ then $x + y$ is singular and so $\widehat{(x + y)} = \hat{x} + \hat{y}$.

CASE 2. If $\lambda \neq 0$ then choose a singular 2-space $S \supseteq \langle x + y + \lambda t \rangle$. By dimension counting, $S \cap (X \oplus Y)$ is one dimensional, so let it be generated by the isotropic vector s. By case 1, $\hat{s} \in \hat{X} \oplus \hat{Y}$. By a similar dimension count, $z^\perp$ is a $(2n)$-dimensional subspace intersecting S in a one dimensional isotropic subspace. This 1-space, together with z, generate an isotropic 2-space which intersects $X \oplus Y$ in an isotropic 1-space. Thus $z^\perp \cap S \subseteq \hat{X} \oplus \hat{Y} \oplus \langle \hat{t} \rangle$, and hence also $\langle x + y + \lambda t \rangle \subseteq \hat{X} \oplus \hat{Y} \oplus \langle \hat{t} \rangle$. □

COROLLARY 7.4.3. $Sp_{2n}(\mathbb{F}_q) \cong O_{2n+1}(\mathbb{F}_q)$ for q a power of 2.

PROOF. The above theorem shows that there is an action of $Sp_{2n}(\mathbb{F}_q)$ on the $(2n + 1)$-dimensional orthogonal space $H_0(\mathcal{F}_M)$, so that the natural map $O_{2n+1}(\mathbb{F}_q) \to Sp_{2n}(\mathbb{F}_q)$ is surjective. On the other hand, if an orthogonal matrix acts trivially on the symplectic quotient space, then it fixes the unique isotropic lift of each vector in this symplectic space. But these lifts generate the orthogonal space. □

COMMENT. Ronan and Smith [223] have proved more generally that for any **minimum weight** module (i.e., one in which there is only one orbit of weights under the action of the Weyl group) except for the natural module for $Sp_{2n}(\mathbb{F}_q)$ in characteristic two, one has $H_0(\mathcal{F}_M) = M$. This condition includes the natural modules for the symplectic and orthogonal groups, as well as the spin modules for the orthogonal groups, the exterior powers of the natural module for the linear groups, the 27-dimensional module for type E_6 and the 56-dimensional module for type E_7.

7.5. Smith's theorem and universal coefficient systems

Much of the representation theory of Chevalley groups in their own characteristic can be understood in terms of the representation theory of the corresponding algebraic group. The interested reader should consult the work of R. Steinberg, J. C. Jantzen, J. E. Humphreys and H. H. Andersen, among others. We shall instead try to see what information we can get out of the approach using local coefficient systems, following Ronan and Smith. The basic input for this theory is the following theorem.

Let k be an algebraically closed field of characteristic p.

THEOREM 7.5.1 (Smith [**240**]). *Suppose G is a Chevalley group in characteristic p, H a parabolic subgroup and M an irreducible kG-module. Then the fixed point space $M^{O_p(H)}$ is an irreducible $k(H/O_p(H))$-module. If $M \not\cong N$ are non-isomorphic irreducible kG-modules then there exists a minimal parabolic subgroup H with $M^{O_p(H)} \not\cong N^{O_p(H)}$.*

PROOF. It would take us too far afield to give the proof of this theorem. The proof is very short, but uses the theory of weight spaces for modules for a Chevalley group, see S. D. Smith [**240**]. An alternative proof may be found in M. Cabanes [**58**].   □

EXAMPLES. (i) Let B be a Borel subgroup (which is the same as a Sylow p-normaliser). Then $B/O_p(B)$ is an abelian p'-group, and so its irreducible modules are one dimensional. So for any irreducible kG-module, the fixed points of a Sylow p-subgroup are one dimensional.

(ii) If $q = 2$ then a Borel subgroup is the same as a Sylow 2-subgroup, and the minimal parabolics $H \supseteq B$ have

$$H/O_2(H) \cong SL_2(2) \cong \mathcal{S}_3,$$

the symmetric group of degree three, and so the simple modules for $H/O_2(H)$ have dimensions one and two. Thus associated to each irreducible module is an assignment of ones and twos to the vertices of the Dynkin diagram, and by Smith's theorem this assignment completely determines the module. It turns out that every possible assignment occurs, and so there are exactly 2^r isomorphism classes of irreducible modules, where r is the Lie rank. The Steinberg module is the one corresponding to putting a 2 on each vertex.

More generally for any value of q, if H is a minimal parabolic then there is a natural map $SL_2(\mathbb{F}_q) \to H/O_p(H)$ whose kernel is central and whose cokernel is an abelian group which acts on the image via diagonal matrices. For $SL_2(p)$ there are p simple modules, namely the symmetric powers of the natural 2 dimensional module with dimensions 1 up to p. Steinberg's tensor product theorem gives the irreducible modules for $SL_2(\mathbb{F}_q)$ when $q = p^e$ in terms of these. It turns out that if two modules have isomorphic fixed points when pulled back to each of these $SL_2(\mathbb{F}_q)$'s then they are related by tensoring with a one dimensional representation. A Chevalley group is of **universal type** if every configuration of pulled back fixed points occurs for some irreducible kG-module. In this case the number of isomorphism classes of irreducible modules is $|G : G'|.q^r$. For any Chevalley group the universal p'-central extension is a Chevalley group of universal type. Thus for example $SL_n(\mathbb{F}_q)$ is of universal type and has q^{n-1} isomorphism classes of irreducible modules. $PGL_n(\mathbb{F}_q)$ is not of universal type, but its universal p'-central extension $GL_n(\mathbb{F}_q)$ is; it has $(q-1)q^{n-1}$ irreducibles. Proofs of all these remarks are outside the scope of this book, but the remarks are included for the orientation of the reader.

COMMENT. Smith's theorem is far from being true for a non-Chevalley group. For example for the alternating group A_7 in characteristic two, there is a 20 dimensional simple module. The fixed points of a Sylow 2-subgroup (which has order eight) are therefore more than one dimensional, whereas the Sylow 2-subgroup is self normalising. The sporadic group J_4 looks more like a Chevalley group in characteristic two, but the 112 dimensional irreducible module has fixed points $1 \oplus 5$ under the action of the maximal 2-local subgroup $2^{10} L_5(2)$.

COROLLARY 7.5.2. *Suppose M is an irreducible kG-module as above. Then the fixed point coefficient system $\mathcal{F}_M$ on the building $\Delta(G)$ has the property that $\mathcal{F}_M(\sigma)$ is irreducible for each simplex σ.*   □

LEMMA 7.5.3. *If M and N are irreducible kG-modules for G a Chevalley group, then*

$$\mathrm{Hom}_{\Delta,G}(\mathcal{F}_M, \mathcal{F}_N) \cong \mathrm{Hom}_{kG}(M, N)$$

where Δ denotes the building of G.

PROOF. Suppose we are given a homomorphism from $\mathcal{F}_M$ to $\mathcal{F}_N$. By Smith's theorem, at each simplex the homomorphism is a homomorphism of irreducible modules. By Schur's lemma it is therefore determined up to multiplication by a scalar. By compatibility with the restriction maps and conjugations, any one scalar determines the rest (since Δ/G is connected). If $M \not\cong N$ then there is some simplex on which the values are not isomorphic, so in this case the scalar has to be zero. If $M \cong N$ then of course any scalar can happen.   □

PROPOSITION 7.5.4. *If M is an irreducible kG-module for G a Chevalley group, then $H_0(\mathcal{F}_M)$ has a unique maximal submodule, and the quotient is naturally isomorphic to M.*

PROOF. Suppose N is a simple kG-module. Then

$$\mathrm{Hom}_{kG}(H_0(\mathcal{F}_M), N) \cong \mathrm{Hom}_{\Delta,G}(\mathcal{F}_M, \kappa_N)$$
$$= \mathrm{Hom}_{\Delta,G}(\mathcal{F}_M, \mathcal{F}_N) \cong \mathrm{Hom}_{kG}(M, N).   □$$

COMMENT. The module $H_0(\mathcal{F}_M)$ should not be confused with the **Weyl module** W_M which also has a unique maximal submodule, which again has quotient isomorphic to M. In fact, both of these modules are quotients of a universal coefficient system constructed as follows.

Let $\mathcal{B}_p^{(1)}(G)$ be the subset of $\mathcal{B}_p(G)$ consisting of the Sylow p-subgroups together with the maximal points lying below them; these correspond to the simplices of the building associated to the Borel and minimal parabolic subgroups, i.e., the simplices of dimension r and $r - 1$. Denote by

$$i_G : \mathcal{B}_p^{(1)}(G) \hookrightarrow \mathcal{B}_p(G)$$

the inclusion.

DEFINITION 7.5.5. *The coefficient system $\mathcal{U}_M = (i_G)_! i_G^{-1} \mathcal{F}_M$ is the **universal coefficient system** associated to the irreducible kG-module M.*

PROPOSITION 7.5.6. *The module $H_0(\mathcal{U}_M)$ has a unique maximal submodule, and the quotient is isomorphic to M. If N is a kG-module such that for a minimal parabolic subgroup H, $N^{O_p(H)} \cong M^{O_p(H)}$, and N is generated by these fixed points, then N is a quotient of $H_0(\mathcal{U}_M)$ (this is true in particular if N is the Weyl module W_M).*

PROOF. Just as in the last proposition, we have

$$\mathrm{Hom}_{kG}(H_0(\mathcal{U}_M), N) \cong \mathrm{Hom}_{\Delta, G}(\mathcal{U}_M, \kappa_N) = \mathrm{Hom}_{\Delta, G}((i_G)_! i_G^{-1}\mathcal{F}_M, \kappa_N)$$

$$\cong \mathrm{Hom}_{\Delta, G}(i_G^{-1}\mathcal{F}_M, i_G^{-1}\kappa_N) = \mathrm{Hom}_{\Delta, G}(i_G^{-1}\mathcal{F}_M, i_G^{-1}\mathcal{F}_N)$$

$$\cong \mathrm{Hom}_{kG}(M, N). \qquad \square$$

AN ALGORITHM FOR CONSTRUCTING THE IRREDUCIBLE MODULES. If M is an irreducible kG-module, we write $\hat{M}$ for $H_0(\mathcal{F}_M)$ and $J(\hat{M})$ for the unique maximal submodule of $\hat{M}$.

LEMMA 7.5.7. *If N is a quotient of $\hat{M}$ then there is a canonical splitting $\mathcal{F}_N \cong \mathcal{F}_M \oplus \mathcal{F}_{J(N)}$.*

PROOF. We have maps

$$\hat{M} = H_0(\mathcal{F}_M) \to N \to M$$

whose composite is the natural map $H_0(\mathcal{F}_M) \to M$, and hence associated maps

$$\mathcal{F}_M \to \mathcal{F}_N \to \mathcal{F}_M$$

whose composite is the identity. The kernel of the second of these maps is just $\mathcal{F}_{J(N)}$. $\qquad \square$

Ronan and Smith had the following idea for turning this into an inductive construction of the irreducible modules. The coefficient system $\mathcal{F}_M$ can be constructed entirely from a knowledge of the representation theory of $H/O_p(H)$ for H parabolic. These groups are Chevalley groups of strictly smaller rank, and so we may assume their representation theory is known. This means we can construct $\hat{M}$. By the above lemma, we have

$$\mathcal{F}_{\hat{M}} = \mathcal{F}_M \oplus \mathcal{F}_{J(\hat{M})}.$$

So the next step is to find the complementary coefficient system $\mathcal{F}_{J(\hat{M})}$. The image of the natural map

$$H_0(\mathcal{F}_{J(\hat{M})}) \to \hat{M}$$

has non-zero image contained in $J(\hat{M})$. Quotienting by this image gives us a closer approximation to M, and we may use it to replace $\hat{M}$ in the above construction and start again. In a finite number of steps (often just one!) this process must terminate.

It is not quite clear from this description that this is really a practical process for calculation. In their paper [223], Ronan and Smith describe in detail how to turn this into an effective algorithm for calculation.

Bibliography

[1] J. F. Adams. *Infinite loop spaces*. Ann. of Math. Studies 90, Princeton 1978.

[2] J. F. Adams. *Graeme Segal's Burnside ring conjecture*. Symposium on Algebraic Topology in honour of José Adem (Oaxtepec, 1981), pp. 9–18, Contemp. Math., 12, Amer. Math. Soc., Providence, R.I., 1982.

[3] A. Adem, J. Maginnis and R. J. Milgram. *Symmetric invariants and cohomology of groups*. Math. Ann. 287 (1990), 391–411.

[4] A. Adem, J. Maginnis and R. J. Milgram. *The geometry and cohomology of the Mathieu group M_{12}*. J. Algebra 139 (1991), 90–133.

[5] J. Aguadé. *The cohomology of GL_2 of a finite field*. Arch. Math. 34 (1980), 509–516.

[6] J. L. Alperin. *Minimal resolutions*. Finite groups '72 (Proc. Gainsville Conf. Univ. of Florida, 1972) 1–2, North-Holland Math. Studies, Vol. 7, North Holland, Amsterdam, 1973.

[7] J. L. Alperin. *Resolutions for finite groups*. Proceedings of the Conference on Finite Groups (Univ. Utah, Park City, Utah 1975) 341–356, Academic Press, New York, 1976.

[8] J. L. Alperin. *Periodicity in groups*. Ill. J. Math. 21 (1977), 776–783.

[9] J. L. Alperin. *Cohomology is representation theory*. Arcata Conf. on Representations of Finite Groups, Proc. Symp. Pure Maths 47 (Part 1), Amer. Math. Soc. 1987, 3–11.

[10] J. L. Alperin. *Weights for finite groups*. Arcata Conf. on Representations of Finite Groups, Proc. Symp. Pure Maths 47 (Part 1), Amer. Math. Soc. 1987, 369–379.

[11] J. L. Alperin and L. Evens. *Representations, resolutions, and Quillen's dimension theorem*. J. Pure Appl. Algebra 22 (1981), 1–9.

[12] J. L. Alperin and L. Evens. *Varieties and elementary abelian subgroups*. J. Pure Appl. Algebra 26 (1982), 221–227.

[13] J. L. Alperin and G. Janusz. *Resolutions and periodicity*. Proc. Amer. Math. Soc. 37 (1973), 403–406.

[14] R. Andrews. *Some periodic modules*. Unpublished manuscript, 1989.

[15] S. Araki. *Steenrod reduced powers in the spectral sequence associated to a fibring, I*. Mem. Fac. Sci. Kyusyu Univ. Series A, Math. 11 (1957), 15–64.

[16] S. Araki. *Steenrod reduced powers in the spectral sequence associated to a fibring, II*. Mem. Fac. Sci. Kyusyu Univ. Series A, Math. 11 (1957), 81–97.

[17] J. E. Arnold, Jr. *Homological algebra based on permutation modules*. J. Algebra 70 (1981), 250–260.

[18] M. F. Atiyah. *Characters and cohomology of finite groups*. Publ. Math. IHES 9 (1961), 23–64.

[19] M. F. Atiyah. *K-theory*. Benjamin, New York/Amsterdam, 1967.

[20] M. F. Atiyah and I. G. Macdonald. *Introduction to commutative algebra*. Addison-Wesley, Reading, Mass. (1969).

[21] M. F. Atiyah and G. B. Segal. *Equivariant K-theory and completion*. J. Diff. Geom. 3 (1969), 1–18.

[22] G. S. Avrunin. *The image of the restriction map on mod 2 cohomology*. Arch. Math. (Basel) 34 (1980), 502–508.

[23] G. S. Avrunin. *Annihilators of cohomology modules*. J. Algebra 69 (1981), 150–154.

[24] G. S. Avrunin and L. L. Scott. *Quillen stratification for modules.* Invent. Math. 66 (1982), 277–286.

[25] A. Babakhanian. *Cohomological methods in group theory.* Marcel Dekker, New York 1972.

[26] J. C. Becker and D. H. Gottlieb. *The transfer map and fiber bundles.* Topology 14 (1975), 1–11.

[27] D. J. Benson. *Some recent trends in modular representation theory.* Proc. Rutgers Group Theory Year, 1983–1984, ed. Aschbacher et al., Cambridge University Press 1984.

[28] D. J. Benson. *Modular representation theory: New trends and methods.* Lecture Notes in Mathematics 1081, Springer-Verlag. Berlin/New York 1984.

[29] D. J. Benson. *Modules for finite groups: representation rings, quivers and varieties.* Representation Theory II, Groups and Orders. Proceedings, Ottawa 1984. Lecture Notes in Mathematics 1178, Springer-Verlag, Berlin/New York 1986.

[30] D. J. Benson. *Representation rings of finite groups.* Representations of Algebras, Durham 1985. L.M.S. Lecture Note Series 116, Cambridge University Press 1986.

[31] D. J. Benson. *Resolutions and Poincaré duality for finite groups.* Proceedings of the June 1990 Barcelona conference on Homotopy and Group Cohomology, Lecture Notes in Mathematics 1509, 10–19, Springer-Verlag, Berlin/New York 1992.

[32] D. J. Benson and J. F. Carlson. *Nilpotent elements in the Green ring.* J. Algebra 104 (1986), 329–350.

[33] D. J. Benson and J. F. Carlson. *Diagrammatic methods for modular representations and cohomology.* Comm. in Algebra 15 (1987), 53–121.

[34] D. J. Benson and J. F. Carlson. *Complexity and multiple complexes.* Math. Zeit. 195 (1987), 221–238.

[35] D. J. Benson and J. F. Carlson. *Projective resolutions and Poincaré duality complexes.* Trans. Amer. Math. Soc. 132 (1994), 447–488.

[36] D. J. Benson and J. F. Carlson. *Cohomology of extraspecial groups.* Bull. London Math. Soc. 24 (1992), 209–235. Erratum: Bull. London Math. Soc. 25 (1993), 498.

[37] D. J. Benson and J. F. Carlson. *Periodic modules with large period.* Quarterly Journal of Mathematics 43 (1992), 283–296.

[38] D. J. Benson, J. F. Carlson and G. R. Robinson. *On the vanishing of group cohomology.* J. Algebra 131 (1990), 40–73.

[39] D. J. Benson and L. Evens. *Group homomorphisms inducing isomorphisms in cohomology.* Comm. in Algebra 18 (1990), 3447–3452.

[40] D. J. Benson and M. Feshbach. *Stable splittings of classifying spaces of finite groups.* Topology 31 (1992), 157–176.

[41] S. Berman, R. Moody and M. Wonenburger. *Certain matrices with null roots and finite Cartan matrices.* Indiana Univ. Math. J. 21 (1971/2), 1091–1099.

[42] C. Bessenrodt and W. Willems. *Relations between complexity and modular invariants and consequences for p-solvable groups.* J. Algebra 86 (1984), 445–456.

[43] J. Boardman and R. Vogt. *Homotopy-everything H-spaces.* Bull. Amer. Math. Soc. 74 (1968), 1117–1122.

[44] I. V. Bogačenko. *On the structure of the cohomology ring of the Sylow subgroup of the symmetric group.* (Russian) Izv. Akad. Nauk. SSSR Ser. Mat. 27 (1963), 937–942.

[45] A. Borel and J. Tits. *Eléments unipotents et sousgroupes paraboliques des groupes réductives I.* Invent. Math. 12 (1971), 97–104.

[46] R. Bott. *The stable homotopy of the classical groups.* Ann. Math. 70 (1959), 313–337.

[47] S. Bouc. *Homologie de certains ensembles ordonnés et modules de Möbius.* Thesis, Paris 1983.

[48] S. Bouc. *Modules de Möbius.* Comptes Rendues Acad. Sci. Paris 299, Série I (1984), 9–12.

BIBLIOGRAPHY 253

[49] S. Bouc. *Homologie de certains ensembles ordonnés.* Comptes Rendues Acad. Sci. Paris 299, Série I (1984), 49–52.

[50] A. K. Bousfield and D. M. Kan. *Homotopy limits, completions and localizations.* Lecture Notes in Mathematics 304, Springer-Verlag, Berlin/New York, 1972.

[51] G. E. Bredon. *Introduction to compact transformation groups.* Academic Press, New York 1972.

[52] K. S. Brown. *Cohomology of groups.* Graduate Texts in Mathematics 87, Springer-Verlag, Berlin/New York 1982.

[53] S. R. Bullett and I. G. Macdonald. *On the Adem relations.* Topology 21 (1982), 329–332.

[54] D. Burghelea. *The cyclic homology of the group rings.* Comment. Math. Helvetici 60 (1985), 354–365.

[55] D. Burghelea and Z. Fiedorowicz. *Cyclic homology and algebraic K-theory of spaces I.* Algebraic K-theory, Boulder (1983). Contemp. Math. vol. 55, part I, Amer. Math. Soc., 1986.

[56] D. Burghelea and Z. Fiedorowicz. *Cyclic homology and algebraic K-theory of spaces II.* Topology 25 (1986), 303–317.

[57] N. Burgoyne and C. Williamson. *On a theorem of Borel and Tits for finite Chevalley groups.* Arch. Math. Basel 27 (1976), 489–491.

[58] M. Cabanes. *Irreducible modules and Levi supplements.* J. Algebra 90 (1984), 84–97.

[59] H. Cárdenas. *El algebra de cohomologia del grupo simétrico de grado p^2.* Bol. Soc. Mat. Mexicana (2) 10 (1965), 1–30.

[60] H. C. Cárdenas and E. Lluis. *On the Chern classes of representations of the symmetric groups.* Group theory (Singapore, 1987), 333–345, de Gruyter, Berlin/New York, 1989.

[61] J. F. Carlson. *Periodic modules over group algebras.* J. London Math. Soc. (2), 15 (1977), 431–436.

[62] J. F. Carlson. *Restrictions of modules over modular group algebras.* J. Algebra 53 (1978), 334–343.

[63] J. F. Carlson. *The dimensions of periodic modules over modular group algebras.* Ill. J. Math. 23 (1979), 295–306.

[64] J. F. Carlson. *Endo-trivial modules over (p,p) groups.* Illinois J. Math. 24 (1980), 287–295.

[65] J. F. Carlson. *The complexity and varieties of modules.* Integral representations and their applications, Oberwolfach 1980. Lecture Notes in Mathematics 882, 415–422, Springer-Verlag, Berlin/New York 1981.

[66] J. F. Carlson. *Complexity and Krull dimension.* Representations of Algebras, Puebla, Mexico 1980. Lecture Notes in Mathematics 903, 62–67, Springer-Verlag, Berlin/New York 1981.

[67] J. F. Carlson. *The structure of periodic modules over modular group algebras.* J. Pure Appl. Algebra 22 (1981), 43–56.

[68] J. F. Carlson. *Dimensions of modules and their restrictions over modular group algebras.* J. Algebra 69 (1981), 95–104.

[69] J. F. Carlson. *The varieties and the cohomology ring of a module.* J. Algebra 85 (1983), 104–143.

[70] J. F. Carlson. *The variety of an indecomposable module is connected.* Invent. Math. 77 (1984), 291–299.

[71] J. F. Carlson. *The cohomology ring of a module.* J. Pure Appl. Algebra 36 (1985), 105–121.

[72] J. F. Carlson. *Cohomology rings of induced modules.* J. Pure Appl. Algebra 44 (1987), 85–98.

[73] J. F. Carlson. *Varieties and transfers.* J. Pure Appl. Algebra 44 (1987), 99–106.

[74] J. F. Carlson. *Varieties for modules.* Arcata Conf. on Representations of Finite Groups, Proc. Symp. Pure Maths 47 (Part 1), Amer. Math. Soc. 1987, 37–44.

[75] J. F. Carlson. *Products and projective resolutions.* Arcata Conf. on Representations of Finite Groups, Proc. Symp. Pure Maths 47 (Part 1), Amer. Math. Soc. 1987, 399–408.

[76] J. F. Carlson. *Exponents of modules and maps.* Invent. Math. 95 (1989), 13–24.

[77] G. Carlsson. *Equivariant stable homotopy and Segal's Burnside ring conjecture.* Annals of Math. 120 (1984), 189–224.

[78] H. Cartan and S. Eilenberg. *Homological Algebra.* Princeton University Press, 1956.

[79] G. R. Chapman. *The cohomology ring of a finite abelian group.* Proc. London Math. Soc. (3) 45 (1982), 564–576.

[80] L. Charlap and A. Vasquez. *The cohomology of group extensions.* Trans. Amer. Math. Soc. 124 (1966), 24–40.

[81] L. Chouinard. *Projectivity and relative projectivity over group rings.* J. Pure Appl. Algebra 7 (1976), 278–302.

[82] E. Cline, B. Parshall and L. Scott. *Cohomology of finite groups of Lie type, I.* Publ. Math. IHES (1974), 169–191.

[83] A. Connes. *Non-commutative differential geometry.* Publ. Math. I.H.E.S., 62 (1985), 257–360.

[84] A. Connes. *Cohomologie cyclique et foncteurs* Ext^n. Comptes Rendues Acad. Sci. Paris 296, Série I (1983), 953–958.

[85] C. W. Curtis. *Homology representations of finite groups.* Proc. ICRA II (Ottawa 1979), Lecture Notes in Mathematics 832 (1980), 177–194, Springer-Verlag, Berlin/New York 1980.

[86] E. C. Dade. *Endo-permutation modules over p-groups, I.* Ann. of Math. 107 (1978), 459–494.

[87] E. C. Dade. *Endo-permutation modules over p-groups, II.* Ann. of Math. 108 (1978), 317–346.

[88] C. DeConcini. *The mod 2 cohomology of the orthogonal and symplectic groups over a finite field.* Advances in Math. 27 (1978), 191–229.

[89] T. tom Dieck. *Transformation groups and representation theory.* Lecture Notes in Mathematics 766, Springer-Verlag, Berlin/New York 1979.

[90] T. tom Dieck. *Transformation groups.* De Gruyter Studies in Mathematics 8, Walter de Gruyter, Berlin/New York 1987.

[91] T. Diethelm. *The mod p cohomology rings of the nonabelian split metacyclic p-groups.* Arch. Math. (Basel) 44 (1985), 29–38.

[92] T. Diethelm and U. Stammbach. *On the module structure of the mod p cohomology of a p-group.* Arch. Math. 43 (1984), 488–492.

[93] P. W. Donovan. *Spectral duality for block cohomology.* J. Algebra 88 (1984), 330–345.

[94] P. W. Donovan. *A criterion for a modular representation to be projective.* J. Algebra 117 (1988), 424–436.

[95] J. Duflot. *Depth and equivariant cohomology.* Comm. Math. Helvetici 56 (1981), 617–637.

[96] J. Duflot. *The associated primes of* $H_G^*(X)$. J. Pure Appl. Algebra 30 (1983), 137–141.

[97] W. G. Dwyer, M. J. Hopkins and D. M. Kan. *The homotopy theory of cyclic sets.* Trans. A.M.S. 291 (1985), 281–289.

[98] E. Dyer and R. Lashof. *Homology of iterated loop spaces.* Amer. J. Math. 84 (1962), 35–88.

[99] S. Eilenberg and J. A. Zilber. *Semi-simplicial complexes and singular homology.* Ann. of Math. 51 (1950), 499–513.

[100] D. Eisenbud. *Homological algebra on a complete intersection, with an application to group representations.* Trans. Amer. Math. Soc. 269 (1980), 35–64.

[101] L. Evens. *The cohomology ring of a finite group.* Trans. Amer. Math. Soc. 101 (1961), 224–239.

[102] L. Evens. *A generalization of the transfer map in the cohomology of groups.* Trans. Amer. Math. Soc. 108 (1963), 54–65.

[103] L. Evens. *On the Chern classes of representations of finite groups.* Trans. Amer. Math. Soc. 115 (1965), 180–193.

[104] L. Evens. *The spectral sequence of a finite group extension stops.* Trans. Amer. Math. Soc. 212 (1975), 269–277.

[105] L. Evens and M. Feshbach. *Carlson's theorem on varieties and transfer.* J. Pure & Appl. Algebra 57 (1989), 39–45.

[106] L. Evens and S. Jackowski. *A note on the subgroup theorem in cohomological complexity theory.* J. Pure Appl. Algebra 65 (1990), 25–28.

[107] L. Evens and D. Kahn. *Chern classes of certain representations of symmetric groups.* Trans. Amer. Math. Soc. 245 (1978/9), 309–330.

[108] L. Evens and D. Kahn. *An integral Riemann-Roch formula for induced representations of finite groups.* Trans. Amer. Math. Soc. 245 (1978/9), 331–347.

[109] L. Evens and S. Priddy. *The cohomology of the semi-dihedral group.* Conf. on Algebraic Topology in honour of Peter Hilton, ed. R. Piccinini and D. Sjerve, Contemp. Math. 37, Amer. Math. Soc., 1985.

[110] Z. Fiedorowicz and S. Priddy. *Homology of classical groups over finite fields and their associated infinite loop spaces.* Lecture Notes in Mathematics 674, Springer-Verlag, Berlin/New York 1978.

[111] P. Fleischmann. *The complexities and rank varieties of the simple modules of $(^2A_2)(q^2)$ in the natural characteristic.* J. Algebra 121 (1989), 399–408.

[112] E. M. Friedlander. *Computations of K-theories of finite fields.* Topology 15 (1976), 87–109.

[113] W. Fulton and R. MacPherson. *Characteristic classes of direct image bundles for covering maps.* Annals of Math. 125 (1987), 1–92.

[114] D. Gluck. *Idempotent formula for the Burnside algebra with applications to the p-subgroup simplicial complex.* Ill. J. Math. 25 (1981), 63–67.

[115] T. G. Goodwillie. *Cyclic homology, derivations, and the free loopspace.* Topology 24 (1985), 187–215.

[116] D. Grayson (after D. Quillen). *Higher algebraic K-theory: II.* Algebraic K-theory, Evanston 1976. Lecture Notes in Mathematics 551, Springer-Verlag, Berlin/New York, 1976.

[117] A. Grothendieck. *Classes de Chern des representations linéaires des groupes discrets.* In Dix exposés sur la cohomologie étale des schémas, North Holland, 1968.

[118] K. W. Gruenberg. *Cohomological topics in group theory.* Lecture Notes in Mathematics 143, Springer-Verlag, Berlin/New York 1970.

[119] J. H. C. Gunawardena. *Stiefel-Whitney classes for representations of groups.* J. London Math. Soc. (2) 35 (1987), 539–550.

[120] J. H. C. Gunawardena, B. Kahn and C. Thomas. *Stiefel-Whitney classes of real representations of finite groups.* J. Algebra 126 (1989), 327–347.

[121] J. H. Gunawardena, J. Lannes and S. Zarati. *Cohomologie des groupes symétriques et application de Quillen.* Advances in homotopy theory (Cortona 1988), 61–68. L.M.S. Lecture Note Series 139, Cambridge Univ. Press, 1989.

[122] N. Habegger. *Hypercohomology varieties for complexes of modules, the realizability criterion, and equivalent formulations of a conjecture of Carlsson.* Arcata Conf. on Representations of Finite Groups, Proc. Symp. Pure Maths 47 (Part 1), Amer. Math. Soc. 1987, 431–437.

[123] D. Happel, U. Preiser and C. M. Ringel. *Vinberg's characterization of Dynkin diagrams using subadditive functions with applications to DTr-periodic modules.* Representation theory II, Ottawa 1979. Lecture Notes in Mathematics 832, Springer-Verlag, Berlin/New York 1980.

[124] M. Harada and A. Kono. *On the integral cohomology of extraspecial 2-groups.* J. Pure Appl. Algebra 44 (1987), 215–219.

[125] R. Hartshorne. *Algebraic geometry.* Graduate Texts in Mathematics 52, Springer-Verlag, Berlin/New York 1977.

[126] P. J. Hilton and U. Stammbach. *A course in homological algebra.* Graduate Texts in Mathematics 4, Springer-Verlag, Berlin/New York, 1971.

[127] G. Hochschild and J.-P. Serre. *Cohomology of group extensions.* Trans. Amer. Math. Soc. 74 (1953), 110–134.

[128] S.-T. Hu. *Elements of general topology.* Holden-Day, San Francisco 1964.

[129] J. Huebschmann. *Automorphisms of group extensions and differentials in the Lyndon–Hochschild–Serre spectral sequence.* J. Algebra 72 (1981), 296–334.

[130] J. Huebschmann. *The cohomology of $F\Psi^q$. The additive structure.* J. Pure Appl. Algebra 45 (1987), 73–91.

[131] J. Huebschmann. *Perturbation theory and free resolutions for nilpotent groups of class 2.* J. Algebra 126 (1989), 348–399.

[132] J. Huebschmann. *Cohomology of nilpotent groups of class 2.* J. Algebra 126 (1989), 400–450.

[133] J. E. Humphreys. *Introduction to Lie algebras and representation theory.* Graduate Texts in Mathematics 9, Springer-Verlag, Berlin/New York 1972.

[134] J. E. Humphreys. *Linear algebraic groups.* Graduate Texts in Mathematics 21, Springer-Verlag, Berlin/New York 1975.

[135] N. H. V. Hung. *The mod 2 cohomology algebras of the symmetric groups.* Acta Math. Vietnamica, 6 (1981), 41–48.

[136] N. H. V. Hung. *Algèbre de cohomologie du groupe symmétrique infini et classes caractéristiques de Dickson.* Comptes Rendues Acad. Sci. Paris 297, Série I (1983), 611–614.

[137] D. Husemoller. *Fibre bundles.* Graduate Texts in Mathematics 20, Springer-Verlag, Berlin/New York 1975.

[138] I. M. James and G. B. Segal. *On equivariant homotopy type.* Topology 17 (1978), 267–272.

[139] D. Jeandupeux. *Sur la cohomologie entière du groupe linéaire général et du groupe linéaire spécial sur un corps fini.* Comptes Rendues Acad. Sci. Paris 308, Série I (1989), 71–73.

[140] D. L. Johnson. *On the cohomology of finite 2-groups.* Invent. Math. 7 (1969), 159–173.

[141] D. L. Johnson. *A transfer theorem for the cohomology of a finite group.* Invent. Math. 7 (1969), 174–182.

[142] J. D. S. Jones. *Cyclic homology and equivariant homology.* Invent. Math. 87 (1987), 403–424.

[143] D. M. Kan. *Abstract homotopy I, II.* Proc. Nat. Acad. Sci. USA, 41 (1955), 1092–1096.

[144] D. M. Kan. *On c.s.s. complexes.* Amer. J. Math. 79 (1957), 449–476.

[145] D. M. Kan. *A combinatorial definition of homotopy groups.* Ann. of Math. 67 (1958), 282–312.

[146] D. M. Kan and W. P. Thurston. *Every connected space has the homology of a $K(\pi, 1)$.* Topology 15 (1976), 253–258.

[147] M. Karoubi. *Homologie cyclique des groupes et des algèbres.* Comptes Rendues Acad. Sci. Paris 297 (1983), 381–384.

[148] M. Karoubi. *Homologie cyclique et K-théorie algébrique I.* Comptes Rendues Acad. Sci. Paris 297 (1983), 447–450.

[149] M. Karoubi. *Homologie cyclique et K-théorie algébrique II.* Comptes Rendues Acad. Sci. Paris 297 (1983), 513–516.

[150] C. Kassel. *K-théorie algébrique et cohomologie cyclique bivariantes.* Comptes Rendues Acad. Sci. Paris 306, Série I (1988), 799–802.

[151] H. Kawai. *On module varieties and quotient groups.* J. Algebra 121 (1989), 248–251.

[152] S. N. Kleinerman. *The cohomology of Chevalley groups of exceptional Lie type.* Memoirs of the Amer. Math. Soc. 268, 1982.

[153] R. Knörr and G. R. Robinson. *Some remarks on a conjecture of Alperin.* Preprint, 1988.

[154] C. Kratzer et J. Thévenaz. *Fonctions de Möbius d'un groupe fini et anneau de Burnside.* Comment. Math. Helvetici 59 (1984), 425–438.

[155] O. Kroll. *Complexity and elementary abelian subgroups.* Ph. D. thesis, Univ. of Chicago, 1980.

[156] O. Kroll. *Complexity and elementary abelian p-groups.* J. Algebra 88 (1984), 155–172.

[157] O. Kroll. *An algebraic characterisation of Chern classes of finite group representations.* Bull. London Math. Soc. 19 (1987), 245–248.

[158] O. Kroll. *The cohomology of the finite general linear group.* J. Pure Appl. Algebra 54 (1988), 95–115.

[159] O. Kroll. *A representation theoretical proof of a theorem of Serre.* Århus preprint, Maj 1986.

[160] O. Kroll. *An algebraic construction of Chern classes of finite group representations.* Århus preprint, 1987.

[161] T. Kudo. *A transgression theorem.* Mem. Fac. Sci. Kyusyu Univ. (A) 9 (1956), 79–81.

[162] T. K. Kuo. *On the exponent of $H^n(G, \mathbb{Z})$.* J. Algebra 7 (1967), 160–167.

[163] K. Lamotke. *Semisimpliziale algebraische Topologie.* Grundlehren in der mathematischen Wissenschaften in Einzeldarstellungen, Band 147. Springer-Verlag, Berlin/New York 1974.

[164] D. S. Larson. *The integral cohomology rings of split metacyclic groups.* U. of Minnesota preprint, 1987.

[165] G. Lewis. *The integral cohomology rings of groups of order p^3.* Trans. Amer. Math. Soc. 132 (1968), 501–529.

[166] L. G. Lewis, J. P. May and J. E. McClure. *Classifying G-spaces and the Segal conjecture.* Current Trends in Algebraic Topology. CMS Conference Proc. 2 (1982), 165–179.

[167] J.-L. Loday and D. Quillen. *Cyclic homology and the Lie algebra homology of matrices.* Comment. Math. Helvetici 59 (1984), 565–591.

[168] A. T. Lundell and S. Weingram. *The topology of CW complexes.* Van Nostrand Reinhold, 1969.

[169] G. Lusztig. *The discrete series of GL_n over a finite field.* Ann. of Math. Studies 81, P.U.P. 1974.

[170] S. Mac Lane. *Homology.* Springer-Verlag, Berlin/New York 1974.

[171] S. Mac Lane. *Origins of the cohomology of groups.* Enseign. Math. (2) 24 (1978), 1–29.

[172] B. M. Mann. *The cohomology of the symmetric groups.* Trans. Amer. Math. Soc. 242 (1978), 157–184.

[173] B. M. Mann. *The cohomology of the alternating groups.* Mich. Math. Journ. 32 (1985), 267–277.

[174] H. R. Margolis. *Spectra and the Steenrod algebra.* North-Holland, 1983.

[175] J. Martino and S. Priddy. *The complete stable splitting for the classifying space of a finite group.* Topology 31 (1992), 143–156.

[176] W. Massey. *Exact couples in algebraic topology, I, II.* Ann. of Math. 56 (1952), 363–396.

[177] W. Massey. *Exact couples in algebraic topology, III, IV, V.* Ann. of Math. 57 (1953), 248–256.

[178] W. Massey. *Products in exact couples.* Ann. of Math. 59 (1954), 558–569.

[179] H. Matsumura. *Commutative algebra*. W. A. Benjamin Co., New York, 1969.

[180] J. P. May. *Simplicial objects in algebraic topology*. University of Chicago Press, 1967.

[181] J. P. May. *A general algebraic approach to Steenrod operations*. The Steenrod Algebra and its Applications: a conference to celebrate N. E. Steenrod's sixtieth birthday, Lecture Notes in Mathematics 168, Springer-Verlag, Berlin/New York, 1970, 153–231.

[182] R. McCarthy. *Morita equivalence and cyclic homology*. Comptes Rendues Acad. Sci. Paris 307, Série I (1988), 211–215.

[183] J. McCleary. *User's guide to spectral sequences*. Mathematics Lecture Series 12, Publish or Perish, Inc., Delaware 1985.

[184] J. C. McConnell and J. C. Robson. *Noncommutative Noetherian rings*. J. Wiley and Sons, 1988.

[185] R. J. Milgram and S. B. Priddy. *Invariant theory and $H^*(GL_n(\mathbb{F}_p); \mathbb{F}_p)$*. J. Pure Appl. Algebra 44 (1987), 291–302.

[186] J. W. Milnor. *Construction of universal bundles, I*. Ann. of Math. 63 (1956), 272–284.

[187] J. W. Milnor. *Construction of universal bundles, II*. Ann. of Math. 63 (1956), 430–436.

[188] J. W. Milnor. *The Steenrod algebra and its dual*. Ann. of Math. 67 (1) (1958), 150–171.

[189] J. W. Milnor *On spaces having the homotopy type of a CW-complex*. Trans. Amer. Math. Soc. 90 (1959), 272–280.

[190] J. W. Milnor. *Axiomatic homology theory*. Pacific J. Math. 12 (1962), 337–341.

[191] J. W. Milnor. *Morse theory*. Ann. of Math. Studies 51, Princeton 1963.

[192] J. W. Milnor and J. D. Stasheff. *Characteristic classes*. Ann. of Math. Studies 76, Princeton 1974.

[193] P. A. Minh and H. Mùi. *The mod p cohomology algebra of the group $M(p^n)$*. Acta Math. Vietnamica 7 (1982), 17–26.

[194] G. Mislin. *On group homomorphisms inducing mod-p cohomology isomorphisms*. Comment. Math. Helvetici 65 (1990), 454–461.

[195] R. E. Mosher and M. C. Tangora. *Cohomology operations and applications in homotopy theory*. Harper and Row, 1968.

[196] D. Mostow. *Equivariant embeddings in euclidean space*. Ann. of Math. 65 (1957), 432–446.

[197] H. Mùi. *Modular invariant theory and the cohomology algebras of the symmetric groups*. J. Fac. Sci. Univ. Tokyo, sec. IA, 22 (1975), 319–369.

[198] H. J. Munkholm. *Mod 2 cohomology of $D2^n$ and its extensions by Z_2*. Conference on Algebraic Topology, Univ. of Illinois at Chicago Circle, June 17–28 1968, p. 234–252.

[199] M. Nakaoka. *Cohomology theory of a complex with a transformation of prime period and its applications*. J. Inst. of Polytechnics, Osaka City Univ. Ser. A, 7 (1956), 51–102.

[200] M. Nakaoka. *Cohomology mod p of symmetric products of spheres, II*. J. Inst. of Polytechnics, Osaka City Univ. Ser. A, 10 (1959), 67–89.

[201] M. Nakaoka. *Decomposition theorem for homology groups of symmetric groups*. Ann. of Math. 71 (1960), 16–42.

[202] M. Nakaoka. *Homology of the infinite symmetric group*. Ann. of Math. 73 (1961), 229–257.

[203] T. Niwasaki. *On Carlson's conjecture for cohomology rings of modules*. J. Pure Appl. Algebra 59 (1989), 265–278.

[204] T. Okuyama and H. Sasaki. *Evens' norm map and Serre's theorem on the cohomology algebra of a p-group*. Archiv der Math. 54 (1990), 331–339.

[205] C. Picaronny. *Sur un théorème de Carlson*. Comptes Rendues Acad. Sci. Paris 299, Série I (1984), 899–902.

[206] S. Priddy. *Transfer, symmetric groups, and stable homotopy theory.* In Algebraic K-theory I—Higher K-theories. Lecture Notes in Mathematics 341, Springer-Verlag, Berlin/New York, 1973.

[207] C. Procesi. *Rings with polynomial identities.* Marcel-Dekker, New York, 1973.

[208] D. Quillen. *A cohomological criterion for p-nilpotence.* J. Pure Appl. Algebra 1 (1971), 361–372.

[209] D. Quillen. *The spectrum of an equivariant cohomology ring, I.* Ann. of Math. 94 (1971), 549–572.

[210] D. Quillen. *The spectrum of an equivariant cohomology ring, II.* Ann. of Math. 94 (1971), 573–602.

[211] D. Quillen. *The mod 2 cohomology rings of extra-special 2-groups and the spinor groups.* Math. Ann. 194 (1971), 197–212.

[212] D. Quillen. *On the cohomology and K-theory of the general linear groups over a finite field.* Ann. of Math. 96 (1972), 552–586.

[213] D. Quillen. *The Adams conjecture.* Topology 19 (1971), 67–80.

[214] D. Quillen. *Higher algebraic K-theory: I.* Algebraic K-theory I—Higher K-theories. Proceedings, Battelle Institute 1972. Lecture Notes in Mathematics 341, Springer-Verlag, Berlin/New York, 1973.

[215] D. Quillen. *Finite generation of the groups K_i of rings of algebraic integers.* Algebraic K-theory I—Higher K-theories. Proceedings, Battelle Institute 1972. Lecture Notes in Mathematics 341, Springer-Verlag, Berlin/New York, 1973.

[216] D. Quillen. *Characteristic classes of representations.* In Algebraic K-theory, Evanston, 1976. Lecture Notes in Mathematics 551, Springer-Verlag, Berlin/New York, 1976.

[217] D. Quillen. *Homotopy properties of the poset of nontrivial p-subgroups of a finite group.* Adv. in Math. 28 (1978), 101–128.

[218] D. Quillen and B. B. Venkov. *Cohomology of finite groups and elementary abelian subgroups.* Topology 11 (1972), 317–318.

[219] D. C. Ravenel. *Complex cobordism and stable homotopy groups of spheres.* Academic Press, 1986.

[220] I. Reiner. *Nilpotent elements in rings of integral representations.* Proc. Amer. Math. Soc. 17 (1966), 270–274.

[221] I. Reiner. *Integral representation algebras.* Trans. Amer. Math. Soc. 124 (1966), 111–121.

[222] M. A. Ronan. *Duality for presheaves on chamber systems, and a related chain complex.* J. Algebra 121 (1989), 263–274.

[223] M. A. Ronan and S. D. Smith. *Sheaves on buildings and modular representations of Chevalley groups.* J. Algebra 96 (1985), 319–346.

[224] M. A. Ronan and S. D. Smith. *Universal presheaves on group geometries, and modular representations.* J. Algebra 102 (1986), 135–154.

[225] D. J. Rusin. *The mod-2 cohomology of metacyclic 2-groups.* J. Pure Appl. Algebra 44 (1987), 315–328.

[226] D. J. Rusin. *The cohomology of groups of order 32.* Math. Comp. 53 (1989), 359–385.

[227] J. Sawka. *Odd primary operations in first-quadrant spectral sequences.* Trans. Amer. Math. Soc. 273 (2) (1982), 737–752.

[228] M.-T. Schmidt. *Beziehungen zwischen Homologie-Darstellungen und der Hauptserie endlicher Chevalley-Gruppen.* Diplomarbeit, Bonn, 1984.

[229] G. B. Segal. *Classifying spaces and spectral sequences.* Publ. Math. IHES 34 (1968), 105–112.

[230] G. B. Segal. *The multiplicative group of classical cohomology.* Quart. J. Math. Oxford (3), 26 (1975), 289–293.

[231] J.-P. Serre. *Homologie Singulière des espaces fibrés.* Ann. of Math. 54 (3) (1951), 425–505.

[232] J.-P. Serre. *Sur la dimension cohomologique des groupes profinis.* Topology 3 (1965), 413–420.

[233] J.-P. Serre. *Algèbre locale—multiplicités.* Lecture Notes in Mathematics 11, Springer-Verlag, Berlin/New York, 1965.

[234] J.-P. Serre. *Une relation dans la cohomologie des p-groupes.* Comptes Rendues Acad. Sci. Paris 304, Série I (1987), 587–590.

[235] J. M. Shapiro. *On the cohomology of the orthogonal and symplectic groups over a finite field of odd characteristic.* Algebraic *K*-theory, Evanston 1976. Lecture Notes in Mathematics 551. Springer-Verlag, Berlin/New York 1976.

[236] W. M. Singer. *Steenrod operations in spectral sequences, I.* Trans. Amer. Math. Soc. 175 (1973), 327–336.

[237] W. M. Singer. *Steenrod operations in spectral sequences, II.* Trans. Amer. Math. Soc. 175 (1973), 337–353.

[238] L. Smith. *Homological algebra and the Eilenberg–Moore spectral sequence.* Trans. Amer. Math. Soc. 129 (1967), 58–93.

[239] L. Smith. *Lectures on the Eilenberg–Moore spectral sequence.* Lecture Notes in Mathematics 134, Springer-Verlag, Berlin/New York, 1970.

[240] S. D. Smith. *Irreducible modules and parabolic subgroups.* J. Algebra 75 (1982), 286–289.

[241] S. D. Smith. *Sheaf homology and complete reducibility.* J. Algebra 95 (1985), 72–80.

[242] S. D. Smith. *Constructing representations from group geometries.* Arcata Conf. on Representations of Finite Groups, Proc. Symp. Pure Maths 47 (Part 1), A.M.S. 1987, 303–313.

[243] S. D. Smith. *On decomposition of modular representations from Cohen–Macaulay geometries.* J. Algebra 131 (1990), 598–625.

[244] W. Smoke. *Dimension and multiplicity for graded algebras.* J. Algebra 21 (1972), 149–173.

[245] E. Snapper. *Cohomology of permutation representations. I. Spectral sequences.* J. Math. Mech. 13 (1964), 133–161.

[246] E. Snapper. *Cohomology of permutation representations. II. Cup product.* J. Math. Mech. 13 (1964), 1047–1064.

[247] E. H. Spanier. *Algebraic topology.* McGraw-Hill, New York 1966.

[248] L. Solomon. *The Steinberg character of a finite group with a BN-pair.* Theory of Finite Groups (Harvard Symposium), Benjamin, New York 1969, 213–221.

[249] U. Stammbach. *Homology in group theory.* Lecture Notes in Mathematics 359, Springer-Verlag, Berlin/New York 1973.

[250] U. Stammbach. *On the principal indecomposables of a modular group algebra.* J. Pure Appl. Algebra 30 (1983), 69–84.

[251] R. P. Stanley. *Invariants of finite groups and their applications to combinatorics.* Bull. A.M.S. 1 (3) (1979), 475–511.

[252] N. E. Steenrod. *Homology groups of symmetric groups and reduced power operations.* Proc. Nat. Acad. Sci. U.S.A. 39 (1953), 213–217.

[253] N. E. Steenrod and D. Epstein. *Cohomology operations.* Ann. of Math. Studies 50, Princeton 1962.

[254] R. G. Swan. *Induced representations and projective modules.* Ann. of Math. 71 (3) (1960), 552–578.

[255] R. G. Swan. *The nontriviality of the restriction map in cohomology of groups.* Proc. Amer. Math. Soc. 11 (1960), 885–887.

[256] R. G. Swan. *Minimal resolutions for finite groups.* Topology 4 (1964), 193–208.

[257] R. G. Swan. *Groups with no odd dimensional cohomology.* J. Algebra 17 (1971), 453–461.

[258] R. M. Switzer. *Algebraic topology — homotopy and homology.* Grundlehren der mathematischen Wissenschaften 212, Springer-Verlag, Berlin/New York 1973.

[259] O. Talelli. *On the minimal resolutions for metacyclic groups with periodic cohomology.* Commun. in Algebra 12 (1984), 1343–1360.

[260] M. Tezuka and N. Yagita. *The varieties of the mod p cohomology rings of extra special p-groups for an odd prime p.* Math. Proc. Camb. Phil. Soc. 94 (1983), 449–459.

[261] M. Tezuka and N. Yagita. *The cohomology of subgroups of $GL_n(\mathbb{F}_q)$.* Proc. Northwestern Homotopy Theory Conf., Contemp. Math. 19 (1983), 379–396.

[262] M. Tezuka and N. Yagita. *The mod p cohomology ring of $GL_3(\mathbb{F}_p)$.* J. Algebra 81 (1983), 295–303.

[263] J. Thévenaz. *Permutation representations arising from simplicial complexes.* J. Combinatorial Theory Ser. A, 46 (1987), 121–155.

[264] J. Thévenaz and P. J. Webb. *Homotopy equivalence of posets with a group action.* J. Combin. Theory Ser. A 56 (1991), 173–181.

[265] C. B. Thomas. *Chern classes and metacyclic p-groups.* Mathematica 18 (1971), 196–200.

[266] C. B. Thomas. *The integral cohomology ring of S_4.* Mathematica 21 (1974), 228–232.

[267] C. B. Thomas. *Modular representations and the cohomology of finite groups.* Topology Appl. 25 (1987), 193–201.

[268] C. B. Thomas. *Characteristic classes and the cohomology of finite groups.* Cambridge Studies in Advanced Mathematics 9, Cambridge University Press 1986.

[269] C. B. Thomas. *Chern classes of representations.* Bull. London Math. Soc. 18 (1986), 225–240.

[270] C. B. Thomas. *On the subring of $H^*(G, \mathbb{F}_2)$ generated by Stiefel–Whitney classes.* J. Algebra, to appear.

[271] C. B. Thomas. *Characteristic classes and 2-modular representations of some sporadic simple groups.* Algebraic Topology (Evanston, IL, 1988), 303–318, Comtemp. Math. 96, Amer. Math. Soc., Providence, RI, 1989.

[272] B. L. Tsygan. *Homology of matrix algebras over rings and the Hochschild homology.* Uspekhi Mat. Nauk. 38 (1983), 217–218 (= Russ. Math. Surveys 38: 2 (1983), 198–199).

[273] B. B. Venkov. *Cohomology algebras for some classifying spaces.* (Russian) Dokl. Akad. Nauk. SSSR 127 (1959), 943–944.

[274] B. B. Venkov. *Characteristic classes for finite groups.* (Russian) Dokl. Akad. Nauk. SSSR 137 (1961), 1274–1277.

[275] J. B. Wagoner. *Delooping classifying spaces in algebraic K-theory.* Topology 11 (1972), 349–370.

[276] C. T. C. Wall, Ed. *Homological group theory.* London Math. Soc. Lecture Note Series 36, Cambridge University Press 1979.

[277] C. T. C. Wall. *Resolutions for extensions of groups.* Proc. Camb. Phil. Soc. 57 (1961), 251–255.

[278] P. J. Webb. *Restricting $\mathbb{Z}G$-lattices to elementary abelian subgroups.* Integral representations and their applications, Oberwolfach 1980. Lecture Notes in Mathematics 882, 423–429, Springer-Verlag, Berlin/New York 1981.

[279] P. J. Webb. *Bounding the ranks of $\mathbb{Z}G$-lattices by their restrictions to elementary abelian subgroups.* J. Pure Appl. Algebra 23 (3) (1982), 311–318.

[280] P. J. Webb. *Complexes, group cohomology, and an induction theorem for the Green ring.* J. Algebra 104 (1986), 351–357.

[281] P. J. Webb. *A local method in group cohomology.* Comment. Math. Helvetici 62 (1987), 135–167.

[282] P. J. Webb. *Subgroup complexes.* Arcata Conf. on Representations of Finite Groups, Proc. Symp. Pure Maths 47 (Part 1), Amer. Math. Soc. 1987, 349–365.

[283] E. Weiss. *Cohomology of groups.* Academic Press, New York 1969.

[284] G. W. Whitehead. *Elements of homotopy theory.* Graduate Texts in Mathematics 61, Springer-Verlag, Berlin/New York, 1978.

[285] C. Wilkerson. *A primer on Dickson invariants*. Proc. Northwestern Homotopy Theory Conf., Contemp. Math. 19 (1983), 421–434.

[286] O. Zariski and P. Samuel. *Commutative Algebra, Volume II*. Graduate Texts in Mathematics 29, Springer-Verlag, Berlin/New York, 1975.

FURTHER REFERENCES (ADDED SINCE THE FIRST EDITION):

[287] A. Adem. *On the geometry and cohomology of finite simple groups*. Algebraic topology (San Feliu de Guíxols, 1990), 1–9, Lecture Notes in Math. 1509, Springer-Verlag, Berlin/New York 1992.

[288] A. Adem. *Cohomology and actions of finite groups*. Differential topology, foliations, and group actions (Rio de Janeiro, 1992), 123–141, Contemp. Math., 161, Amer. Math. Soc., Providence, RI, 1994.

[289] A. Adem and D. Karagueuzian. *Essential cohomology of finite groups*. Comment. Math. Helv. 72 (1997), 101–109.

[290] A. Adem and R. J. Milgram. Invariants and cohomology of groups. Papers in honor of José Adem. Bol. Soc. Mat. Mexicana 37 (1992), 1–25.

[291] A. Adem and R. J. Milgram. A_5-*invariants, the cohomology of* $L_3(4)$ *and related extensions*. Proc. London Math. Soc. 66 (1993), 187–224.

[292] A. Adem and R. J. Milgram. *The cohomology of finite groups*. Grundlehren der Math. Wissenschaften 309. Springer-Verlag, Berlin/New York 1994.

[293] A. Adem and R. J. Milgram. *The cohomology of the Mathieu group* M_{22}. Topology 34 (1995), 389–410.

[294] A. Adem and R. J. Milgram. *The subgroup structure and mod* 2 *cohomology of O'Nan's sporadic simple group*. J. Algebra 176 (1995), 288–315.

[295] A. Adem and R. J. Milgram. *The mod* 2 *cohomology rings of rank* 3 *simple groups are Cohen–Macaulay*. Prospects in topology (Princeton, NJ, 1994), 3–12, Ann. of Math. Stud., 138, Princeton Univ. Press, Princeton, NJ, 1995.

[296] A. Adem and R. J. Milgram. *The cohomology of the McLaughlin group and some associated groups*. Math. Zeit. 224 (1997), 495–517.

[297] J. L. Alperin. *A Lie approach to finite groups*. Groups—Canberra 1989, 1–9, Lecture Notes in Mathematics 1456, Springer-Verlag, Berlin/New York, 1990.

[298] T. Asai and H. Sasaki. *The mod* 2 *cohomology algebras of finite groups with dihedral Sylow* 2-*subgroups*. Comm. Algebra 21 (1993), 2771–2790.

[299] M. Aschbacher and P. B. Kleidman. *On a conjecture of Quillen and a lemma of Robinson*. Arch. Math. (Basel) 55 (1990), 209–217.

[300] M. Aschbacher and S. D. Smith. *On Quillen's conjecture for the p-groups complex*. Ann. of Math. 137 (1993), 473–529.

[301] G. S. Avrunin and J. F. Carlson. *Nilpotency degree of cohomology rings in characteristic two*. Proc. Amer. Math. Soc. 118 (1993), 339–343.

[302] A. M. Bajer. *The May spectral sequence for a finite p-group stops*. J. Algebra 167 (1994), 448–459.

[303] L. J. Barker. *Möbius inversion and the Lefschetz invariants of some p-group complexes*. Comm. Algebra 24 (1996), 2755–2769.

[304] D. J. Benson. *The image of the transfer map*. Archiv der Mathematik 61 (1993), 7–11.

[305] D. J. Benson. *Cohomology of modules in the principal block of a finite group*. New York Journal of Mathematics 1 (1995), 196–205.

[306] D. J. Benson. *Decomposing the complexity quotient category*. Math. Proc. Camb. Phil. Soc. 120 (1996), 589–595.

[307] D. J. Benson. *Complexity and varieties for infinite groups, I*. J. Algebra 193 (1997), 260–287.

[308] D. J. Benson. *Complexity and varieties for infinite groups, II*. J. Algebra 193 (1997), 288–317.

[309] D. J. Benson and J. F. Carlson. *Products in negative cohomology.* J. Pure & Appl. Algebra 82 (1992), 107–129.

[310] D. J. Benson, J. F. Carlson and J. Rickard. *Complexity and varieties for infinitely generated modules.* Math. Proc. Camb. Phil. Soc. 118 (1995), 223–243.

[311] D. J. Benson, J. F. Carlson and J. Rickard. *Complexity and varieties for infinitely generated modules, II.* Math. Proc. Camb. Phil. Soc. 120 (1996), 597–615.

[312] D. J. Benson, J. F. Carlson and J. Rickard. *Thick subcategories of the stable module category.* Fundamenta Mathematicæ 153 (1997), 59–80.

[313] D. J. Benson and M. Feshbach. *On the cohomology of split extensions.* Proc. Amer. Math. Soc. 121 (1994) 687–690.

[314] D. J. Benson and J. Greenlees. *The action of the Steenrod algebra on Tate cohomology.* J. Pure & Appl. Alge a 85 (1992), 21–26.

[315] D. J. Benson and P. Kropholler. *Cohomology of Groups.* Survey article, "Handbook of Algebraic Topology" North Holland (1995), 917–950.

[316] C. Bessenrodt. *Some new block invariants coming from cohomology.* Astérisque 181–182 (1990), 247–262.

[317] C. Broto and H.-W. Henn. *Some remarks on central elementary abelian p-subgroups and cohomology of classifying spaces.* Quart. J. Math. Oxford 44 (1993), 155–163.

[318] D. G. Brown. *Relative cohomology of finite groups and polynomial growth.* J. Pure Appl. Algebra 97 (1994), 1–13.

[319] J. F. Carlson. *Projective resolutions and degree shifting for cohomology and group rings.* Representations of algebras and related topics (Kyoto, 1990), 80–126, L.M.S. Lecture Note Series 168, Cambridge Univ. Press, 1992.

[320] J. F. Carlson. *Varieties and modules of small dimension.* Arch. Math. (Basel) 60 (1993), 425–430.

[321] J. F. Carlson. *Systems of parameters and the structure of cohomology rings of finite groups.* Topology and representation theory (Evanston, IL, 1992), 1–7, Contemp. Math. 158, Amer. Math. Soc., Providence, RI, 1994.

[322] J. F. Carlson. *Depth and transfer maps in the cohomology of groups.* Math. Zeit. 218 (1995), 461–468.

[323] J. F. Carlson. *Transfers and the structure of cohomology rings.* Research into algebraic combinatorics (Kyoto, 1993). Sūrikaisekikenkyūsho Kōkyūroku No. 896 (1995), 1–7.

[324] J. F. Carlson. *The decomposition of the trivial module in the complexity quotient category.* J. Pure Appl. Algebra 106 (1996), 23–44.

[325] J. F. Carlson. *Modules and group algebras.* Notes by Ruedi Suter. Lectures in Mathematics ETH Zürich. Birkhäuser Verlag, Basel, 1996.

[326] J. F. Carlson. *The cohomology of groups.* Handbook of algebra, Vol. 1, 581–610, North-Holland, Amsterdam, 1996.

[327] J. F. Carlson. *Varieties and induction.* Bol. Soc. Mat. Mexicana 2 (1996), 101–114.

[328] J. F. Carlson. *Modules over group algebras.* Representation theory of groups, algebras and orders (Constanţa, 1995). An. Ştiinţ. Univ. Ovidius Constanţa Ser. Mat. 4 (1996), 31–42.

[329] J. F. Carlson, P. W. Donovan and W. W. Wheeler. *Complexity and quotient categories for group algebras.* J. Pure Appl. Algebra 93 (1994), 147–167.

[330] J. F. Carlson and H.-W. Henn. *Depth and the cohomology of wreath products.* Manuscripta Math. 87 (1995), no. 2, 145–151.

[331] J. F. Carlson and H.-W. Henn. *Cohomological detection and regular elements in group cohomology.* Proc. Amer. Math. Soc. 124 (1996), 665–670.

[332] J. F. Carlson and C. Peng. *Relative projectivity and ideals in cohomology rings.* J. Algebra 183 (1996), 929–948.

[333] J. F. Carlson and G. R. Robinson. *Varieties and modules with vanishing cohomology.* Math. Proc. Cambridge Philos. Soc. 116 (1994), 245–251.

[334] J. F. Carlson and W. W. Wheeler. *Varieties and localizations of module categories.* J. Pure Appl. Algebra 102 (1995), 137–153.

[335] L. S. Charlap and A. T. Vasquez. *Characteristic classes for modules over groups.* Trans. Amer. Math. Soc. 137 (1969), 533–549.

[336] A. Chin. *The integral cohomology rings of certain p-groups.* Comm. Algebra 23 (1995), 3003–3023.

[337] A. Chin. *The cohomology rings of finite groups with semi-dihedral Sylow 2-subgroups.* Bull. Austral. Math. Soc. 51 (1995), 421–432.

[338] A. Chin. *The cohomology rings of some p-groups.* Publ. Res. Inst. Math. Sci. 31 (1995), 1031–1044.

[339] J. Clark. *Mod-2 cohomology of the group* $U_3(4)$. Comm. Algebra 22 (1994), 1419–1434.

[340] F. R. Cohen, J. Harper and R. Levi. *On the homotopy theory associated to certain finite groups of 2-rank two.* Homotopy theory and its applications (Cocoyoc, 1993), 65–79, Contemp. Math. 188, Amer. Math. Soc., Providence, RI, 1995.

[341] J. Dietz, J. Martino and S. Priddy. *Cohomology of groups with metacyclic Sylow p-subgroups.* Proc. Amer. Math. Soc. 124 (1996), 2261–2266.

[342] J. Dietz and S. Priddy. *The stable homotopy type of rank two p-groups.* Homotopy theory and its applications (Cocoyoc, 1993), 93–103, Contemp. Math. 188, Amer. Math. Soc., Providence, RI, 1995.

[343] L. Evens. *Steenrod operations and transfer.* Proc. Amer. Math. Soc. 19 (1968), 1387–1388.

[344] L. Evens. *The cohomology of groups.* Oxford University Press, 1991.

[345] L. Evens and S. Priddy. *The ring of universally stable elements.* Quart. J. Math. Oxford (2) 40 (1989), 399–407.

[346] L. Evens and S. F. Siegel. *Generalized Benson–Carlson duality.* J. Algebra 179 (1996), 775–792.

[347] D. J. Green. *On the cohomology of the sporadic simple group* J_4. Math. Proc. Camb. Phil. Soc. 113 (1993), 253–266.

[348] D. J. Green. *The 3-local cohomology of the Mathieu group* M_{24}. Glasgow Math. J. 38 (1996), 69–75.

[349] D. J. Green and I. J. Leary. *Chern classes and extraspecial groups.* Manuscripta Math. 88 (1995), 73–84.

[350] D. J. Green and I. J. Leary. *The spectrum of the Chern subring.* Darstellungstheorietage Jena 1996, 47–53, Sitzungsber. Math.-Naturwiss. Kl., 7, Akad. Gemein. Wiss. Erfurt, Erfurt, 1996

[351] J. P. C. Greenlees. *Commutative algebra in group cohomology.* J. Pure Appl. Algebra 98 (1995), 151–162.

[352] H.-W. Henn and S. Priddy. *p-nilpotence, classifying space indecomposability, and other properties of almost all finite groups.* Comment. Math. Helv. 69 (1994), 335–350.

[353] J. Huebschmann. *The mod p cohomology rings of metacyclic groups.* J. Pure Appl. Alg. 60 (1989), 53–103.

[354] J. Huebschmann. *Chern classes for metacyclic groups.* Arch. Math. (Basel) 61 (1993), 124–136.

[355] K. Inoue and A. Kono. *Nilpotency of a kernel of the Quillen map.* J. Math. Kyoto Univ. 33 (1993), 1047–1055.

[356] S. Jackowski. *Group homomorphisms inducing isomorphisms in cohomology.* Topology 17 (1978), 303–307.

[357] S. Jackowski and J. McClure. *Homotopy decomposition of classifying spaces via elementary abelian subgroups.* Topology 31 (1992), 113–132.

[358] S. Jackowski, J. McClure and R. Oliver. *Homotopy classification of self-maps of BG via G-actions, I.* Ann. of Math. (2) 135 (1992), 183–226.

[359] S. Jackowski, J. McClure and R. Oliver. *Homotopy classification of self-maps of BG via G-actions, II.* Ann. of Math. (2) 135 (1992), 227–270.

[360] B. Kahn. *The total Stiefel–Whitney class of a regular representation.* J. Algebra 144 (1991), 214–247.

[361] H. Kawai. *On the cohomology of finite groups and the applications to modular representations.* Osaka Math. J. 27 (1990), 937–945.

[362] B. Keller. *Invariance of cyclic homology under derived equivalence.* CMS Conference Proceedings, Vol. 18 (1996), 353–361.

[363] O. Kroll. *On Chern classes of finite group representations.* Enseign. Math. 35 (1989), 363–374.

[364] S. Lang. *Topics in cohomology of groups.* Translated from the French original by the author. New chapter X based on letters written by John Tate. Lecture Notes in Mathematics, 1625. Springer-Verlag, Berlin/New York 1996

[365] I. J. Leary. *The mod-p cohomology rings of some p-groups.* Math. Proc. Camb. Phil. Soc. 112 (1992), 63–75.

[366] I. J. Leary. *A differential in the Lyndon–Hochschild–Serre spectral sequence.* J. Pure Appl. Algebra 88 (1993), 155–168.

[367] I. J. Leary. *3-groups are not determined by their integral cohomology rings.* J. Pure Appl. Algebra 103 (1995), 61–79.

[368] I. J. Leary. *p-groups are not determined by their integral cohomology groups.* Bull. London Math. Soc. 27 (1995), 585–589. Erratum: Bull. London Math. Soc. 29 (1997), 368.

[369] I. J. Leary and N. Yagita. *Some examples in the integral and Brown–Peterson cohomology of p-groups.* Bull. London Math. Soc. 24 (1992), 165–168.

[370] J.-L. Loday. *Cyclic homology.* Grundlehren der Math. Wissenschaften 310, Springer-Verlag, Berlin/New York 1992.

[371] J. Maginnis. *The cohomology of the Sylow 2-subgroup of J_2.* J. London Math. Soc. 51 (1995), 259–278.

[372] J. Maginnis. *Local control of group cohomology.* J. Pure Appl. Algebra 103 (1995), 81–90.

[373] J. R. Martino. *Classifying spaces of p-groups with cyclic maximal subgroups.* Topology and representation theory (Evanston, IL, 1992), 157–174, Contemp. Math. 158, Amer. Math. Soc., Providence, RI, 1994.

[374] J. R. Martino. *Classifying spaces and their maps.* Homotopy theory and its applications (Cocoyoc 1993), 161–198. Contemp. Math. 188, Amer. Math. Soc., Providence, RI, 1995.

[375] J. R. Martino and S. Priddy. *On the dimension theory of dominant summands.* Adams Memorial Symposium on Algebraic Topology, 1 (Manchester, 1990), 281–292. L.M.S. Lecture Note Series 175, Cambridge Univ. Press, 1992.

[376] J. R. Martino and S. Priddy. *Unstable homotopy classification of $BG_p^{\wedge}$.* Math. Proc. Camb. Phil. Soc. 119 (1996), 119–137.

[377] J. R. Martino and S. Priddy. *Stable homotopy classification of $BG_p^{\wedge}$.* Topology 34 (1995), 633–649.

[378] R. J. Milgram and M. Tezuka. *The geometry and cohomology of M_{12}. II.* Bol. Soc. Mat. Mexicana 1 (1995), 91–108.

[379] J. W. Milnor. *Introduction to algebraic K-theory.* Ann. of Math. Studies 72, Princeton Univ. Press, 1971.

[380] P. A. Minh. *On the mod p cohomology groups of extra-special p-groups.* Japan. J. Math. (N.S.) 18 (1992), 139–154.

[381] P. A. Minh. *A cohomological approach to theory of groups of prime power order.* Workshop on Geometry and Topology (Hanoi, 1993). Kodai Math. J. 17 (1994), 571–584.

[382] P. A. Minh. *On the restriction map in the mod-2 cohomology of groups.* J. Pure Appl. Algebra 102 (1995), 67–73.

[383] P. A. Minh. *Transfer map and Hochschild–Serre spectral sequences.* J. Pure Appl. Algebra 104 (1995), 89–95.

[384] P. A. Minh. *The mod p cohomology group of extra-special p-group of ord$_\ell$ p^5 and of exponent p^2.* Math. Proc. Camb. Phil. Soc. 120 (1996), 423–440.

[385] P. A. Minh. *Group homomorphisms inducing mod-p cohomology monomorphisms.* Proc. Amer. Math. Soc. 125 (1997), 1577–1578.

[386] P. A. Minh. *On a conjecture of Kahn for the Stiefel–Whitney classes of the regular representation.* J. Algebra 188 (1997), 590–609.

[387] P. A. Minh. *Proper singularities of extraspecial p-groups for the mod-p cohomology functor.* Comm. Algebra 25 (1997), 965–971.

[388] G. Mislin. *Cohomologically central elements and fusion in groups.* Algebraic topology (San Feliu de Guíxols, 1990), 294–300, Lecture Notes in Math. 1509, Springer-Verlag, Berlin/New York 1992.

[389] T. Niwasaki. *On codimensions of maximal ideals in cohomology rings.* Hokkaido Math. J. 22 (1993), 211–223.

[390] Y. Ogawa. *On the subring of universally stable elements in a mod-2 cohomology ring.* Tokyo J. Math. 15 (1992), 91–97.

[391] Y. Ogawa. *On a Noether normalization for a mod-2 cohomology ring.* J. Algebra 178 (1995), 343–373.

[392] T. Okuyama and H. Sasaki. *Periodic modules of large periods for metacyclic p-groups.* J. Algebra 144 (1991), 8–23.

[393] K. Pearson. *Integral cohomology and detection of w-basic 2-groups.* Math. Comp. 65 (1996), 291–306.

[394] J. Rickard. *Idempotent modules in the stable category.* J. London Math. Soc., to appear.

[395] J. Rosenberg. *Algebraic K-theory and its applications.* Graduate Texts in Mathematics 147, Springer-Verlag, Berlin/New York 1994.

[396] D. J. Rusin. *Kernels of the restriction and inflation maps in group cohomology.* J. Pure Appl. Algebra 79 (1992), 191–204.

[397] H. Sasaki. *The mod 2 cohomology algebras of finite groups with semidihedral Sylow 2-subgroups.* Comm. Algebra 22 (1994), 4123–4156.

[398] H. Sasaki. *The mod p cohomology algebras of finite groups with metacyclic Sylow p-subgroups.* J. Algebra 192 (1997), 713–733.

[399] Y. Segev. *Quillen's conjecture and the kernel on components.* Comm. Algebra 24 (1996), 955–962.

[400] Y. Segev and P. Webb. *Extensions of G-posets and Quillen's complex.* J. Austr. Math. Soc. Ser. A 57 (1994), 60–75.

[401] S. F. Siegel. *The spectral sequence of a split extension and the cohomology of an extraspecial group of order p^3 and exponent p.* J. Pure Appl. Algebra 106 (1996), 185–198.

[402] S. F. Siegel. *On the cohomology of split extensions of finite groups.* Trans. Amer. Math. Soc. 349 (1997), 1587–1609.

[403] S. D. Smith and K. L. Umland. *Stability of cohomology via double-coset products and suborbit diagrams.* J. Algebra 182 (1996), 627–652

[404] P. Symonds. *The orbit space of the p-subgroup complex is contractible.* Preprint, 1996.

[405] M. Tezuka and N. Yagita. *Calculations in mod p cohomology of extra special p-groups. I.* Topology and representation theory (Evanston, IL, 1992), 281–306, Contemp. Math. 158, Amer. Math. Soc., Providence, RI, 1994.

[406] M. Tezuka and N. Yagita. *On some results of the cohomology of extra special p-groups.* Representation theory of finite groups and algebras (Kyoto, 1993). Sūrikaisekikenkyūsho Kōkyūroku No. 877 (1994), 122–132.

[407] M. Tezuka and N. Yagita. *On odd prime components of cohomologies of sporadic simple groups and the rings of universal stable elements.* J. Algebra 183 (1996), 483–513.

[408] C. B. Thomas. *Modular representations and the cohomology of finite Chevalley groups.* Papers in honor of José Adem (Spanish). Bol. Soc. Mat. Mexicana 37 (1992), 535–543.

[409] B. Totaro. *Cohomology of semidirect product groups.* J. Algebra 182 (1996), 469–475.

[410] L. G. Townsley Kulich. *Investigations of the integral cohomology ring of a finite group.* Ph. D. Thesis. Northwestern University, Evanston, 1988.

[411] P. J. Webb. *Graded G-sets, symmetric powers of permutation modules, and the cohomology of wreath products.* Algebraic topology (Oaxtepec, 1991), 441–452, Contemp. Math. 146, Amer. Math. Soc., Providence, RI, 1993.

[412] N. Yagita. *Note on the spectral sequence converging to cohomology of an extra special p-group for odd prime p.* Math. Proc. Camb. Phil. Soc. 119 (1996), 35–41. Corrigenda: Math. Proc. Camb. Phil. Soc. 121 (1997), 575.

Index

⋈, 87
≃, 2
$\simeq_G$, 220

$\mathbb{A}^n(k)$, 161
$\mathcal{A}_p(G)$, 217
$a(G)$, 218, 227
$a(G)$, nilpotent elements in, 191
absolute Hurewicz theorem, 10
acyclic Hochschild complex, 73, 74
Adams operations, 45
additive K-theory, 80
Adem relations, 138, 144
affine
    polynomial identity algebra, 177
    space, 161
    variety, 163, 228
        homogeneous, 165
algebra
    commutative, 161
    Lie, 229
algebraic
    geometry, 161
    group
        connected, 228
        linear, 228
    K-theory, 68
Alperin
    's conjecture, 233
    –Evens theorem, 160
alternating group
    A_6, 201
    A_7, 247
Andrews' theorem, 192
annihilation of cohomology, 189
approximation, cellular, 15
Atiyah
    –Hirzebruch spectral sequence, 114
    completion theorem, 48
    spectral sequence, 114
attaching
    cells, 12

map, 12
augmentation map, 216
augmented chain complex, 216
Avrunin
    –Carlson example, 172
    –Scott theorem, 179
axiom
    exactness, 45
    excision, 45
    homotopy, 45
axioms, Eilenberg–Steenrod, 45

$B_*(\Lambda)$, 84
BC, 27
$\mathcal{B}_p(G)$, 231
$\mathcal{B}_p^{(1)}(G)$, 248
$\mathcal{B}_p(G)$, 217
b'_n, 73
$b(G', G)$, 59
$b(G)$, 58, 216
b_n, 74
bar resolution, 30
barycentric subdivision, 27, 215, 217
base space, 16
basepoint, 1
β, $\hat{\beta}$, 132
$\hat{\beta}$, 135
Becker–Gottlieb transfer, 53
BG, 38
$BGL_n(\mathbb{C})$, 42
$BGL_n(\mathbb{F}_q)$, 65
$BGL_n(\mathbb{R})$, 41
bilinear form, symmetric, 169, 245
bimodule, vi
BO, 46
$BO(n)$, 41
Bockstein
    higher, 134
    homomorphism, 132
    spectral sequence, 134
    twisted, 135
Borel